SCRUTINIZING SCIENCE

SCRUTINIZING

SCIENCE

Empirical Studies of Scientific Change

Edited by

ARTHUR DONOVAN

U.S. Merchant Marine Academy, New York

LARRY LAUDAN and RACHEL LAUDAN

University of Hawaii

THE JOHNS HOPKINS UNIVERSITY PRESS

BALTIMORE AND LONDON

First published in English in 1988.
Copyright 1988 Kluwer Academic Publishers.
All rights reserved
Printed in the United States of America on acid-free paper

Johns Hopkins Paperbacks edition, 1992

The Johns Hopkins University Press
701 West 40th Street
Baltimore, Maryland 21211-2190
The Johns Hopkins Press Ltd., London

Library of Congress Cataloging-in-Publication Data

Scrutinizing science : empirical studies of scientific change / edited
  by Arthur Donovan, Larry Laudan, and Rachel Laudan.
      p.   cm.
    Originally published: Dordrecht ; Boston : Kluwer Academic
  Publishers, c1988. (Synthese library ; v. 193).
    Includes bibliographical references and index.
    ISBN 0-8018-4517-3 (pbk. : alk. paper)
    1. Science—Philosophy—Case studies.   2. Science—History—Case
  studies.   3. Change—Case studies.   4. Hypothesis—Case studies.
  I. Donovan, Arthur L.   II. Laudan, Larry.   III. Laudan, Rachel.
  1944–
  [Q175.S4245   1992]
  509—dc20   92-23964

A catalog record for this book is available from the British Library

TABLE OF CONTENTS

PART I

PART II: CASE STUDIES

ACKNOWLEDGEMENTS

This project would have been quite impossible without the generous participation of numerous scholars from history, philosophy, and the natural and social sciences. Since the support for interdisciplinary science studies is limited, we were particularly fortunate to be able to count on the good graces of all those who took time out from their own research programs to assist in this project. Harold Brown, Jarrett Leplin, Paul Thagard, Stephen Wykstra and Peter Barker co-authored with us the *Synthese* monograph that formed the first stage of the project. Early drafts of that monograph were circulated widely within the science studies community. Our ideas were much clarified by the comments and criticisms we received from skeptics and supporters alike, particularly Steve Brush, Geoffrey Cantor, I. B. Cohen, Jim Cushing, Maurice Finocchiaro, Alan Gabbey, Loren Graham, Fred Gregory, Jon Hodge, Larry Holmes, Rod Home, Homer LeGrand, Tim Lenoir, David Lindberg, Andrew Lugg, Jane Maienschein, Sy Mauskopf, Alan Musgrave, Tom Nickles, Ilkka Niiniluoto, Ronald Numbers, Mary Jo and Bob Nye, David Oldroyd, Carl Perrin, Fritz Rehbock, Alan Rocke, Nils Roll-Hansen, Dorothy Ross, Michael Ruse, Warren Schmaus, Kristin Schrader-Frechette, George Stocking, Stephen Turner, Richard Westfall and Robert Westman.

Following the initial stage of the project, Alan Gabbey, David Hull, Nils Roll-Hansen and Homer LeGrand presented papers on particular case studies. Discussions with them allowed us to articulate our methodology for testing the claims of the theorists of scientific change.

Cristina Bicchieri, Richard Burian, Donald Campbell, Jim Cushing, Stephen Fuller, Ronald Giere, Scott Kleiner, Lorenz Kruger, Peter Lamal, Jarrett Leplin, Andrew Lugg, Jane Maienschein, Bert Moyer, Tom Nickles, Joe Pitt, Margaret Schabas, Warren Schmaus, Sam Schweber, Dudley Shapere, Stephen Turner, and Sjerp Zeldenrust, as well as all the authors in this volume, gave papers or comments at a conference on the topic in Fall 1986.

Finally, we should thank Donald Campbell for his willingness to discuss methodological issues with us and for his encouragement at various stages in the development of the project.

Blacksburg, Virginia
June 1987

INTRODUCTION TO THE JOHNS HOPKINS EDITION

1. INITIAL RESPONSES TO *SCRUTINIZING SCIENCE*

Four years separate the original publication of *Scrutinizing Science* from the appearance of this paperback edition, so it is too early to arrive at a definitive appraisal of its reception and impact. Indeed, the book is being reissued because its editors and the Johns Hopkins University Press are convinced it deserves a wider audience. The original edition was reviewed at length in a number of scholarly journals and some of the issues the reviewers have raised are discussed below.[1] We hope that with the book available in a more accessible format, many more scientists and students of science will become familiar with and contribute to its research program.

That research program has elicited a wide range of responses. In this introduction we do not summarize the reviews that have come to our attention or answer specific points of criticism, but instead take these initial reactions as representative of current patterns of thought in the small but lively field of scholarly specialization called science studies.[2] When we began the project that led to the conference upon which this book is based, one of our goals was to encourage a greater sense of coherence and direction among scholars interested in the interdisciplinary study of scientific change. We remain convinced that this is a pressing and worthwhile goal and that the research program we devised provides a well-conceived and promising strategy for achieving it.

Students of science—that is, those who take science itself rather than nature as their subject of study—have for the most part been trained in one or another of two very different realms of culture. Many early students of science were scientists who brought to the task considerable backgrounds in classical learning as well. During the past half-century, however, the study of science has been increasingly dominated by academic specialists who, like most of the contributors to this volume, were trained primarily as historians or philosophers. One might expect scientists and humanists to begin with shared assumptions when addressing a common subject, yet it often turns out that they hold radically different images of science.

Scientists understand that it is fundamentally important to test theoretical claims against empirical evidence. *Scrutinizing Science* addresses diverse and frequently conflicting claims about how science changes. It seeks to test these claims against well-researched historical cases. Such a program is fraught with methodological difficulties, many of which are briefly discussed in part one of the book (and many of which we also review later in this introduction), but these difficulties did not strike us then, nor do they strike us now, as insurmountable. Scientists who encounter similar difficulties in their day-to-day work have not been troubled by our use of empirical evidence to appraise the theories of interest to us, a procedure that is hardly unique to science.

Many historians and philosophers are made uneasy by this use of evidence, however, for they see it as an attempt to create a science of science. Yet as we made plain in 'Testing Theories of Scientific Change' (see p. 8 and notes 6 and 8), we do not draw analogies between science and science studies because we think science can be adequately understood as a rule-governed game played by decontextualized minds. But having granted that contexts play a role, we nonetheless incline to the view that inquiry, like every other human activity, exhibits more and less successful instances. We begin with the natural assumption that some strategies of inquiry are more effective than others, and we engage the task of finding out, by an examination of particular cases, which are which. Specifically, we seek to determine which cognitive strategies have repeatedly played a role in effecting major theory changes in science. The theses tested represent various popular *strategies of inquiry* (e.g., 'ignore anomalies', 'seek surprising predictions', etc.). We find ourselves alternately bemused and perplexed by those critics of the project who suppose that there are no such strategies to be found, even while taking us to task because, by their lights, the strategies of research that we use are less than optimal! Such a contradictory reaction to our modest proposal says more about the uncertainties and insecurities that afflict certain areas of contemporary culture than about the research program we proposed.

While science is characterized by vigorous controversy, great intellectual freedom, and revolutionary upheavals, it is also remarkably successful in bringing controversies to closure and in establishing cognitive consensus. Although students of science have long sought to explain how this progressive dynamism is sustained, they have not yet managed to make it a central feature of their own scholarly pursuits. Science studies is currently characterized by extreme freedom in the formulation of hypotheses and intense sectarianism in debate. The parties seeking to establish hegemony commonly ignore all accounts of science except their own. Partisans fortify their positions by citing case studies that have been carefully selected and fashioned to support their claims. This struggle of all against all within science studies is encouraged by the ideological posturing that is currently widespread in the humanities. Scientists understandably find it quite baffling.

Still worse, this sectarian feuding about science appears to the outsider as extraordinarily parochial in that it ignores completely certain theories of science that the wider public, including the scientific public, thinks have considerable explanatory power. One need not be a disciple of Karl Popper or Thomas Kuhn to acknowledge that their theories have shaped the informed public's image of science for nearly two generations and have rendered intelligible many aspects of scientific belief and change that were ignored or misunderstood by earlier positivists.[3] The research program of *Scrutinizing Science*, while not designed to falsify or verify these or any other accounts of scientific change as a whole, treats the theories it examines with respect. Rather than simply dismissing prominent theories of science, it proposes that they be analyzed into their component theses and that those claims be tested against the historical record. This procedure does not, of course, demonstrate the truth of those theses that survive scrutiny; quite apart from the fact that we are all fallibilists, intellectual honesty requires us to acknowledge that the existing 'data base' is too partial to sustain more than judgments of plausibility. Our goal is simply to initiate an ongoing discourse that addresses widely accepted theories and brings reliable evidence to bear in comparative evaluations. We have not attempted to predict or determine where that discourse might lead.

Scrutinizing Science has attracted rather selective attention among historians of science. This is not surprising, for few of the historians we approached during the early stages of this project greeted it with enthusiasm. This is not the place to explore the increasingly tenuous relations between the history and the philosophy of science.[4] It is sufficient to note that all but a few Anglo-American historians of science are either resolutely atheoretical in practice or model their narratives on accounts of social and cultural change not primarily concerned with science. The traditional dialogue between the history and philosophy of science no longer commands center stage in the history of science, and few historians are inclined to commit time to evaluating theories of scientific change. Perhaps if scholars more concerned with theoretical issues exhibited a greater sense of collective purpose and a higher regard for empirical evidence, some of these defecting historians might be brought back into the discourse of science studies.

2. MAJOR CRITICISMS OF THE PROJECT

Some of the criticisms raised in reviews and discussion of this project can be dealt with quite briefly, whereas others call for more extended replies. Let us begin with the question,

Why, in devising strategies for testing, fasten on the hypothetico-deductive method of empirical inquiry?

This question is motivated by the realization, which has been thoroughly explored by philosophers for many centuries, that the method of hypothesis inevitably yields fallible results. One of the editors has written several articles and a book on this subject, and thus we were hardly unaware of this problem when we began this project.[5] Two major considerations recommended this approach, however, despite its limitations. First, because widespread public acceptance of the conflicting claims of Popper, Kuhn, and other theorists of science make it incumbent on the science-studies community to address the empirical warrant for these theories, we decided that the hypothetico-deductive method offered the most convenient way to proceed. Second, we were encouraged by the fact that this method, despite its flaws, had been used to good effect by scientists for many centuries. We were also swayed by a third, more heuristic consideration. We expected, as was indeed the case, that nuanced discussion of the theses would result not in simple rejection or acceptance, but rather in their modification and refinement in light of new evidence.

We could, of course, have begun in other ways as well. Rather than focusing on older theories that are clearly flawed, why not simply start afresh and construct new theories designed to interpret the novel empirical evidence available in recent studies of scientific change? We considered this strategy in 'Testing Theories of Scientific Change' (pp. 12–13) but found it wanting. Theoretical novelty is much rarer than empirical novelty, and one would be hard-pressed to identify much that is truly new in the accounts of science that go in and out of fashion today. Of course, it is possible that one of the current factions in science studies will achieve a dominant position and succeed in imposing its image of science on the field as a whole. Were this to happen, and were a tradition of rational critique and empirical testing to be established as well, then science studies might begin to make real strides towards understanding its subject. However, we continue to favor beginning with the testing of current theories, because we do not wish to prolong the current state of dissensus in science studies. As Bacon observed long ago, knowledge is more likely to emerge from error than from confusion. We also believe it is both politic and wise to begin with an image of science that scientists find accessible and informative, and to critically evaluate this image rather than rejecting it wholesale.

We have also been asked,

Why do you rely simply on the published record of science? Doesn't this commit you to the discredited discovery/justification distinction?

The fact remains, however, that most theorists of scientific change have been principally concerned with explaining how whole communities of specialists react to new theories. Such appraisal can only take place *after* the discovery phase and in response to a series of published debates and exchanges concerning the merits and deficiencies of rival theories. The public record, and especially the published one, therefore provides the historian with his or her principal point of access and is crucial to any understanding of the reception accorded to scientific theories.

Other critics have insisted that

The authors of many of the theses under scrutiny formulated them as normative claims about science. To claim that empirical evidence can have a bearing on claims about standards is a classic instance of the naturalistic fallacy.

The short answer to this criticism is that we do not believe the so-called naturalistic fallacy is a fallacy. We believe that standards of inquiry, like substantive claims about nature, ought to be exposed to and be obliged to survive empirical scrutiny. Our arguments for this view of norms have been developed at length elsewhere.[6]

Yet another line of criticism asks,

Are you not misrepresenting the claims of the authors of the theses under scrutiny when you strip them of the rich contexts of meaning in which they were originally embedded?

We previously acknowledged (p. 9) that this way of proceeding would appear to diminish the range and power of the theories being examined, yet we believe that competing claims must be made comparable if they are to be tested empirically. Perhaps the charge, made in several reviews, that we have radically misunderstood the meaning of the claims being assessed is a serious one; to determine if it is, one would have to return to the texts in question to see if the meanings we extracted constitute reasonable readings. But criticism of this sort, no matter how justified in detail, leaves unaddressed the larger questions raised by this project. Would anyone deny that today's most prominent theories of scientific change contradict one another on many important points? Must we be limited to juxtaposing competing and incompatible systems of interpretation while being denied access to the only source that can settle the issues before us?

Some critics have bluntly asked,

Is it possible to make any reliable generalizations whatsoever about scientific change?

In the current climate of opinion in science studies, this criticism is often thought to pose the ultimate challenge to the research program of *Scrutinizing Science*. Is science in all its manifestations so profoundly shaped by local and contingent factors that all generalizations about what science is and how it changes are hopelessly misconceived? The question at issue is obviously one of degree, for no one who has taken even the briefest look at a recent history of science can claim that contemporary students of science are ignoring the force of personality, interest, and extra-scientific belief. Everyone now knows that scientific theories are not a direct reflection of nature, just as everyone, or almost everyone, has the good sense to realize that science is more than a projection onto an undifferentiated nature of the attitudes and relationships that structure the social and psychological milieu of the scientist. What is in question is the extent to which science can lay claim to a distinctive cultural tradition, one that in significant and ineliminable ways is self-conscious, continuous, and consequential.

Many contemporary theorists believe that the very cognitive content of science is determined by the immediate circumstances in which its theories are formulated. Students of politics have long insisted that 'all politics is local', and students of science seem to be coming to a similar conclusion. Historians, many of whom luxuriate in the specific, find this emphasis on the particular and the conditioned congenial. If one blocks out the noise of sectarian battle, however, it appears we are hearing only half of what the scientific texts are trying to tell us. Who would deny that scientists, like not only poets, artists, generals, and statesmen, but also philosophers and historians, speak to one another across the ages and across national and cultural boundaries?[7] How is it possible to carry on a universalized scientific discourse, as scientists clearly do, when local cultural resources cannot be meaningfully deployed and local interests cannot be effectively invoked? Surely modern science has a degree of autonomy that should be attended to and that can be rendered intelligible. Contemporary scientists obviously believe that when acting professionally they have as much freedom and are as self-determining as anyone else, and one is inclined to say they are right. One need accept nothing more to refute extreme nominalism in science studies and provide a foundation for the research program proposed in *Scrutinizing Science*.

Should scholars studying science concentrate solely on that which is individual in each case? We remain unpersuaded and note, without claiming to be original, that students of science can be loosely divided into generalizers and particularizers. The former look for reiterated patterns of regularity in their data; the latter revel in the richly textured uniqueness of the particular. Both approaches illuminate the subjects being studied. The generalizers make the valid point that all collections of phenomena have numerous points in common. Yet because every event is unique, there are also numerous respects in which each event differs from every other event. Both interpretive approaches are correct; neither is exhaustive. If one looks hard at any two episodes of scientific change, one can easily list both differences and similarities.

Scrutinizing Science is an attempt to subject certain general claims about science to empirical testing. Although particularizers may consider this attempt doomed from the start, they have failed to demonstrate that the several case studies presented in this volume have nothing in common. Their repudiation is a dismissal based on two improbable general claims about science. The first, which seeks to marginalize and discredit whatever remains of the positivist account of science, asserts that the common methodological and epistemic elements identified in the cases studied are either 'trivial' or 'vacuous'. The second, which reflects the powerful influence of the new historicism in science studies, asserts that using general properties to explain particular episodes ignores the methodologically privileged position of narrative in contemporary cultural studies. According to this view, truth is revealed by employing actors' categories rather than those of the observer and by focusing on the local and contingent rather than the general and the causal.

The charge of vacuity and triviality is unfounded, as one can see by examining several of the theses tested in this volume. Consider, for instance, the claim that scientists convert from one world view to another 'all at once'. If this claim is true, it has profound implications for our understanding of the logic and psychology of scientific inference. For the historian, it would entail that narrative descriptions of how scientists come to hold particular views are nothing more than fictions concocted by the historians themselves. Consider also the claim that scientists generally ignore anomalies that challenge their theories. Charles Sanders Peirce and Karl Popper certainly did not think this thesis is trivially true, for each constructed a comprehensive philosophy of inquiry based on the assumption that it is false. Nor can one simply flip-flop and assert that this claim is trivially false, for Thomas Kuhn, Imre Lakatos, and a generation of British sociologists of knowledge have argued that one of science's core features is its ability to ignore counter-instances.

Clearly the claim that certain generalizations about science are trivial is grounded on equally general, if unstated, beliefs about science that are themselves held to be non-trivial. As has often been remarked, those who insist they are not committed to any theory are most heavily committed to deeply entrenched theories. The particularists' insistence that certain general claims about science are obvious and trivial reveals that they themselves are guided by general hunches about how the scientific enterprise works. These hunches give purpose and direction to the narratives they construct to explain particular events. We do not fault them for holding guiding assumptions; we merely urge them to make these assumptions explicit and subject them to empirical appraisal.

Are the theses under review vacuous, in the sense of having no explanatory power? Consider an example. One thesis asserts that scientists do not ignore anomalies that challenge their theories. If this is true, then we have reason to expect that scientists who encounter anomalous evidence will cast about for ways to accommodate that evidence within their theories. Because apparent anomalies crop up frequently in science, this thesis indicates that scientists spend much of their time coping with anomalies. Clearly this thesis, if true, is far from vacuous. Virtually all the other theses under review have equally consequential implications.

Let us now turn to the second general criticism, which is that generalizations about scientific change lack credibility because they ignore crucial local and contingent factors. The particularists draw a sharp contrast between the type of account they favor, which is that of narrative history, and the generalized accounts provided by theories of scientific change. They are correct, of course, in noting that historical narratives acquire persuasive power through the accumulation of discrete bits of evidence rather than through the invocation of general laws. But are they correct in insisting that such *pointillist* accounts should be the major or even the only way of organizing the data collected by students of science?

In short, we are not impressed by the hushed piety with which some historians speak of the mystical powers of the narrative. After all, narration is but one among many ways of rendering historical phenomena intelligible. Economic and social historians have frequently demonstrated that historical events can often best be represented as data entries in a tabulation or as recurrent episodes in a general process or cycle. Most of the social sciences, except, perhaps, cultural anthropology, and all of the natural sciences have rejected narration as their primary mode of representation or interpretation.

Perhaps students of cultural reflexivity will someday trace out the dialectical convolutions that have led so many historians of science to convert quite suddenly to the descriptivist, anti-theoretical style of narrative anthropology and to adopt its commitment to radical contextualization and its nominalist hostility to generalization. This move appears especially bizarre in the history of science, for it is science, more than any other aspect of modern culture, that has advanced by insisting on the primacy of the general over the particular. The progress of modern science confirms the wisdom of those early investigators who dared to ignore immediate and obvious differences while reaching for general truths. For Galileo, it was the general pattern, rather than the minor variations, he could detect in the speed of falling bodies that mattered. The narrative method seldom serves the purposes of science, whose characteristic quest is for general truths. Those who assert the self-reflexively incoherent thesis that generalizations have no place in science studies must therefore expect considerable opposition from those who have been attracted to this subject by the undeniable historic and epistemic success of science.

It is ironic that many of those who objected to our concentration on general theories nonetheless commended the historical case studies included in *Scrutinizing Science*. Historians routinely dismiss case studies motivated by philosophical questions as misinformed and hopelessly partisan. However, if the authors of the cases in our volume have, in the course of testing a small set of theses about scientific change, written informative historical studies, is it not reasonable to conclude that their essays are interesting precisely because they address general theoretical claims? Surely context matters here, too. It would be exceedingly tendentious to insist the case studies are interesting *despite* their common concern or that they might just as well have been written in the way in which they appear here had they not been composed as tests of general claims about science. At the very least, these case studies demonstrate that an interest in the general nature of science need not be fatal to the writing of well-grounded and informative histories.

3. HOW THE PROJECT CAN BE ADVANCED

In the years since embarking on this project we have reflected frequently on what we have learned and how it might be developed further. The difficulties involved in constructing a coherent and comprehensive image of modern science are daunting, for the subject is vast and ever-changing. The case studies in *Scrutinizing Science* are drawn from the physical sciences, but the theses examined also need to be tested against cases drawn from the biological and social sciences. We hoped, and still hope, that scholars who are experts in these subjects will organize conferences of their own and publish volumes in which the research program of *Scrutinizing Science* is adapted to their topics and needs. The theories whose claims we set out to test certainly were not designed for the physical sciences alone, and it is worth recalling in this regard how powerful an impact Thomas Kuhn's account of scientific revolutions has had among social scientists (see pp. 6–7). Contemporary science studies appears chaotic because its reach so clearly exceeds its grasp, but only the faint-hearted would suggest that order should be achieved by reducing its aspirations and abandoning the search for a comprehensive image of science.

Some of the more interesting and unanticipated comments on our project came from scholars who share our conviction that science studies should focus on identifying testable generalizations about science. For example, both the psychologist Dean Keith Simonton and the historian of science Frank Sulloway agree that historical evidence can help us move from individual narratives toward well-grounded general conclusions.[8] The approaches they propose differ somewhat from those pursued in *Scrutinizing Science*, but we suspect that some of the differences arise from the particular empirical questions asked rather than from deep methodological disagreements. In science itself strategies of investigation are constantly being adapted to the particular questions being addressed and one should expect the same kind of adaptation to occur in science studies. But because selecting a research strategy is serious business, we would like to give some indication as to why we have not made use of the particular techniques employed by these two investigators.

Questionnaires: Frank Sulloway has made extensive use of questionnaires in his attempt to assess various factors, ranging from birth order to degree of religious commitment, that affect attitudes toward scientific innovation. He has collected a vast amount of information from experts on controversies in a wide variety of sciences from the seventeenth century to the present. We were at first tempted to adopt his technique and send a large number of specialists a questionnaire on the theses we were examining. Such a procedure would have enabled us rapidly to accumulate a large data base, but we nonetheless rejected this approach. We suspected that even experts would find it difficult to reconceptualize the cases they know well in ways that would enable them to make meaningful assessments of the theses under examination. Our experience at the conference confirmed this suspicion. We also worried that if we presented those completing a questionnaire with the stark choice of 'yes', 'no', or 'not applicable', their answers might reflect little more than their preconceptions about the theories being tested rather than a disinterested conjecture about the cases they know best. Finally, because we were seeking only a first-cut indication of the veracity of theses that deserve further analysis, we did not feel compelled to gather a large number of samples. In the end, therefore, we decided to collect our empirical data in the form of a small number of specially prepared case studies.

Statistical Analysis: Both Sulloway and Simonton have collected large samples and subjected them to elaborate statistical analysis. David Hull also employed a quantitative approach in his pioneering study of the ages at which members of Darwin's scientific community accepted his new theory. We applaud the efforts of these authors even though we did not choose to emulate them. Our choice was a result of two factors. The first was a simple matter of finance. Collecting the research of hundreds of scholars, rather than holding a conference for the three dozen who joined us, was simply beyond our limited resources. The second factor is that we do not concede that large samples of ill- or quasi-digested data are necessarily to be preferred to smaller samples in which the data are carefully groomed. The theses under test often make reference to subtle concepts with nuanced meanings. One key function of the conference and the exchange of papers that followed was to insure—so far as this was possible—that the philosophical vocabulary being invoked was used in a univocal fashion by all scholars appraising the same thesis. We admit to some skepticism about collaborative research efforts in this area in which the bulk of the 'collaborators' fail to interact with one another with respect to defining the key concepts of appraisal being used.

Finally, those who wish to bring order and direction to science studies need to find forms of institutional support that will enable them to link together the diverse research groups currently at work. The importance of formulating more adequate theories of science is beyond question, as is the need to render this singular cultural enterprise intelligible to non-specialists. Yet at present we lack the institutional and intellectual means needed to do so. *Scrutinizing Science* is the product of a university-based group created to launch a new research and graduate program in science studies. Such undertakings are rare, difficult to initiate, and difficult to maintain in the best of times, and even before the original edition of *Scrutinizing Science* had appeared, the research group that conceived it and brought it to life had been disrupted and dispersed. Progress in complex projects of this sort cannot be achieved without suitable institutional support. This is another fact of modern research that is a truism in science but that students of science have not yet made central to their own research efforts.

When drawing up the research program for *Scrutinizing Science*, we purposefully suppressed the temptation to formulate yet another new theory of scientific change, our goal being rather to bring into a common discourse several existing views of science. The case studies that form such an important part of this project demonstrate that historical evidence can be made probative when weighing the merits of competing theoretical claims about how science changes. We continue to believe we have asked the right questions, and we are pleased that other students of science are taking up the challenge.[9] Some have done so unwittingly, but their contributions can be useful nonetheless. Consider, for instance, a recent article on anomalies.[10] After discussing several carefully researched and instructive case studies, the authors conclude that 'an important class of scientific anomalies are recognized as anomalies only after they are given compelling explanations within a new conceptual framework'. In other words, they have provided historical evidence relevant to one of the theses in *Scrutinizing Science*, namely Guiding Assumption 2.7: 'When a set of guiding assumptions runs into empirical difficulties, those difficulties become acute only if a rival theory explains them'. As it turned out, none of the case studies in *Scrutinizing Science* provided clear evidence on the truth or falsity of this thesis (cf. p. 30). Had the authors of this article connected their findings to the more comprehensive project of *Scrutinizing Science*, the significance of their research would have been considerably enlarged. We look forward to seeing connections of this sort being made in the future.

The Editors
June 1992

NOTES

[1]Two of the editors of this book have described its purpose and method at public panels on the testing of theories of scientific change, one at an annual meeting of the American Association for the Advancement of Science, the other at an annual meeting of the History of Science Society. Frank Sulloway is currently editing a selection of papers on this topic for publication.

[2]Essay reviews include Thomas Nickles, 'Historicism and scientific practice', *Isis* **80** (1989), 665–669; David Gooding, 'How to be a good empiricist', *British Journal for the History of Science* **22** (1989), 419–427; Ron Curtis, 'Scrutinizing science', *Philosophy of the Social Sciences* **20** (1990), 376–384; Colin Howson, 'The poverty of historicism', *Studies in History and Philosophy of Science* **21** (1990), 173–179; and Paul Hoch, 'An historical philosophy of science', *History of Science* **28** (1990), 211–219.

[3]As a recent book review makes clear, disputes over the interpretation of Kuhn's account of scientific change continue to inform significant controversies concerning the nature of science and its public impact; see Virginia E. Nolan and Edmund Ursin, 'Science and tort law', *Science* **254** (13 Dec. 1991), 1663–64.

[4]See Thomas Nickles, 'Philosophy of science and history of science', in the pre-circulated papers of the Conference on Critical Problems and Research Frontiers, History of Science Society and Society of the History of Technology, Madison, Wisc., 1991, pp. 349–373.

[5]Larry Laudan, *Science and Hypothesis* (Dordrecht: D. Reidel, 1981).

PART I

TESTING THEORIES OF SCIENTIFIC CHANGE

RACHEL LAUDAN, LARRY LAUDAN, AND ARTHUR DONOVAN

TESTING THEORIES OF SCIENTIFIC CHANGE

1. PREFACE

Science is accorded high value in our culture because, unlike many other intellectual endeavors, it appears capable of producing increasingly reliable knowledge. During the last quarter century a group of historians and philosophers of science (known variously as 'theorists of scientific change', the 'post-positivist school' or the 'historical school') has proposed theories to explain progressive change in science. Their concepts and models have received such keen attention that terms like 'paradigm' have passed from obscurity to common speech. In this volume, we subject key claims of some of the theorists of scientific change to just that kind of empirical scrutiny that has been so characteristic of science itself. Certain claims emerge unscathed – the existence and importance of large-scale theories (guiding assumptions) in the physical sciences for example. Others, such as the supposed importance of novel predictions or the alleged insignificance of anomalies, seem to be without foundation. We conclude that only by engaging in testing of this sort will the study of science be able to make progress.

In the following section of this essay, we explain why we think testing so urgent a matter. Following that, we lay out one possible program for testing. The final and most substantial section of the introduction discusses the results that we obtained.

2. THEORIES OF SCIENTIFIC CHANGE: A CALL FOR TESTING

Because science is an icon of our time and our culture, the image of science that we possess is immensely important. It determines the attitude of the general public toward science, it sways the politicians' willingness to support pure and applied research, and perhaps most important of all, it bears on the scientist's conception of the appropriate goals and methods for his own research. These observations are familiar clichés, but they drive home the importance of ensuring that our image of science is not egregiously wide of the mark.

3

During the past twenty-five years, the so-called 'historical school' has been attempting to re-draw the image of science, replacing the positivist or logical empiricist image that dominated for the first half of the century. According to the older picture (hammered out by scientists and philosophers such as Duhem, Carnap, Bridgman, Reichenbach, Popper and Hempel and widely accepted by the lay community) science can be sharply delimited from other intellectual activities. It is value-free, above the vagaries of theological belief, metaphysical commitment or political involvement, and hence hostage to no non- or extra-scientific influences. Science has a method, *the* scientific method, which provides unity to the disparate areas of inquiry which go by the name 'science'. The assessment of scientific laws and theories is simply a matter of confrontation between them and the data. If the right kind and quantity of positive evidence can be collected, the theory is acceptable. If, on the other hand, the evidence is negative, then the proposed law or theory has to be rejected. The key process of comparing theory and evidence is, at least in principle, uncomplicated because it is thought that a clear distinction can be drawn between 'observational' and 'theoretical' statements. The latter are speculative and tentative; the former by contrast are bedrock. Finally, on this view, science is cumulative, i.e., later scientific theories encapsulate all the true consequences (the 'empirical successes') of earlier theories. Thus, through time, the scope of successful science becomes steadily wider, and nothing important is lost in the process.

During the late 1950s and 1960s, a number of serious cracks began to appear in the positivist picture of an objective, distinct, value-free and cumulative science. Some thorny difficulties were philosophical in origin – the result of technical obstacles in working out the logic of confirmation, problems about the viability of the theory-observation distinction, or logical puzzles about the ambiguities involved in testing individual scientific claims in isolation. But even more important were the difficulties created by the growing recognition that the record of past science is radically at odds with the positivist image of science. Investigation of the origin of, evidential warrant for, and subsequent career of, Galilean and Newtonian mechanics, Lavoisian chemistry, the Darwinian theory of evolution, the Bohr atom, or relativity theory (to name only the more prominent) showed the then-current account of science to be hopelessly wide of the mark. As it became evident that the most successful and prominent scientists of the past had acted in ways that failed totally to square with the positivist picture of how they ought to have acted, the historical school concluded that the image of science was badly in need of overhaul.

Accordingly, new theories of scientific change proliferated. Thomas Kuhn's *Structure of Scientific Revolutions*, first published in 1962 and now a seemingly mandatory reference for every scientist who believes that his discipline is in need of, or actually in the throes of, some radical reconstruction, is the best known of these efforts. But other influential theories of science have been advanced by Imre Lakatos, Paul Feyerabend, Donald Campbell, Stephen Toulmin, Wolfgang Stegmueller, I. B. Cohen, Gerald Holton, Dudley Shapere and Larry Laudan, among others. While these theories differ one from another, they all deny many of the central theses of the positivist image of science.

Moreover, and herein lies the rationale for this volume, the members of the historical school insist that their alternative theories evolved from a careful examination of past and present science. Common to most members of this school is the conviction that any theory of scientific change needs to accommodate the impressive body of evidence assembled by historians from the career of science itself, while simultaneously circumventing the philosophical and logical problems besetting positivism.

Yet whatever the provenance of these post-positivist theories, the sad truth is that most theories of scientific change – including those which have won widespread acceptance – have not yet been extensively or systematically tested against the empirical record. The historical examples liberally scattered throughout the writings of these theorists are *illustrative* rather than *probative* and none is developed in sufficient detail to determine whether the analysis fits the case in hand. Furthermore, detailed studies of examples favored by the historical school – the chemical revolution, Bohr's atomic theory, or the Copernican revolution – have undermined conclusions based on the scantier evidence available in the 1960s and 1970s, when so many of the new theories were formulated. But for a variety of reasons, that increase in detailed knowledge has not been brought to bear on the assessment of theoretical claims about scientific change.

True, there are a handful of case studies which seek to apply various theories of scientific change to selected episodes in the natural sciences.[1] The results of these 'tests' – if tests they be – are decidedly mixed, with at least as many disconfirming as confirming results. But even if the results were not so mixed, and even if there were significantly more empirical studies, one would still be wary about saying that any extant theory of scientific change was well confirmed. There are several reasons for this judgment: (a) most of the case studies to date attempt to apply some theory *in toto* to a particular episode, so that points of agreement and disagreement between the theory and the episode are difficult to disentangle; (b) thus far, virtually none of the 'tests' has attempted to assess the *relative adequacy* of *rival* theories of science, so that we simply do not know whether apparent confirmations of one theory might not also support theories that differ in very important respects; and (c) many of the case studies have been written by authors who have drawn chiefly on secondary or tertiary sources whose reliability is, at best, suspect.[2] In short, most of the avowed 'tests' would not pass muster on even the most tolerant view of robust experimental or quasi-experimental design.[3]

In the absence of evidence that can decisively adjudicate between extant theories of scientific change, the student of science can simply 'pay his money and take his choice'. Typically personal preference, institutional setting or philosophical inclination, rather than evidence, determines which theory a particular scholar espouses. In recent years the historical school of theorists of scientific change – so active in the 1960s and 1970s – has shown signs of losing momentum, largely because no serious attempt has been made to determine the extent to which relevant evidence supports the claims made by the various theories that have been proposed – an ironic situation in light of the importance this school attaches to the empirical grounding of theories of scientific change. Hence, a number of us decided that, if the historical school was to flourish, the time was ripe to take seriously the project of the testing of theories of scientific change.

But more was at stake than simply forwarding the program of the theorists of scientific change. During the past couple of decades, the conclusions of the historical school have become widely accepted as sound, particularly by outsiders to history and philosophy of science. The eagerness, not to say haste, with which these still-speculative accounts of science have been embraced indicates that a public discussion of their empirical credentials is urgently required. After all, the fact that scientists and social scientists find the theories plausible on first encounter is no guarantee of their correctness. Yet natural scientists have made ample use of these models when discussing recent developments in their specialties.[4] Their impact has been even more dramatically felt across the entire spectrum of the social sciences. Sociologists, political scientists, economists, policy specialists, geographers, anthropologists and marketers have pounced with glee on the theories of the historical school, in part because they find non-positivist approaches to the structure of science appealing and refreshing.[5]

The widespread belief (except among the narrow circle of specialists in history and philosophy of science) that Kuhn – or some other theorist – has delivered the last word on the nature and evolution of science results in large measure from the failure of historians and philosophers of science to point out in accessible publications (and not just in the final footnotes of technical monographs) the implications of historical research for theories of scientific change. In the absence of such readily-accessible critiques, others have taken at face value the claim that members of the historical school base their models on a serious examination of the historical record.

The fact that these theories have not been seriously tested against a broad range of scientific examples does not, of course, establish that they are wrong. Nor, even if they turn out to be seriously inadequate for one stage or discipline within science, does it follow that they are wholly invalid or useless. But there is a serious epistemic problem which needs to be faced squarely. Although the new theories of science are interesting and provocative, the evidence for their soundness is much too flimsy to warrant accepting them as a tentative, let alone a definitive, template for describing whole disciplines or sub-disciplines. Yet this is becoming increasingly widespread. Serious scholars straight-facedly ask "Does sociology count as a science?", where counting or not counting depends on compliance with the formulae which one or another of the theorists of scientific change have laid down. Others, even more adventurous, ask how to re-structure their disciplines so as to make them more like a science. All these projects rest on the assumption that recent accounts of the character of natural science are sound, or at least pretty close to the mark. Yet we have no reason whatsoever for making that assumption, either about Kuhn's theory or about its more recent rivals.

In this project we seek to establish a dialogue between the theoretical and the empirical investigation of science, parallel in many ways to the dialogue between the theoretical and experimental scientist. As scientific practice shows, the soundness of theoretical claims and the significance of empirical data can be determined only by constantly moving back and forth between theory on the one hand and observation and experiment on the other. We take the analogy between science and science studies seriously and see no reason why science itself should not be studied scientifically.[6] Empirical investigations do not exhaust the activities of the scientist but they do form an ineliminable counterpoint to the articulation of theories. Of late, the study of scientific change has been suffering a surfeit of the latter at the expense of the former. Testing is, of course, no simple matter, for we are simultaneously trying to refine our understanding of the methods of science itself.[7] But only such a self-referential and constantly iterated investigation is likely to give us an empirically well-grounded picture of the workings of science. In contrast to the individualistic pattern of scholarship typical of the humanities that has thus far predominated in science studies, this approach necessitates large-scale, collaborative work. There are all kinds of institutional barriers to a shift in emphasis of this kind but there is no reason to think that the serious study of science should not require the same kind of support that, say, is required for the study of business, the family, government or any of the other major societal institutions.[8]

3. A PROGRAM FOR TESTING

3.1. *Formulating Testable Theses*

From the outset we had decided to test the claims of the theorists of scientific change *comparatively* since we believe *all* theory appraisal is comparative. Thus the first step in the comparative empirical testing of theories of scientific change involved rendering the claims made by the various theories comparable. We and our co-authors addressed this task in a monograph 'Scientific Change: Philosophical Models and Historical Research' published in *Synthese* (Laudan *et al.*, 1986).

We set ourselves the following aims: (1) to extract from some major theories of scientific change specific empirical claims, claims which we called 'theses'; (2) to formulate these theses so as to make it relatively easy to compare the claims of different theorists; (3) to render the theses intelligible to readers unfamiliar with the technical language of philosophy or the specific terminology of different theorists; and (4) to express the theses in a 'neutral' vocabulary that did not presuppose the underlying assumptions of any of the theories.

Any rigorous attempt to make the claims of diverse theories comparable inevitably involves some violation of the holistic character of those theories. We recognized from the outset that we would not provide an account that would explicate all their different facets without incurring any conceptual loss. Consider, for instance, the normative language in which these theories were originally formulated. Consider too, the attention most theorists devote to conceptual philosophical issues such as the theory of meaning. Since such concerns about how science should function or about how certain philosophical positions should be articulated have no direct bearing on claims about how science *in fact* functions, we ignored them when extracting the theses.

Since we intended to explicate the empirical claims with as little distortion as possible, we on occasion included multiple readings of ambiguous texts rather than presenting only the one which appeared to be the most likely. We made no attempt either to emphasize or to hide internal inconsistencies, for our plan was not to judge the theories as wholes, but rather to facilitate the evaluation of their discrete parts (Laudan *et al.*, 1986: 144–145).

To render these descriptive claims accessible and comparable, we extracted them from the theoretical discussions in which they originally appeared and redescribed them in a common neutral language. Clearly, this entire project would have been fruitless had we, for example, described Kuhn's claims about science in terms of 'paradigms', 'normal science' and 'incommensurability' while describing Lakatos's in terms of 'research programmes', 'negative heuristics' and 'monster-barring stratagems'. Of course, the terms in which we have redescribed the central concepts of the original theories are not neutral in any absolute sense, and certainly not in the sense of being 'theory-free' concepts that presuppose nothing about science. But they are neutral with respect to the theories being scrutinized. The three terms used in the theses examined in this volume are as follows (Laudan *et al.*, 1986: 161–162):

empirical difficulty (anomaly): An empirical difficulty confronting a theory is an experimental result or observation for which the theory yields inaccurate predictions. Also included are cases in which a theory fails to make a prediction in circumstances in which it would be expected to do so.

guiding assumptions: Some theories are, over a substantial historical period, well established, relatively insulated against empirical refutation, wide-ranging in application, and highly influential across a variety of scientific fields. They include both substantive assumptions about the world and guidelines for theory construction and theory modification. Such theories are variously called paradigms (Kuhn), research programmes (Lakatos), and research traditions (Laudan). Although theorists of scientific change say conflicting things about such theories, they agree in their identification. Here such theories are called 'guiding assumptions' to distinguish them from more specific, narrower, more readily datable, and more directly testable theories.

revolution: The replacement, abrupt or gradual, of one set of guiding assumptions by another.

We presented the theses that emerged from this process of selection and abstraction in two different formats. Those derived from the works of the main theorists under consideration (viz., Kuhn, Feyerabend, Lakatos and Laudan) are listed seriatum in sections devoted to each individual. In addition, in order to facilitate comparative evaluation, we also regrouped the theses under a series of thematic headings, an arrangement that makes it possible to juxtapose claims made by the main authors and by several other theorists as well. The textual source for every thesis is given in both listings.

3.2. *Designing the Test*

The formulation of testable theses was only the first phase of this project, for until such time as specific theses had been tested against the evidence, it was impossible to say whether the testing project could produce credible and interesting results. We therefore organized a conference to compare a selected set of theses with specific historical evidence.

Deciding to test the theses was one thing, actually developing a methodology for testing them was quite another. The two groups who had thus far worked on the project, historians and philosophers, were unaccustomed to developing testing methodologies. And the testing procedure that seemed the obvious choice – the appeal to historical case studies – has long been the Cinderella of methods among social scientists. It would be disingenuous to suggest that we fully clarified our ideas before pressing on. Some decisions were already implicit in the way we had designed the *Synthese* monograph. On other issues, our understanding of the opportunities and limitations of testing grew as the project proceeded. We felt that a decision to solve all methodological problems before proceeding with testing would postpone, perhaps forever, the evaluation of the theories in question. Hence, we addressed the methodological challenge in tandem with the testing itself. During the academic year prior to the conference we consulted with several historians and philosophers of science about how testing might best be structured. Several conference sessions also focused on how to relate cases to theory and on the difficulties that attend the use of historical evidence to appraise philosophical theories. We make no apologies for our procedure. One aim of the exercise was to refine what Donald Campbell calls "strategies for an incomplete science", even as we were attempting to develop that science. For our readers, though, we feel that at this point some explanation and justification of our methodology will help explain why our conclusions should be taken seriously.

We deliberately chose a historical case-study (or more accurately a multi-case-study) method rather than experiments, surveys or eth-nomethodological studies (that is, the participant observation of scientists in the laboratory).[9] Our reasons for rejecting an experimental method are so obvious as to scarcely need explaining. Given our lack of control over the events that constitute scientific change and the impossibility of creating a situation in which we could manipulate such events, an experimental study of scientific change was out of the question. In this respect the study of science is on a par with certain physical sciences (such as astronomy) and with most of the social sciences.

Perhaps less immediately obvious is our rationale for rejecting certain popular techniques of the social sciences, particularly the use of surveys and of ethnomethodological techniques, both of which usually concentrate on contemporary science. Scientists, in particular, have questioned why we turn to history instead of to contemporary scientific practice. We would be the last to deny that much can be learnt from the study of contemporary science. But although the study of contemporary science has an important role to play, it has significant limitations.

For one thing, the apparently abundant additional evidence often turns out to be less impressive than it might appear at first sight. Responses to surveys or to interviews have to be carefully interpreted in the light of written records precisely because respondents' memories are notoriously unreliable. Laboratory studies are ill-suited for detecting the factors that determine the acceptance and rejection of theories (which constitute the subject of theories of scientific change) since these generally are not primarily the result of acts of negotiation at the laboratory bench. The analyst of laboratory discourse has no way of following the private deliberations that lead to belief changes; nor can he chart broad shifts in opinion across an entire discipline whose scientists are scattered across the globe. Once again the written record has to be taken into account. Indeed, in some cases we actually have better access to the thoughts of dead scientists than living ones as a result of the records they kept – think of the evidence available about the changing beliefs of a Newton, a Darwin or a Faraday over their lives. Furthermore, theories of scientific change deal with long-term as well as short-term shifts. Guiding assumptions often persist for centuries, revolutions more typically take decades (the Copernican revolution, the Newtonian revolution) than years. History alone can give us any access to those.

Even those who grant the soundness of the historical case-study method raise further questions that need to be addressed, even though we do not think any of them create crushing difficulties for the project.[10] Some skeptics doubt whether *anything* can be said at the general level about how science works. Perhaps the process is to complex, the role of adventitious circumstances too great, the variations from science to science or from period to period too large, to permit one to say anything which is both general and non-trivial. Science may not be a natural kind, and its features may be so various that no general 'laws' govern its history. Like other human institutions science presumably learns from its past and changes itself deeply in the process. Perhaps the sciences are 'expert systems' whose rationale and way of going on is so deeply tacit and implicit that it will forever elude being made explicit. Any or all of these things *may* be so. But such a conclusion would be warranted only *after*, not before, a sustained effort to identify the rule and rhythm of scientific change.

Nor do we presume, as some of our critics charge, that science is sharply demarcated from other intellectual pursuits or that there is a unity of method within science. Indeed our strong suspicion is that both these propositions are false. One excellent reason for empirical studies is to accumulate evidence that can decide such questions.

Some have asked why we chose to test theoretical claims already extant in the literature rather than to inspect the past of science in an attempt to generate some inductive generalizations. After all, the hypothetico-deductive method is not the only scientific method nor is it without serious problems. But our reading of the history of methodology suggests that hypothetical methods have been more successful in science than inductive ones. Add to that (a) our belief that there is a need to offer an informed public assessment of extant theories of change and (b) that we see no immediate prospect of inductive generalizations emerging from the historical scholarship of the last couple of decades and it will be clear why we decided on this as the best course of action. Many of the theorists' claims are couched as universals and, hence, even single case-studies (Duhem–Quine worries notwithstanding for the moment) can bear decisively on them. Further, given enough cases, evidence bearing on even statistical claims can be compounded.

More specifically, we are open-minded about the question of whether the concepts used by the theorists, particularly the concept of a guiding assumption, are sound and if sound, correctly specified. Those who looked at cases from the physical sciences had no trouble identifying units of change at the level of guiding assumptions. There was more confusion so far as the biological sciences were concerned and it is conceivable (though we still think unlikely) that biologists operate exclusively at the level of the specific theory most of the time. Again, we do not believe that the results we obtained are viciously theory-laden, for the authors of the case studies did not have to assume the very theories to be tested.

We were sustained in our belief that testing was possible by the analogy with science itself. For at least the last few hundred years, scientists have been successfully testing theories despite all the familiar ambiguities of testing. While granting that some theses are very tricky to test, there are others that seem readily amenable to empirical scrutiny. Claims such as: "Scientists are prepared to accept only those theories which embody all the results of earlier theories", or "Scientists systematically ignore negative evidence" seem straightforwardly testable. And we believe the results of the conference, described below, justify our optimism.

Following a call for papers, thirty-two historical case studies were presented at the conference, the topics addressed ranging in time from the sixteenth century to the present and in subject across the physical, biological and social sciences. Unfortunately it has not been possible to include all the papers from the conference in this volume. Faced with the unenviable task of having to determine which of the papers should be revised for publication, we selected case studies drawn from the physical sciences. This was not an arbitrary decision, for the theorists of scientific change likewise depended primarily on evidence drawn from the history of the physical sciences.

The conference was designed to test a limited set of thematically-grouped theses rather than the more comprehensive theories from which they were drawn. Participants were asked not to act as advocates for any one theory in its entirety and to focus on a limited number of the theses pertinent to the historical episode they knew best. We selected the short list of theses in light of their centrality to two topics that have dominated recent discussions of scientific change, namely the character and assessment of high-level guiding assumptions and of more specific theories.

4. FINDINGS

4.1. *The Prevailing Wisdom about Guiding Assumptions*

According to most theories of scientific development, broad-scale scientific change (i.e., a 'revolution') consists in the replacement of one set of guiding assumptions (paradigms/research programs/research traditions) by a quite different set. On one familiar interpretation (Kuhn's), a single set of guiding assumptions typically dominates a given science in any epoch. That dominant set directs inquiry towards particular phenomena and leads to the formulation of specific theories which particularize and render those guiding assumptions more precise and determinate. In the course of testing such theories, so the story continues, scientists will encounter situations which suggest an apparent failure of the guiding assumptions (Kuhn's category of 'unsolved puzzles'). Scientists will not construe them as failures of the guiding assumptions however. Instead, they will blame themselves rather than their tools and will, in effect, 'shelve' these apparent anomalies as problems for further investigation, especially if the reigning guiding assumptions still manage occasionally to make novel predictions (Lakatos).

Eventually though, these apparent anomalies accumulate in such numbers that scientists begin to raise serious doubts about the credentials of their guiding assumptions. This leads to a period of 'crisis' when, allegedly for the first time, scientists are willing actively to consider alternative sets of guiding assumptions. Once a new and rival set is on the scene, *inconclusive* debate takes place between advocates of the new and the old assumptions. Neither side fully communicates with the other since the rival sets are incommensurable. As a result, no telling arguments can be marshalled for either set of guiding assumptions. Instead of rational debate determining the outcome, generational change does: specifically, younger scientists tend to opt for the newer assumptions while their older colleagues typically soldier on, tinkering with the old set, hoping to bring them into line with experience. 'Converts' to the newer set are quite literally just that. The acceptance of a new set of guiding assumptions (according to Kuhn) (a) happens all of a piece, (b) involves a leap of faith, and (c) occurs abruptly and totally or not at all. If the new set of guiding assumptions is very effective at winning converts and in quick order everyone in the field comes to accept the new assumptions, we say that a scientific revolution has occurred. If not, everyone reverts to the old. But in either case, normal science (with one set of guiding assumptions enjoying hegemony in the field) quickly reasserts itself.

4.2. Theses about Guiding Assumptions

Virtually every element of the perspective just described is under scrutiny in the papers in this volume which address the following specific theses:[11]

Guiding assumptions 1: Acceptability
GA1. The acceptability of a set of guiding assumptions is judged largely on the basis of:

1.1 empirical accuracy; [3.1] (Kuhn, 1977: 323)
1.2 the success of its associated theories at solving problems; [3.3] (Laudan, 1977: 82, 124)
1.3 the success of its associated theories at making novel predictions; [3.4] (Lakatos, 1978: 185–186)
1.4 its ability to solve problems outside the domain of its initial success; [3.5] (Kuhn, 1970: 206, 208; 1977: 322; Lakatos, 1978: 39, 69)
1.5 its ability to make successful predictions using its central assumptions rather than assumptions invented for the purpose at hand. [3.6] (Lakatos, 1978: 185–186)

Guiding assumptions 2: Anomalies
GA2. When a set of guiding assumptions runs into empirical difficulties:

2.1 scientists believe this reflects adversely on their skill rather than on inadequacies in the guiding assumptions; [6.1] (Kuhn, 1970: 35, 80; 1977: 362–363)

2.2 scientists are prepared to leave the difficulties unresolved for years; [6.2] (Kuhn, 1970: 81; Fleck, 1979: 30–31)

2.3 scientists often refuse to change those assumptions; [6.3] (Kuhn, 1977: 288; Lakatos, 1978: 111, 126, 128)

2.4 scientists ignore the difficulties as long as the guiding assumptions continue to anticipate novel phenomena successfully; [6.4] (Lakatos, 1978: 111, 126)

2.5 scientists believe those difficulties become grounds for rejecting the guiding assumptions only if they persistently resist solution; [6.5] (Kuhn, 1970: 69; 1977, 272; Lakatos, 1978: 16, 72, 76, 86, 111)

2.6 scientists introduce hypotheses which are not testable in order to save the guiding assumptions; [6.6] (Lakatos, 1978: 126)

2.7 those difficulties become acute only if a rival theory explains them. [Kuhn (35): change in wording] (Kuhn, 1970: 84, 86–7)

Guiding assumptions 3. Innovation
GA3. New sets of guiding assumptions are introduced only when the adequacy of the prevailing set has already been brought into question. [11.1] (Kuhn, 1970: 67, 74–75, 97; 1977: 235; 1963: 349, 365)

Guiding assumptions 4. Revolutions
GA4. During a change in guiding assumptions (i.e., a scientific revolution):

4.1 scientists associated with rival guiding assumptions fail to communicate; [16.3] (Kuhn, 1970: 109, 147–149)

4.2 a few scientists accept the new guiding assumptions, which fosters rapid change, but resistance intensifies when change appears imminent; [16.4] (Feyerabend, 1981b: 146–147; Laudan, 1977: 137; Cohen, 1985: 35)

4.3 guiding assumptions change abruptly and totally; [16.5 & 16.6] (Kuhn, 1970: 92. 103, 106, 108–109, 147–149, 150–151; Stegmueller, 1978: 243)

4.4 the entire scientific community changes its allegiance to the new guiding assumptions; [16.7] (Kuhn, 1970: 166–167)

4.5 younger scientists are the first to shift and then conversion proceeds rapidly until only a few elderly holdouts exist. [15.2] (Kuhn, 1970: 158, 159; Toulmin, 1967: 469).

4.3. Findings about Guiding Assumptions

4.3.1. *How They Come to be Accepted* (GA1.1–GA1.5). Virtually all theorists of scientific change (except the most extreme cognitive relativists) agree that the acceptability of a set of guiding assumptions is dependent, in some way or other, on the empirical support for those guiding assumptions and their associated theories and models. Science, they say, would not be 'scientific' unless decisions about core assumptions were based, at least in part, on the available evidence. However, there has been substantial disagreement among theorists of scientific change about precisely what constitutes positive evidence for a set of guiding assumptions. Some (such as Kuhn and Laudan) have claimed that a set of guiding assumptions is positively supported (and thus worthy of acceptance) so long as its associated theories are able to explain a broad range of phenomena (GA1.2). Others (such as Popper and Lakatos) have held that the ability of a global theory to explain a broad range of problems is neither here nor there so far as empirical support is concerned. Rather, a set of guiding assumptions establishes its empirical credentials *only to the extent that it can successfully generate predictions of novel effects* (GA1.3).

This dispute has roots deep within the philosophy of science. More than a century ago, Whewell and Mill were already attempting to thrash it out. Mill's position was close to (GA1.2) in spirit, while Whewell adumbrated a third alternative, one which finds echoes in several recent theories of scientific change. Specifically, Whewell insisted that although a theory need not make surprising predictions, it must nonetheless exhibit its explanatory potential by allowing one to explain phenomena from domains very different from those it was initially introduced to explain (GA1.4). Such predictions need not be novel in the sense intended by Popper or Lakatos, since the support might derive from the explanation of things already known; but Whewell believed that we should be impressed with a global theory only insofar as its promised extensions of explanatory range had been delivered on, as a result of successful applications to domains different from those of its initial problematic. Finally, Lakatos and several of his followers have offered an even more demanding version of the successful prediction requirement. Specifically, they have said that a set of guiding assumptions, even if it makes surprising predictions successfully, should not be regarded as well confirmed unless those predictions are drawn from the guiding assumptions themselves rather than from collateral and *ad hoc* doctrines invented for the purposes at hand (GA1.5). Only in this way, they claim, can we avoid the ever-present possibility of *ad hocness*.

Several of the papers in this volume address these claims about positive evidence for guiding assumptions. Thus, Maurice Finocchiaro examines Galileo's lifelong flirtation with heliocentrism, his purpose being to understand what sorts of positive evidence Galileo sought before endorsing the Copernican worldview. He finds substantial evidence that Galileo insisted on empirical accuracy (GA1.1), general problem-solving success (GA1.2), and the ability of a theory to explain phenomena outside its initial domain of application (GA1.4). However, he finds no evidence that Galileo – at any stage of his career – demanded or expected that the Copernican hypothesis should be judged on the strength of its ability to make novel predictions. Quite to the contrary, Galileo was willing to accept and campaign for the Copernican system even though he knew of few, if any, confirmed surprising predictions that it made.

Moreover, Finocchiaro directs our attention to a salient fact about Galileo's involvement with the Copernican system which is probably of quite general applicability and which exhibits vividly how testing of specific theses of this sort can lead to a refinement of the analytic categories which we bring to an understanding of science. Specifically, Finocchiaro argues that Galileo did not begin his career as a full-fledged Copernican. On the contrary, his early stance towards the heliocentric system was neither one of rejection nor of acceptance. It was, instead, what Finocchiaro calls partial pursuit or what R. Laudan has elsewhere called 'entertainment' (Laudan, 1987). Even before his telescopic discoveries, Galileo was intrigued with the Copernican system, prepared to engage in thought experiments concerning its consequences, and puzzling over what sort of terrestrial and celestial physics might sustain it. Prior to about 1612, Galileo was pretty clearly a fellow-traveller of Copernicus, though not yet a fully-fledged Copernican.

Finocchiaro's account indicates that we need not invoke hankerings after Pythagorean mysticism nor a metaphysical preoccupation with the metaphorical centrality of the sun (of the sort Copernicus himself exhibited) to explain why Galileo seriously entertained the Copernican hypothesis. Rather, even at this early stage, Galileo was struck by the empirical success of his own system of mechanics, a system which harmonized quite neatly with Copernican astronomy. Thus, Galileo's early pursuit of the Copernican system appears to bear out the claim (GA1.2) that a set of guiding assumptions is judged largely on the basis of the success of its associated theories. Yet, as Finocchiaro tells the story, this sort of evidence was insufficient to convince Galileo that he should accept or believe the Copernican system. Only in the mid-1610s in the immediate aftermath of his telescopic researches did Galileo fully accept the Copernican system. What Galileo then had in hand was evidence which exhibited that the guiding assumptions of heliocentrism were empirically accurate (GA1.1), that its associated theories were successful problem-solvers (GA1.2) and that those assumptions could be utilized to solve problems outside the original domain (kinematic astronomy) which generated the assumptions (GA1.4). As noted, Finocchiaro finds *no* evidence to bear out the claim that Galileo ever judged the Copernican assumptions in light of their ability to generate surprising predictions (GA1.3). Moreover, Galileo's gradual, decade-long conversion to the heliocentric system gives the lie to those theories of scientific change which insist that when new guiding assumptions are accepted, the transition is quick and total (GA4.3).

The next detailed examination of this cluster of theses about positive support is James Hofmann's study of the appraisal of the guiding assumptions of Ampère's electrodynamics during the late 1810s and early 1820s. Through much of the later part of that period, Biot was pursuing a rival set. Ampère's task, in such circumstances, was to show the greater merits of his approach. According to Hofmann, he did so chiefly by making two sorts of argument: (1) an argument that his theory could explain all the known facts of electrodynamics, particularly those which lay outside the initial problem domain (GA1.4); and (2) an argument that his guiding assumptions (especially the assumption of centrally-directed forces) were simple and coherent. Despite the fact that Ampère's approach actually made some novel predictions, Hofmann claims that Ampère himself attached no particular epistemic weight to its ability to do so, thus contra-indicating (GA1.3). Like his disciple Savary, Ampère stressed the greater explanatory scope of his guiding assumptions over those of Biot rather than their differential ability to make novel predictions.

Henk Zandvoort pursues many of the same themes in his study of the history of nuclear magnetic resonance in the decades after World War II. He finds that the impressive empirical accuracy (GA1.1) of nuclear

magnetic resonance and the problem-solving success of its associated
theories (GA1.2) do much to explain the enthusiastic reception accorded that program during its early years. But Zandvoort's most
important finding is that surprising or novel predictions played no role in
the reception of nuclear magnetic resonance. We have already seen that
Finocchiaro and Hofmann found no evidence for Popper–Lakatos style
proclivities on the part of their respective subjects. As telling as these
cases are, they could be partially neutralized by a defender of the
novelty requirement who could claim that the sciences in which they
were working were still immature.[12] That move is not possible in the
case of nuclear magnetic resonance, one of the best known and most
influential post-war programs of research in theoretical chemistry and
physics. Yet in this case, as with Copernican astronomy and Ampèrean
electrical theory, scientists put no premium on the ability of a theory to
make surprising predictions. What is especially helpful about Zandvoort's analysis is the attention he pays to giving a clear and perspica-
cious formulation of various versions of the novelty requirement.
Indeed, his characterizations of several different forms of heuristic and
novel success represent precisely the sort of nuanced clarification
needed to move this project further ahead.

The last author to examine how guiding assumptions come to be
accepted is Richard Nunan. Nunan focusses not on the role of novel
predictions but rather on whether, as Laudan suggests, rival sets of
guiding assumptions can be appraised in terms of the problem-solving
success of their associated theories (GA1.2). Nunan holds that *individual* theories (with*in* a particular research program) typically are assessed
in terms of their comparative problem-solving success. However, when
it comes to the assessment of theories associated with rival sets of
guiding assumptions or to the assessment of those guiding assumptions
themselves, Nunan maintains that 'success' is decidedly in the eye of the
beholder. The particular historical case which leads to his skepticism
about the possibility of comparative judgements between rival sets of
guiding assumptions involves the rarely-studied encounter between the
advocates of plate tectonics and the partisans of the theory of an
expanding earth. Nunan argues that certain problems which advocates
of the expansionist approach regarded as glaring anomalies for plate
tectonics were regarded as minor by its proponents and *vice-versa*. The
scientist, faced with two rival sets of guiding assumptions, each of which
exhibits certain apparent successes and failures (of both an empirical

and a conceptual sort), has no objective decision criterion – in part because advocates of rival guiding assumptions weight problems in significantly different ways. Nunan thus disputes Laudan's claim that a non-question-begging technique for weighting the empirical successes and failures of rival sets of guiding assumptions can be formulated.

4.3.2. *The Impact of Empirical Difficulties* (GA2.1–GA2.7). The role of anomalies was initially given great emphasis by Karl Popper and his school. The well-known cornerstone of his philosophy of science was that all scientific doctrines (whether specific theories or what we are here calling guiding assumptions) which encounter refuting instances should be abandoned without further ado. Popper's claim flew in the face of an older doctrine, associated with Pierre Duhem, to the effect that global theories can always be retained in the face of apparent refutations by introducing suitable modifications in the auxiliary assumptions. In the early 1960s Kuhn entered the fray squarely on the side of Duhem (and Quine), insisting that scientists regarded apparent anomalies for guiding assumptions simply as unsolved puzzles, challenges that reflected more on the abilities of the experimentalist than on the core assumptions at stake (GA2.1; GA2.2). Kuhn went on to argue that unsolved puzzles become seriously threatening to a set of guiding assumptions only when several such anomalies accumulate and only if they remain unresolved for a long time (evidently he had in mind several decades) (GA2.5). Once recognized as genuinely anomalous for the reigning paradigm, however, such recalcitrant phenomena (according to Kuhn) generate a *crisis* in the scientific community, a state characterized by the proliferation of alternatives and the abandonment of the other patterns of normal science.

Imre Lakatos, writing in the early 1970s, agreed with Kuhn that scientists typically ignore apparent anomalies. But Lakatos held that they were entitled to do so (a) only insofar as those anomalies had already been anticipated in advance of their appearance, and (b) only so long as the guiding assumptions in question continued to generate novel predictions, at least some of which were successful (GA2.4). In the mid-1970s Larry Laudan argued that apparent anomalies became significant *only if* there existed a rival set of guiding assumptions which could give a plausible alternative account of the anomalous phenomena (GA2.7). Anomalies unaccounted-for by any extant set of guiding assumptions are wholly non-threatening. Paul Feyerabend held that many anomalies could not even be discovered in the absence of a rival set of guiding assumptions. These disagreements about the relationship between anomalies and guiding assumptions have ramifications for many other features of scientific change. For instance, if Laudan or Feyerabend is correct, then Kuhnian crises (which are, in Kuhn's view, a necessary prelude to scientific revolution) could never occur, since Laudan and Feyerabend believe that the existence of rival guiding assumptions occasions the recognition of the anomalies as serious. Kuhn, by contrast, maintains that the emergence of rivals only occurs *after* the acknowledgement of the existence of serious anomalies for the prevailing point of view. Thus, here we have a direct clash between quite divergent theories of scientific change, a clash that ought to be readily decidable.

Discussion of the role of anomalies in the assessment of guiding assumptions begins in this volume with the complementary papers of Brian Baigrie and Betty Jo Dobbs, which deal with reactions to the realization that (as Newton would have it) the fact that planets move according to Keplerian laws is incompatible with the existence of a resisting space-filling medium. In particular, Newton held that if such a medium existed, the motions of the planets should be perceptibly slowed and thus should be neither exactly elliptical nor sweep out equal areas in equal times. Yet the assumption of a space-filling aether of subtle matter was a lynch pin of 17th-century mechanical philosophy.

In the wake of the initial discovery of the anomaly, neither Newton nor the more traditional mechanists abandoned mechanical principles. This reluctance to change was not merely stubborn adherence to the familiar. On the contrary, as Dobbs shows, Newton himself felt that there was considerable experimental evidence for the existence of a subtle resisting fluid, even in apparent vacua. For instance, quite early on, Newton had performed experiments involving the motion of a pendulum bob in vacuo and in air. The degree of damping of the bob's motion in vacuo, while not as great as in air, was nonetheless striking – a fact that Newton, like other mechanical philosophers, had taken as evidence for the existence of resisting matter even in apparently empty space.

Newton's initial problem was to reconcile these experimental results (appearing to support a resisting aether in vacua) with the Keplerian motions (which tended to speak against the existence of resisting matter in space). As Dobbs points out, this conflict between two rival lines of experimental evidence prompted Newton to design more discriminating experiments on the motion of pendula, experiments which, once performed, seemed to bear out the claim that space is not full of resisting matter. Newton's initial cautious reaction to the discovery of this apparent anomaly thus seems to illustrate the claim that scientists often respond to anomaly by doubting their own experimental techniques rather than the assumptions under test (GA2.1). But this was only Newton's preliminary response. Once he had assured himself that the (now more sophisticated) evidence from pendula bore out the evidence from planetary kinematics, he developed two distinct lines of argument against this core assumption of the mechanical philosophy. With those arguments in mind, Newton abandoned this tenet of mechanistic physics, even though he had no fully-articulated replacement. By abandoning this particular guiding assumption in the absence of a clear alternative, Newton would appear to be violating Laudan's claim that anomalies become acute only if a rival can give a plausible account of them (GA2.7). Indeed, Newton's response to these anomalies runs against most of the claims here under test about the relations between guiding assumptions and anomalies (specifically GA2.2–GA2.5).

But, as Baigrie reminds us, Newton's response to the anomalous Kepler motions was neither the only possible, nor the most characteristic, reaction. For several decades after Newton's description of these anomalies in the *Principia*, most of Newton's fellow mechanical philosophers (Leibniz and Huygens, for instance) soldiered on with ethereal, vorticular physics. At first blush, this would appear to be evidence for Kuhn and Lakatos's claims about the dogmatic dismissal of the significance of apparent anomalies (GA2.2–GA2.5). But Baigrie goes to some pains to argue that such would be a misleading interpretation of the behavior of late-17th-century natural philosophers.

Indeed, to see the full force of Baigrie's argument, one needs to recognize that prevailing accounts of anomaly have been predicated on a false dilemma. They have assumed that the only possible responses to an anomaly involve either (as Popper would have it) immediate rejection of the doctrines generating the anomaly or (as Kuhn would have it) 'shelving' the anomaly and thereby ignoring indefinitely its potential for threatening one's core assumptions. Baigrie's analysis shows that neither of these descriptions fits what was going on in his case. As he describes it, prominent defenders of vortex theory (especially Huygens and Leibniz) neither denied the importance of Kepler motions nor saw them as the occasion for a wholesale abandonment of the mechanical philosophy. With Newton, they granted that the Kepler motions were genuinely anomalous for Descartes's version of vortex theory. They also acknowledged that the mechanistic approach now labored under a *burden of proof presumption*. Specifically, they appeared to believe that if a vorticular explanation for Keplerian motions were not found in quick order the mechanistic program would have to be abandoned. As Baigrie puts it, most influential mechanistic philosophers acknowledged the challenge posed by the Keplerian anomalies and granted the obligation to show how those motions were reconcilable with their guiding assumptions. Where (GA2.2) suggests that anomalies are merely ignored, Baigrie holds that scientists often actively engage the apparent anomalies facing their theories. Where (GA2.5) suggests that one need not be worried about anomalies until they have been around for a long time, Baigrie insists that many mechanical philosophers began responding to the problem of Kepler motion immediately on learning about Newton's arguments.

Another twist to the story Baigrie tells is that, in the interest of retaining their general commitment to mechanistic physics, many natural philosophers of the period were quite prepared to modify some of their core assumptions while retaining others. Huygens, for instance, was willing to allow the existence of vacuities within the aether if that would enable him to reconcile Keplerian motions with a general mechanical world-picture. This point is important because both Kuhn and Lakatos have denied that scientists ever permit tinkering with any of the component elements of their guiding assumptions. As Kuhn sees it, scientists hold a paradigm sacrosanct; to give up any of its key ingredients is to repudiate it. Lakatos similarly maintains that the 'hard cores' of research programs are retained at all costs. Responses to anomalies (insofar as Lakatos allows for any response to anomalies) involve tinkering, not with the core assumptions, but only with the ancillary ones. Baigrie's analysis makes it quite clear, however, that at least in this episode, scientists were willing to jettison some of their central theoretical commitments in the interest of reconciling the rest with the evidence.

One final point deserves to be made about Baigrie's treatment of this case. He maintains that the Keplerian motions became intensely threatening to the mechanistic approach because Newton had an alternative non-mechanical explanation for them (in terms of gravitational action-at-a-distance).[13] The ability of Newton's theory to handle Keplerian motions so readily and the apparent inability of mechanistic physics to handle them at all posed the real challenge for the mechanists. This would appear to bear out the general claim that anomalies become acute only if a rival already explains them (GA2.7).

The anomalies cluster is also central to William Bechtel's case from early-20th-century biochemistry. Among the assumptions guiding biochemical research at that time were: (1) that biological systems are chemical machines for extracting energy from food via standard chemical reactions; (2) that such reactions are catalyzed by enzymes; and (3) that the chemical reactions productive of energy follow identifiable linear pathways. Bechtel identifies two apparent anomalies for an early-20th-century model of alcoholic fermentation grounded on these guiding assumptions: the non-fermentation of methylglyoxal in living systems and the curious role of phosphates in the fermentation process. In examining the response to each of these anomalies, he notes that scientists, at least initially, disputed the basic experimental findings (GA2.1), and that they were prepared to hang on to their guiding assumptions despite apparent anomalies (GA2.2). But, like Baigrie, Bechtel adds an important qualification: to wit, that scientists did not ignore these anomalies or dismiss them as trivial or non-threatening. On the contrary, much work in the biochemistry of the period was directed toward finding some way of reconciling the anomalies with the preferred guiding assumptions. The scientists involved did not ignore the anomalies (contra-indicating GA2.4); nor did they give up their guiding assumptions even when the anomalies had persistently resisted solution (contra-indicating GA2.5); nor finally does Bechtel see any evidence that they introduced untestable hypotheses deliberately designed to save their guiding assumptions (contra-indicating GA2.6). The problems were finally resolved, interestingly, by the abandonment of the linearity assumption, while retaining the rest of the guiding assumptions. As in the Baigrie case, we have here clear evidence that sets of guiding assumptions can change on a piecemeal basis.

On another front, Bechtel reminds us that it is often difficult to know whether a particular anomaly indicts a set of guiding assumptions or only a specific theory based upon them. Our understanding of the differential response to anomaly will often depend on how we answer this key question. If one thinks of theories as specific *versions* of a set of guiding assumptions, then an anomaly for the theory presumably leaves open whether *other* versions based on those guiding assumptions would fare equally badly. If, on the other hand, one can show that experimental results impugn not merely the theory directly under test but a whole family of theories involving a certain fundamental claim, then the anomaly is far more general in scope. Thus, Newton saw the Keplerian motions as indicting not merely Descartes's version of vortex theory, but all mechanical, ethereal explanations of planetary motions. Others (such as Leibniz and Huygens) saw those same anomalies as indicating weaknesses in one particular family of ethereal explanations. In the case Bechtel analyses, most biochemists evidently agreed that the anomalies indicted specific theories or models of fermentation (e.g., Neuberg's) without raising fundamental challenges to the underlying guiding assumptions. (The irony here, of course, is that these anomalies were finally resolved only after significant modifications had been made in the guiding assumptions themselves.)

4.3.3. Innovation and Revolution (GA3 and GA4.1–4.5). Since discussions of the character of scientific revolutions have loomed large in the literature on scientific change ever since Whewell and Duhem, it is no surprise to find that more than half of the papers in this volume address one or another of the theses concerning scientific revolutions. The best known theory of scientific revolutions is, of course, Thomas Kuhn's, and most of the individual theses about revolutions under investigation here derive from his *Structure of Scientific Revolutions*. Among Kuhn's unorthodox views on this topic are: scientific revolutions involve paradigm change; a new paradigm is introduced only *after* its predecessor is in a state of crisis (GA3); scientists holding rival paradigms fail to communicate (GA4.1) because their respective paradigms are *incommensurable*. Paradigm change does not happen gradually or in a piecemeal fashion; on the contrary, it is abrupt and involves replacement of *all* the formerly prevalent guiding assumptions (GA4.3). It also involves a shift of allegiance on the part of the entire relevant scientific community (GA4.4). On the threshold of a revolution, it will be the younger scientists who shift to the new paradigm first (GA4.5). As the new paradigm wins more converts, resistance to it from advocates of the older paradigm will intensify (GA4.2). The results of this struggle will be either the triumph of the new paradigm (a scientific revolution) or the reassertion of the hegemony of the older paradigm.

The first paper directly germane to this set of claims is Carl Perrin's study of the chemical revolution, an episode which has been often used as an example by theorists of scientific change.[14] His is a rich paper and we shall touch here only on some of its more salient conclusions. Perrin's central concern is to ascertain whether, as Kuhn maintains, a change of guiding assumptions is a process that is both abrupt and total. In making this claim, Kuhn is saying several things at the same time. At the level of the individual scientist, he is maintaining that the acceptance of a new paradigm is like a gestalt switch; rather than taking place gradually and piece by piece, as it were, Kuhn believes that the individual scientist's acceptance of a new paradigm takes place all at once or not at all. To accept a new paradigm is to accept *all* its component elements, and to do so all at once. On the group or community level, these phrases take on a rather different meaning. Kuhn holds that the community-wide acceptance of a new paradigm, if it is to occur, will occur quite quickly – a matter of a few years at most. More importantly for Kuhn, a scientific revolution involves the conversion of virtually an entire community of scientists to a new way of working (GA4.4), with the possible exception of a few isolated and elderly hold-outs (GA4.5).

Perrin's research decisively refutes most of these claims. As his evidence clearly reveals, Lavoisier's new guiding assumptions did not emerge full-blown but piece by piece over a decade or more. Indeed, Lavoisier's earliest work on weight gain was conducted within a generally phlogistic framework. Lavoisier's new assumptions were both developed and publicized sequentially. By the time Lavoisier's chemistry had emerged as a mature alternative to its predecessors, Lavoisier's chemical contemporaries had already begun both to digest it and to respond to it. Some chemists accepted the new chemistry quickly; others slowly; some not at all. The slow and gradual conversion to Lavoisier's approach raises grave doubts about whether community-wide change is a rapid process (contra-GA4.3). More significantly, Perrin's analysis makes it clear that many chemists endorsed one or another of the core elements of Lavoisier's system before accepting others; contrary to another element of (GA4.3), acceptance of this new paradigm was not an 'all-or-nothing' affair. Furthermore, Perrin shows that there was extensive discussion between phlogistonists and oxygen theorists; these exchanges exhibit none of the incommensurability often assumed to make inter-paradigmatic communication exceedingly difficult (contra-GA4.1).[15] If Perrin's case is typical, it is clear that scientific revolutions are much more gradual and piece-meal affairs (both for individuals and for the community) than Kuhn claims.

Perrin does, however, find positive evidence for two of Kuhn's theses about scientific revolutions. Specifically, he notes that resistance to the new chemistry of Lavoisier intensified as that approach appeared on the verge of winning large numbers of converts (GA4.2), and he found that younger scientists tended to be more receptive to the new chemistry than their older counterparts (GA4.5).[16]

Quite a different revolution is the subject of Seymour Mauskopf's essay. He focusses on the early-19th-century reaction to the set of guiding assumptions Haüy formulated for crystallography. These had been very widely accepted in the late 18th century, dominating much of the work in crystallography. According to prevailing historiography, they fell into disrepute in the early 19th century. However, Mauskopf argues that Haüy's guiding assumptions, although indeed abandoned by German and British scientists, persisted in France through most of the 19th century. Not only did they persist, they were modified in very promising ways by a succession of French natural philosophers, especially Ampère. Mauskopf's point is that the abandonment of sets of guiding assumptions is not always (as is typically assumed) an affair that is discipline-wide, contra-indicating (GA4.3) and (GA4.4). There are, he maintains, national styles and traditions which can make it reasonable to retain a particular program of research, even when that program has been repudiated by most workers in the field. Mauskopf's analysis poses as many questions as it answers since the issue of national styles in science is a vexatious one. But for purposes of this volume, his central point is that, contrary to Kuhn's claim that scientific revolutions involve the entire scientific community in a discipline, pockets of resistance to such change are sometimes able to sustain themselves as 'scientists' (without being relegated to the lunatic fringe) for generations, particularly when such pockets are language or culture specific.

With John Nicholas' study of the early quantum theory we leap ahead almost a century and focus on the related issues of innovation and the *generation of crisis* in science. Kuhn has maintained that revolutionary new approaches never emerge in science until older approaches are already perceived as being in a state of crisis. That crisis state, in turn, depends (according to Kuhn) on the prior emergence of multiple anomalies confronting the older views. Nicholas's paper directly tackles the vexed question of the relationship between anomalies, crisis and the emergence of rival approaches. He believes that the history of early quantum mechanics completely inverts the Kuhnian sequence just described.

Planck has often been called a 'reluctant revolutionary', and Nicholas's study offers an intriguing justification for that label. Specifically, Nicholas argues (contra-Kuhn) that mere persistence through time is insufficient to promote an unsolved, but initially non-threatening, anomaly into a paradigm-defeating anomaly. More constructively, Nicholas shows that it was the ability of quantum theory to account for blackbody radiation that transformed the radiation distribution laws into decisive anomalies for *classical* theory. Of course, those peculiarities had been there all along and had been noted by many 19th-century physicists. But none of them regarded the classical theory's inability to give a coherent account of the radiation problem as a debilitating weakness; not, that is, until a rival (specifically Planck's theory, especially as elucidated by Einstein and Ehrenfest) emerged which gave a plausible account of those phenomena. If Nicholas's account is correct, it provides substantial grounds for abandoning the view, associated particularly with Kuhn but much more widely held, which insists that the emergence of novel theories is always preceded by the recognition of the crisis-inducing character of recalcitrant phenomena (contra-GA2.7). The introduction of a new set of guiding assumptions is typically a pre-condition for, rather than a sequel to, basic doubts being raised about the prevalent guiding assumptions (contra-GA3.1). As we have had occasion to observe earlier, anomaly-crisis-emergence of new theory is not (as Kuhn and others would have it) the normal sequence. Rather, as Feyerabend and Laudan have insisted, the type is new theory-recognition of anomaly-crisis.

4.3.4. *Summary.* The cases examined in this volume offer support for the following tentative, general conclusions concerning the role of guiding assumptions in the physical sciences.

With respect to *acceptability*:
- guiding assumptions are accepted, not on the strength of leaps of faith, but because they have exhibited an ability to generate theories which possess considerable demonstrated problem-solving success, particularly in domains outside those of their initial successes (GA1.2 and GA1.4);
- guiding assumptions are not generally expected to make successful surprising predictions before they are accepted (contra GA1.3 and GA1.5).

With respect to *empirical difficulties* (*anomalies*):
- when guiding assumptions encounter anomalies, those anomalies are neither ignored nor do they typically occasion immediate abandonment of the guiding assumptions. Rather they place those holding the guiding assumptions for which they are anomalous under an obligation to find some non-*ad-hoc* way of accommodating them within the framework (contra GA 2.2 and GA2.4);
- failing that, the guiding assumptions in question are apt to be abandoned, particularly if a rival set can successfully handle the anomalies (contra GA2.3).

With respect to *innovation* and *revolutions*:
- the generation of new sets of guiding assumptions is not deferred until the prevailing set is perceived as confronted by acute anomalies (contra GA 3);
- scientists advocating rival sets of guiding assumptions understand one another's work and arguments (contra GA 4.1);
- a scientific revolution does not consist in the wholesale replacement of one set of guiding assumptions by a wholly different set. Rather, scientists work their way gradually and in piecemeal fashion from one framework to the other so that guiding assumptions change neither rapidly nor in holistic bundles (contra GA4.3);
- scientific revolutions are not invariably or even usually community-wide (contra GA4.4).

About other theses there was either no consensus or an insufficient number of authors addressing them for even tentative conclusions to emerge.

4.4. *The Prevailing Wisdom about Theories*

There continues in the recent literature on scientific change – as in the 'positivist' literature that preceded it – an abiding concern with specific theories, their structure, their interrelations and their evidential base. In a sense, there is no *prevailing* wisdom about these aspects of theories. Instead, there are a variety of quite divergent accounts of the character of theories. At the risk of some over-simplification, one can divide them into two quite distinct families, the positivist and the post-positivist.

According to the older positivist account, scientific theories are judged entirely in terms of the empirical support they enjoy; metaphysical or other ideological factors should have no role to play in theory assessment. Moreover, on this view, theories must be abandoned immediately they encounter an anomalous or refuting instance (Popper, Mill). Good theories, on this view, are theories which, can explain everything explained by their predecessors and rivals (Popper, Reichenbach, Whewell). Beyond that, a theory is not really acceptable unless it has successfully made novel predictions (Popper, assorted Bayesians).

Alongside the positivist account of theories is a rival. According to this, theories are judged in part in relation to the larger world views with which they are associated (Kuhn, Feyerabend, Laudan, Lakatos). The most important empirical phenomena which support a theory (according to this perspective) are those instances which were formerly apparent counterexamples and those instances quite distinct from the problems the theory was designed to explain. Almost all post-positivist theorists reject the idea that successful theories must be able to solve all the problems solved by their rivals and predecessors.

4.5. Theses about Theories

Theories 1. Inter-theory relations
T1. When confronted with rival theories, scientists prefer a theory which:

    1.1. can solve some of the empirical difficulties confronting its rivals; [20.1] (Laudan, 1977: 18, 27; Kuhn, 1979: 148)

    1.2. can solve problems not solved by its predecessors; [20.4] (Kuhn, 1970: 97, 153; Lakatos, 1978: 66–70; Laudan, 1984: 100)

    1.3. can solve all the problems solved by its predecessors plus some new problems. [20.5] (Lakatos, 1978: 32)

Theories 2. Appraisal
T2. The appraisal of a theory:

    2.1. depends on its ability to turn apparent counter-examples into solved problems; [20.2] (Laudan, 1977: 31)

    2.2. depends on its ability to solve problems it was not invented to solve; [20.3] (Laudan, 1984: 100; Lakatos, 1978: 32)

    2.3. is based on the success of the guiding assumptions with which the theory is associated; [21.3] (Lakatos, 1978: 33–35, 47; Laudan, 1977: 107; Feyerabend, 1975: 181–182)

2.4. is based entirely on those phenomena gathered for the express purpose of testing the theory and which would not be recognized but for that theory; [21.4] (Lakatos, 1978: 38)

2.5. is based on phenomena which can be detected and measured without using assumptions drawn from the theory under evaluation; [21.6] (Laudan, 1977: 143)

2.6. is usually based on only a very few experiments, even when those experiments become the grounds for abandoning the theory; [21.7] (Lakatos, 1978: 38)

2.7. is sometime favorable even when scientists do not fully believe the theory (specifically when the theory shows a high rate of solving problems); [21.8] (Laudan, 1977: 22–23. 110, 119, 125)

2.8. is relative to prevailing doctrines of theory assessment and to rival theories in the field; [21.9] (Laudan 1977: 1–3, 124; 1984: 27–28)

2.9. occurs in circumstances in which scientists can usually give reasons for identifying certain problems as crucial for testing a theory; [21.10] (Laudan, 1984: 10)

2.10. depends on certain tests regarded as 'crucial' because their outcome permits a clear choice between contending theories; [21.11] (Laudan, 1984: 100)

2.11. depends on its ability to solve the largest number of empirical problems while generating the fewest important anomalies and conceptual difficulties. [120.6] (Laudan, 1977: 5, 13, 66, 68, 119).

4.6. Findings about Scientific Theories

4.6.1. Inter-Theory Relations (T1.1-T1.3). The problem of inter-theory relations, namely the logical and evidential relations between successive theories, forms a recurring theme in the theory of science. Of greatest importance is the question of the *evidential* linkages between successive or rival theories. Authors such as Popper and Lakatos have argued that later theories must be able to explain *all* the phenomena explained by their predecessors (as well as other phenomena besides), a claim that entails strong *evidential continuity* or cumulatively in science (T1.3). By contrast, Kuhn, Feyerabend and Laudan have stressed the non-cumulative character of theory change, both evidentially and logically.

The latter incline to the weaker thesis that a new theory, to acquire *prima facie* plausibility, must be able to solve *some* problems *not* solved by its predecessors (i.e., some of the anomalies that confound its predecessors), along with some new problems besides (T1.3).

The first paper to be discussed dealing with inter-theory relations is Henry Frankel's detailed study of the evaluation of plate tectonic revolution. As he shows, from very early on continental drift was able to explain a number of phenomena which were apparently anomalous for its rivals (e.g., coastline similarities, Permo-Carboniferous glaciation patterns, etc.). This fund of empirical successes for drift (and anomalies for its rivals) was supplemented in the 1950s by studies of polar wandering and, dramatically during the 1960s, by the confirmation of sea-floor spreading and the existence of transform faults – two hypotheses which were closely linked to the drift program. Frankel sees these features of drift theory as supporting the idea that scientists prefer a certain approach if it can solve some of the empirical difficulties confronting its rivals (T1.1). Equally important in Frankel's view is a theory's ability to solve some problems which, although perhaps not refutations of its predecessors, have nonetheless not yet been solved by them (T1.2). Frankel shows a host of such cases where drift theory is concerned.[17]

However, if Frankel's research bears out (T1.1) and (T1.2), he believes it speaks decisively against Popper and Lakatos's requirement that successful theories must solve all the problems mastered by their rivals and predecessors (contra-indicating T1.3). He notes, for instance, that the sea-floor spreading hypothesis (which was at the core of plate tectonic theory) had *no* account of mountain-building when it was accepted, despite the fact that it had been a central and (in most cases, successfully solved) problem for almost all sea-floor spreading's fixist predecessors. Where Frankel approaches the plate tectonic debates with a view to comparing mobilist with earlier fixist theories, Richard Nunan turns the tables and compares plate tectonic theory with one of its contemporary rivals, expansionist geophysics. Nunan maintains that both plate tectonics and expansionist theories are able to solve some empirical problems confronting their rivals (T1.1) and that each can solve some problems not solved by its predecessors (T1.2). As we shall see below, however, Nunan worries that if two conflicting theories can equally well satisfy such demands, we may have to acknowledge the possible nondecidability of the choice between them.

Hofmann considers (T1.1) in the context of the controversy between

Ampère and Biot about electrodynamics. He shows that one of the major perceived strengths of Ampère's theory was that it could readily handle many of the problems which had been obstacles for Biot's. Hones too claims to have positive evidence for (T1.1). There is a similar consensus (Nunan, Zandvoort and Frankel) for the idea that theories get high marks when they can solve problems unsolved by their predecessors (T1.2). But interestingly, none of our authors claims to have any evidence that theories typically solve *all* the problems confronting their rivals and predecessors (T1.3), despite the fact that several prominent theorists of science have made this feature a *sine qua non* for scientific progress.

4.6.2. *Theory Appraisal* (T2.1-T2.11). This is our largest cluster of theses and so it should be, since positivists and post-positivists have been of one mind in holding that the evaluation of scientific theories is the central epistemic problem in the philosophy of science. Some theses deal with questions of empirical support, e.g., the ability of a theory to handle its apparent counter-examples (T2.1), its ability to explain novel phenomena (T2.2), and the kind and range of experiments used to test a theory (T2.4, T2.5, T2.6, T2.10). Others deal with the relation between the theory being appraised and its related guiding assumptions (T2.3), the existence of universal standards of appraisal (T2.8), rival theories (T2.9, T2.10), and conceptual problems (T2.11).

The ability of a theory to turn apparent counter-examples into solved problems (T2.1) figures in several cases in this volume, with virtually all authors who considered this thesis finding it borne out. Thus, Henry Frankel looks at the conversion of anomalies into confirming instances in the acceptance of plate tectonics. When first proposed in the early-20th-century, drift theory faced certain apparent anomalies (e.g., absence of geodetic evidence for drift, apparent counterexamples to supposedly matching coastlines and fauna and flora). Frankel shows how, in case after case, drift theorists transformed counter-examples into confirming instances. In Frankel's view, these transformations helped to keep drift theory alive when it was unfashionable and then to make it predominant by the mid-1960s, thus supporting (T2.1). Similarly, Hofmann finds that Ampère's electrodynamic theory was given a significant epistemic boost when it exhibited its ability to handle what had previously been seriously anomalous phenomena. Finally, Michael Hones, in his investigation of electro-weak unification theory, shows how a significant anomaly (the so-called strangeness non-conserving events) facing the Weinberg–Salam model of gauge-invariant theory was transformed into a solved problem by Glashow's introduction of the concept of charm. Hones sees this transformation as one of the most significant features in the wide acceptance of the Weinberg–Salam model.

Just as scientists evidently expect theories to adapt successfully to their counter-instances (or at least give them strong marks when they do so), so – it has been claimed – must acceptable theories exhibit their fertility by solving problems quite different from those they were invented to explain (21.2 in the *Synthese* monograph). Frankel, for instance, points to several occasions when drift theory was able to do precisely that. (It is important to add, however, that those instances constitute – in Frankel's view – only a small subset of the empirical evidence for drift theory, whereas Popper and Lakatos would argue that such should be the *only* evidence germane to the assessment of drift theory.) Nunan agrees with Frankel's reading of the plate tectonic case here, although he goes to some pains to stress that plate tectonics was not the only geological theory in the mid-1960s exhibiting this feature.

Some theorists of scientific change (especially Laudan) have maintained that the appraisal of a theory is, in part, a function of the larger program of research or guiding assumptions with which it is associated (T2.3). On this view, theories linked with otherwise successful sets of guiding assumptions will be given more credence than theories associated with less successful guiding assumptions – even if their straightforward empirical support appears equivalent. Deborah Mayo, in her study of the history of Brownian motion, considers this claim in some detail. On the whole she finds (T2.3) counter-indicated by the history of ideas about Brownian motion. Prior to the 1870s, several theories were propounded to explain Brownian motion – theories which were not clearly associated with any particular program of research. Thus, (T2.3) can do nothing to illuminate the early disputes about theories concerning Brownian motion. What about the later stages of the controversy, when the fortunes of Brownian theories appear to have been closely linked up with the kinetic theory of gases? Even here, Mayo argues, the partisans and detractors of the kinetic theory did not regard it as particularly decisive to point to weaknesses in the guiding assumptions with which kinetic theory was associated (e.g., an apparent denial of the second law of thermodynamics). In her view, the participants in the later stages of the debate were quite clear about the guiding assumptions with which the kinetic theory was associated, and many had serious reservations about them. But on her reading, those doubts at no point played as significant a role as the direct experimental appraisal of kinetic theory.

Mayo also discusses Lakatos's well-known doctrine that a theory is appraised *exclusively* in light of phenomena expressly gathered to test *that* theory and which would be unrecognized but for that theory (T2.4). On the face of it, the use of Brownian motion to test the kinetic theory of gases is a direct refutation of Lakatos's claim. After all, the phenomenon of Brownian motion had been known for almost a century before it came to be regarded as relevant to the appraisal of statistical mechanics. Moreover, Einstein's theory of Brownian motion was specifically constructed to avoid certain mistaken predictions about Brownian motion associated with earlier versions of statistical mechanics. Thus, on most conventional accounts of the novelty of predictions (e.g., Lakatos's, Zahar's, Worrall's), (T2.4) would appear to be counter-indicated. Zandvoort's discussion of nuclear magnetic resonance leads to a similar conclusion about the merits of (T2.4).

By contrast, Alan Rocke's study of the early fortunes of benzene theory might appear to provide an impressive instance of (T2.4). As he shows, Kekulé's version of benzene theory quickly exhibited an ability to explain all manner of things that it was not invented to explain. In many cases, however, as Rocke observes, these explanations involved phenomena which were already well known prior to the development of the theory. As a result, although the ability of benzene theory to make surprising predictions (and retrodictions) weighed heavily in the minds of late-19th-century chemists, most of the telling confirmations of it did *not* arise from experimental situations created for the express purpose of testing benzene theory; nor were the data such that they could be recognized as important only in light of the benzene theory. Hence, all three of our authors' case studies raise grave doubts about (T2.4).

Thesis (T2.5) is, in a sense, the mirror image of (T2.4), insisting that the experimental evidence used to test a theory must not presuppose that theory. More importantly, (T2.5) is directed against the claim of writers like Kuhn and Feyerabend, who sometimes seem to insist that observations are not only 'theory-laden' but laden with the very theory which those observations are designed to test. If this thesis were true, it would be very difficult to see how a theory could ever accumulate positive evidence in the eyes of its critics. If every piece of evidence cited as positive for a theory had to be interpreted using the conceptual apparatus of that theory, and could not be formulated without it, then a scientist who rejected the theory would never be obliged to acknowledge the existence of evidence on its behalf. Thesis (T2.5) denies that strong sort of parasitism. Numerous authors here address this important thesis. Hones, for instance, claims that although the experiments carried out in recent quantum electrodynamics were often guided and motivated by specific theories under test, those experiments and their interpretation did *not* presuppose the theories under test. Similarly, Zandvoort claims that experim·nts carried out to test theories of nuclear magnetic resonance did not presuppose the conceptual framework of that theory. Diamond likewise finds positive support for (T2.5) in his study of the experiments done to test polywater theories. Mayo also chimes in here. She shows in considerable detail that Perrin's experimental tests of Einstein's version of kinetic theory do not presuppose that theory.

The lone dissenter from the consensus about (T2.5) is Ron Laymon, whose entire paper is devoted to a careful scrutiny of the Michelson–Morley experiments, his purpose being to determine whether they were conceptually dependent on one of the theories under test. Laymon argues that certain theories are so fundamental that the measurements needed to determine the values of the initial conditions in a test situation will presuppose either the very theory to be tested (or, even more troublingly, a denial of the theory under test). Laymon manifestly does not hold that all tests of all theories have this particular character, but he does believe that certain theories are so basic (e.g., theories of space and time) that any test of them will either presuppose them or a contrary to them. Consider the Michelson–Morley case in particular. Their experiment was designed to test the claim that light should travel through the aether with different apparent velocities in different directions. Michelson and Morley's interpretation of the workings of the interferometer they were using to detect aether drag rested on the assumption that the arms of the interferometer were of equal length. But precisely that assumption involves the denial of any aether-induced contraction in the arms! Hence, Michelson and Morley 'test' the contraction hypothesis by adopting an understanding of their measuring instruments which denies the contraction hypothesis. Laymon argues, however, that this procedure is perfectly harmless, *provided* one can

show (as can be shown for the Michelson-Morley hypothesis) that there
will be no computationally significant discrepancy between the observed
results, whether the instruments are interpreted using the theory under
test or one of its rivals. Laymon's message thus is that (T2.5) can be
violated just so long as its violation will not produce significantly
different results than would be obtained if it were honored. In short,
Laymon's piece is not an argument against (T2.5) *tout court* but an
important codicil to it: one need not obey the letter of (T2.5) as long as
one observes its spirit.

It is commonly believed that when it comes to experiments, the more
the better. Thesis (T2.6) explicitly denies that old saw, insisting that
scientists generally decide to accept or reject theories on the strength of
a small handful of experiments. Mayo wholeheartedly disagrees. As she
sees it, the history of the study of Brownian motion is a series of
thousands of experiments designed sequentially to test a variety of
theories. Perrin's efforts to test Einstein's claim that Brownian motion
showed no systematic effects and was genuinely random required, given
the nature of the statistical tests involved, a large body of experimental
data. Similarly, Rocke finds that the appraisal of the benzene theory was
based on literally thousands of experiments, most of which were essential to ascertain whether aromatic compounds as a class constituted
confirming or disconfirming evidence for Kekulé's theory, contra-
indicating (T2.6). Diamond rejects the plausibility of (T2.6) in light of
his study of polywater as well. Here again, full consensus emerges.

Both Rocke and Diamond tackle directly another claim often associated with Kuhn and Feyerabend, specifically the claim that scientists
working in the same field but with different sets of guiding assumptions
often disagree about the appropriate standards to bring to bear in
assessing theories. Laudan and Lakatos, by contrast, have insisted that
scientists in a field generally subscribe to common standards that are not
dependent on particular sets of guiding assumptions (T2.8). Rocke
notes that Kekulé's preferred method was hypothetico-deductive in
character. That method was widely accepted by his contemporaries,
including many who did not share his guiding assumptions. Diamond
likewise notes that pro- and anti-polywater chemists worked with substantially the same standards of chemical analysis. Similarly, Mayo
holds that there were no significant differences in methodology between
the advocates of the kinetic theory and its critics. On her view, they all
agreed about what would be required to establish statistically that the

motion of Brownian particles was random. In fact, none of our authors notes any significant divergences of standards which can be neatly mapped onto differences about underlying guiding assumptions. This is not to suggest that the standards of scientific research are fixed and unchanging. They pretty clearly are not. But there is no substantial evidence in this volume for the proposal that differences in standards of experimental adequacy or theoretical inference can be attributed in any simplistic fashion to divergent programmatic commitments.

One final pair of theses deserves brief mention. Early in this century, Pierre Duhem denied that there could be crucial experiments in physics, in the sense of logically-compelling experiments which simultaneously refuted one theory and proved another. The philosopher Willard Quine later made this doctrine the cornerstone of his epistemology of science. Few would now quarrel with the idea that theories cannot be definitively proved or disproved. But there remain many authors (e.g., Popper, Laudan) who hold that certain experiments can be relatively crucial, i.e., can provide compelling grounds for rejecting one theory and accepting another. That is the sense of (T2.9) and (T2.10). This doctrine, even in its weaker form, is denied by Kuhn and even by Lakatos (as least so far as the latter claims that the 'cruciality' of an experiment can be determined only long after it was performed). All the authors who examined this thesis find it plausible in the case of their chosen historical controversies. For instance, Rocke finds repeated instances where scientists agreed on the centrality and cruciality of certain experiments for benzene theory. Mayo, in her analysis of the reception of Perrin's experiments, finds them to have been perceived, and perceived without significant delay, as crucial by almost all parties. Hones holds, with Mayo and Rocke, that certain recent experiments in high-energy physics (e.g., the neutral-weak-current scattering events) have been regarded as crucial in guiding theory choices; however, Hones maintains that their crucial character is often not recognized until many years after the experiments were performed and reported.

4.6.3. *Summary*. What have we learned from the papers in this volume about theory appraisal and inter-theory relations from the papers? Several points of consensus are quite striking.

With respect to *inter-theory relations*, for instance, there is very broad agreement that:

- theories are expected to solve some of the problems unsolved by their rivals (T1.1) and predecessors (T1.2);
- theories are not required to solve all problems solved by their predecessors, and some loss of problem-solving power occurs in the transition from one theory to another (contra T1.3).

With respect to *theory appraisal*, there is strong agreement that:

- scientists award high marks to theories that can turn apparent counter-examples into confirming instances (T2.1);
- theories are expected to solve problems they were not invented to solve (T2.2);
- the test instances of a theory are not drawn exclusively from phenomena expressly gathered to test the theory (contra T2.4);
- (subject to Laymon's qualifications) the instruments and measurements used to test a theory generally do not presuppose the very theory under test (T2.5);
- theories are usually not tested only against a very small class of experiments (contra T2.6);
- the standards for assessing theories do not generally vary from one paradigm to another (T2.8);
- there are (relatively) crucial experiments that decide between rival theories (T2.9 and T2.10).

About several other theses in the appraisal cluster there was either no consensus or an insufficient number of authors addressing them for even tentative conclusions to be drawn.

5. CONCLUSION

Quite apart from the specific and concrete conclusions which the authors of our case studies have reached, there are important side-effects of studies of this sort, although to call them 'side-effects' is perhaps to underestimate their significance. In attempting to apply these various theses about scientific change, most authors have been forced to engage in a great deal of conceptual clarification and disambiguation of the specific doctrines under test.[18] Several of these theses can be interpreted in significantly different ways, and the tasks of interpretation and application produce a kind of hermeneutic refinement of the analytic categories one brings to the analysis of scientific change. We are still a long way from having an understanding of the dynamics of science which is theoretically integrated and empirically solid. But the picture that is emerging from studies of this sort arguably represents a dramatic improvement over the caricatures associated not only with positivism but also with the first generation of post-positivistic theories of science.

NOTES

[1]A select bibliography of these case studies can be found in Laudan *et al.* (1986) 218–223.

[2]See, for example, Brooke's criticism of a variety of case studies attempting to use Avagadro's hypothesis as a test of the theories of Duhem, Popper and Lakatos (Brooke, 1981). One may disagree with Brooke's generally negative construal of the case study method, while granting his criticisms of its sloppy application.

[3]To make matters worse, those few individual claims that have been subjected to well-designed tests using the history of the natural sciences as a data base – such as the thesis that new paradigms are likely to be accepted by younger scientists but rejected by older ones – have generally shown negative results (Hull, Tessner and Diamond, 1978).

[4]To take but a few examples, Alvin Weinberg takes for granted that Kuhn has shown that "scientific 'progress' is punctuated by 'revolutions' which break existing paradigms" (1970: 615), while James Conant adopts Kuhn's thesis that "scientific revolutions are quite unlike usual scientific progress" (1967: 318). Tuzo Wilson enthusiastically invokes Kuhn's analytic machinery in order to describe developments in geology in the 1960s (Wilson, 1968), while Hallam decides in favor of Lakatos (Hallam, 1983).

[5]Paul Samuelson (1981: 169), for instance, cites Kuhn to justify his belief that "a new and better paradigm may be motivated by criticisms of an old paradigm that are themselves confused and even wrong". H. L. Roitblatt claims that cognitivism in psychology represents a "paradigmatic revolution" (1982: 99); A. W. Coats (1969) propounds the applicability of Kuhn's ideas to economics. J. G. A. Pocock states that Kuhn's theory makes possible the understanding of intellectual revolutions in political science (Heyl, 1975: 65). Sheldon Wolin (1968) shares this view. Within American studies, Bruce Kuklick and Gene Wise speak for Kuhn's ideas (Heyl, 1975: 62). Robert Friedrichs (1970) attests that Kuhn's theory "made all the difference" in his intellectual development, and finds a variety of paradigms lurking in sociology. Michel De Vroey (1975) sees the rise of neoclassical economics as a straightforward Kuhnian revolution. As Heyl has correctly

observed, "Kuhn's ideas penetrated social scientific thought and practice with quite extraordinary speed" (1975: 62).

[6] Had the phrase "the science of science" not been co-opted by Derek Price for one particular branch of science studies (and one very different from this project) we would happily use that expression.

[7] Hence although we share David Bloor's commitment to studying science scientifically, we part company with him in the explication of what such a program amounts to (L. Laudan, 1984; D. Bloor, 1984).

[8] To many readers of this volume, particularly those from the natural and social sciences, it may seem self-evident that the claims of theorists of scientific change should be tested empirically. But for many of the historians and philosophers who commented on the project, the concept of 'testing' comprehensive theories of science caused considerable difficulty. The absence of sustained discussion of the methodology appropriate to developing an empirically-sound theory of science – indeed the skepticism about the necessity and very viability of such a task – indicates the embryonic state of our understanding of science.

[9] See Yin (1986) for a defense of the method in the social sciences.

[10] Some were expressed at the conference. Many can be found summarized in Nickles (1986).

[11] Note that the numbers in square brackets after the theses refer to the original numbering in the *Synthese* monograph and that the references indicate where the passages can be found in the works of the theorists of scientific change. The theses are also printed on the endpapers of this volume for ease of references.

[12] Both Kuhn and Lakatos distinguish science into two phases, an immature (or 'pre-paradigm') state and a mature state. Their claims about scientific change are meant to apply to the latter only, although neither clearly defines the boundary between them.

[13] Where Dobbs argued that Newton abandoned a resisting medium before he had a clearly developed alternative, Baigrie stresses that Newton's continental contemporaries perceived him as having a definite alternative in mind. The two positions can be readily reconciled if one realizes that Dobb's claims are about Newton's conceptual repertoire in the years immediately prior to *Principia*, whereas Baigrie is discussing the state of the debate in the decade(s) after *Principia*.

[14] For instance, it looms large in Kuhn's treatment of revolutions, as it had earlier in Conant's.

[15] See also Mayo's paper which argues against (GA 4.1).

[16] Interestingly, Diamond's paper in this volume reaches quite a different conclusion about the connection between chemists' age and their receptivity to innovation (GA 4.5).

[17] In Hofmann's study of the debate between Biot and Ampère about electrical theory, he notes a similar concurrence of (T1.1) and (T1.2).

[18] Similar processes go on, of course, in virtually any attempt to flesh out theoretical claims empirically.

REFERENCES

Bloor, David (1984), 'The Strengths of the Strong Program', in J. R. Brown (ed.) *Scientific Rationality: The Sociological Turn*, Dordrecht: D. Reidel.

Brooke, John (1981), 'Avagadro's Hypothesis and its Fate: A Case-Study in the Failure of Case-Studies', *History of Science* **19**, 235–273.

Coats, A. W. (1969), 'Is There a Structure of Scientific Revolutions in Economics?', *Kyklos* **22** (2), 289–294.

Cohen, I. B. (1985), *Revolution in Science*, Cambridge, Mass.: Harvard University Press.

Cook, Thomas and Campbell, Donald (1979), *Quasi-Experimentation: Design and Analysis for Field Settings*, Chicago: Rand McNally.

Conant, J. B. (1967), 'Scientific Principles and Moral Conduct', *American Scientist* **55** (3), 311–328.

De Vroey, Michael (1975), 'The Transition from Classical to Neoclassical Economics: A Scientific Revolution', *J. Economic Issues* **9** (3), 415–440.

Feyerabend, Paul (1978), *Against Method*, London: Verso.

Feyerabend, Paul (1981), *Rationalism, Realism and Scientific Method: Philosophical Papers, Volume I*. Cambridge: Cambridge University Press.

Fleck, Ludwig (1979), *Genesis and Development of a Scientific Fact*, Chicago: University of Chicago Press.

Friedrichs, Robert (1970), *A Sociology of Sociology*, New York: Free Press.

Hallam, A. (1983), *Great Geological Controversies*, Oxford: Oxford University Press.

Heyl, John (1975), 'Paradigms in Social Science', *Society* **12** (5), 61–67.

Hull, David, Tessner, P. and Diamond, A. (1978), 'Planck's Principle', *Science* **202**, 717–722.

Kuhn, Thomas 1970, *Structure of Scientific Revolutions*, 2nd enlarged edition, Chicago: University of Chicago Press.

Kuhn, Thomas (1977), *The Essential Tension*, Chicago: University of Chicago Press.

Lakatos, Imre (1978), *Philosophical Papers. Vol. I. The Methodology of Scientific Research Programmes*, Cambridge: Cambridge University Press.

Laudan, L. (1977), *Progress and Its Problems*, Berkeley: University of California Press.

Laudan, L. (1984a) *Science and Values*, Berkeley: University of California Press.

Laudan, L. (1984b) 'The Pseudo-Science of Science?', in J. R. Brown (ed.) *Scientific Rationality: The Sociological Turn*, Dordrecht: D. Reidel.

Laudan, L., Donovan, A., Laudan, R., Barker, P., Brown, H., Leplin, Jarrett, Thagard, Paul, and Wykstra, Steve (1986), 'Scientific Change: Philosophical Models and Historical Research', *Synthese* **69**, 141–223.

Laudan, R. (1987), 'The Rationality of Entertainment and Pursuit', in Marcello Pera and J. Pitt (eds.) *Rational Changes in Science*, Boston Studies in History and Philosophy of Science.

Nickles, Thomas (1986), 'Remarks on the Use of History as Evidence', *Synthese* **69**, 253–266.

Roitblatt, H. L. (1982), 'The Meaning of Repression in Animal Memory', *Behavioral and Brain Sciences* **5** (3), 353–372.

Samuelson, Paul (1981), 'Justice to the Australians', *Quart. J. Economics* **96** (1), 169–170.

Stegmueller, Wolfgang (1976), *The Structure and Dynamics of Theories*, New York: Springer-Verlag.

Toulmin, Stephen (1967), 'The Evolutionary Development of Natural Science', *American Scientist* **55**, 456–470.

Weinberg, Alvin (1970), 'The Axiology of Science', *American Scientist* **58** (6), 612–617.

Wilson, Tuzo (1968), 'Static or Mobile Earth: The Current Scientific Revolution', *Proc. Amer. Philosophical Soc.* **112**, 309–320.

Wolin, Sheldon (1968), 'Paradigms and Political Theories', in P. King and B. Parekh (eds.), *Politics and Experience*, Cambridge: Cambridge University Press.

Yin, Robert K. (1986), *Case Study Research: Design and Methods*, Beverly Hills: Sage.

PART II

CASE STUDIES

1. 17TH-CENTURY MECHANICS

MAURICE A. FINOCCHIARO*

GALILEO'S COPERNICANISM AND THE ACCEPTABILITY OF GUIDING ASSUMPTIONS

1. INTRODUCTION

The aim of this paper is to examine the problem of the appraisal of guiding assumptions on the basis of historical evidence about Galileo's Copernicanism. That is, I plan to determine whether this episode confirms or disconfirms (GA1.1) – (GA1.5). To be more exact, I shall be examining the evolution of Galileo's attitude toward Copernicanism in order to ascertain whether the acceptability of a set of guiding assumptions is judged largely on the basis of: (GA1.1) empirical accuracy, (GA1.2) the success of its associated theories at solving problems, (GA1.3) the success of its associated theories at making novel predictions, (GA1.4) its ability to solve problems outside the domain of its initial success, and (GA1.5) its ability to make successful predictions using its central assumptions rather than assumptions invented for the purpose at hand. Such testing will be impossible to carry out without a number of clarifications, to which we now turn.

2. CONCEPTUAL CONSIDERATIONS

Notice, to begin with, that even without any reference to the background literature, the theses as formulated above embody a distinction between theories and guiding assumptions. In fact, (GA1.2) and (GA1.3) speak of 'associated theories'. This corresponds to the standard distinction between sets of ideas that are more, and sets that are less, general from a methodological point of view; also more or less long-lasting from a historical point of view; and more or less elusive from the point of view of textual documentation. So we are obviously assuming that Copernicanism was a set of guiding assumptions rather than merely a theory.

It might seem that just as the semi-technical meaning of guiding assumptions involves us in presupposing the distinction just mentioned, the specific meaning of 'acceptability' would require us to presuppose another distinction, that between acceptance and pursuit. However, this

49

is not so. If one checks the passages of the works from which these theses are gleaned, it becomes clear not only that the distinction is not being uniformly maintained, but also that some of the appraisal criteria are being put forth explicitly as criteria for pursuit rather than acceptance. This means that our efforts will have to be two-sided. We cannot merely assume that Galileo exhibited toward Copernicanism the well-defined stance of acceptance, and then turn our attention to his works so as to determine what were the criteria on the basis of which he accepted it. Instead, we shall have to ascertain, and indeed to define and identify, the various cognitive stances he adopted, keeping an open mind as to the various possibilities mentioned above, while simultaneously identifying and defining his operative criteria. This second task is facilitated by the fact that our list of five criteria provides us with clues of what to look for. But this is not to say that we are assuming that his motivating reasons were necessarily one or more of the five types in this cluster. Obviously one must be open to the possibility that none of these five would apply. Thus, our assumption is merely that one or more of the five types of criteria may apply to the case of Galileo's Copernicanism. If so, we shall try to determine which one or which ones. If none apply, we shall try to define what his operative criteria are.

Before we proceed to that sort of testing, we need to clarify the individual meaning and the mutual interrelationships of the five criteria. The first, empirical accuracy, does not require any special comment here. Thesis (GA1.2) involves the success of associated theories at solving problems. The only thing to note here is the existence of conceptual and of anomalous problems, in addition to empirical problems. This criterion may be labeled, for short, 'general problem-solving success'. If need be, we may also speak of empirical problem-solving success, conceptual problem-solving success, and comparative problem-solving success. Thesis (GA1.3) speaks of the success of associated theories at making novel predictions, and may be called simply 'predictive novelty'. Thesis (GA1.4) deals with the ability of a set of guiding assumptions to solve problems outside the domain of its initial success. This would seem to be a special case of the general problem-solving success, previously mentioned, and may be labeled 'external problem-solving success'. The last factor (GA1.5) is the ability to make successful predictions using its central assumptions rather than assumptions invented for the purpose at hand. No special emphasis on novel predictions, as distinct from explanations, is meant here. So what is intended is the avoidance of ad hocness. I shall label it either 'anti-ad hocness', or less awkwardly, 'explanatory coherence'.

Now, what sorts of answers can we expect? That is, we cannot yet proceed to the testing, for to do so would be blind empiricism. The fact is that due to the various logical interrelationships among the five factors, only a certain range of answers may turn out to be warranted by the historical evidence. This interrelationship is in turn connected to the exact import of one other crucial term used in the formulation of the theses, namely the word 'largely'. This word obviously does not mean 'only'. Nor does it mean 'primarily'. Presumably it does not mean 'normally'. It probably means 'commonly' or 'frequently', or 'to a large extent'. But how large is large? Let us mention some clear examples of consistent subclusters, some clear examples of inconsistent ones, and some cases of unclear relationships. It is obvious that it may conceivably turn out that guiding assumptions are appraised largely on the basis of predictive novelty and of explanatory coherence, or largely on the basis of general problem-solving success and of external problem-solving success. This yields two consistent subclusters. I believe it is equally obvious that our historical testing could not possibly reveal that guiding assumptions are appraised largely on the basis of empirical accuracy and of problem-solving success of the nonempirical sort (e.g., involving conceptual problems). But here we must be careful. I do not think it would be impossible for guiding assumptions to be appraised in large measure on the basis of some combination of, for example, empirical accuracy plus conceptual problem-solving success. Such a combined large measure could not, however, consist of equal parts, because I believe that a large measure of one would exclude a large measure of the other; but we might have a combined large measure consisting of a large measure of empirical accuracy plus a small part of conceptual problem-solving success. I believe a similar tension exists between empirical accuracy and predictive novelty. In other words, our testing may be said to presuppose several assumptions, each derivable from one of these apparently inconsistent subclusters.

3. HISTORIOGRAPHICAL CONSIDERATIONS

These procedural considerations put us in a better position to conduct our testing. They give us a better idea of what it is we are looking for, what the various possibilities are, and what is unlikely or impossible to happen. However, they only relate to issues of a generally philosophical sort, which is only one side of our methodological situation, the other being the historical data. This material has to be collected before it can be analyzed for the purpose of our test. So a number of historical considerations are needed.

It might seem that, since we are dealing with only one scientist, it would be relatively easy to locate the evidence. However, Galileo's Copernicanism is a commitment that spans his entire career and, hence, we are faced with a large collection of writings, analogous to other historical episodes which may involve large numbers of scientists, but where the individual contributions are limited to such documents as a book or article. Yet in this particular case we are fortunate because of the existence of a critical edition of his collected works.[1]

The next thing called for, logically if not chronologically, is a critical survey of the secondary literature. This, of course, is itself another labyrinth.[2] The fruitful way to proceed here is to examine the secondary literature critically and judiciously with an eye toward collecting the relevant evidence, rather than toward accepting any one particular thesis or even any one approach. In speaking of the relevant evidence, I mean that which is directly relevant, for I am willing to admit that anything in Galileo's complete works is indirectly relevant. Here, of course, I am assuming that it is possible to distinguish direct from indirect evidence.

If we briefly reflect on how it is possible actually to make this distinction, I believe we can come to the identification of another assumption needed in our investigation, namely something about the content of Copernicanism. In fact, I believe one draws the distinction between direct and indirect evidence on the basis of another distinction, that between explicitness and implicitness. That is, directly relevant evidence is that in which Galileo discusses or mentions Copernicanism explicitly. But what does 'explicit' mean? There is a linguistic criterion for this, namely the occurrence of the word Copernicus and cognate terms, but that is obviously not enough. Equally explicit are writings where Galileo discusses any of the following topics: the state of rest or motion and the location in the universe of the earth and of the other heavenly bodies, the arrangement of such bodies, and their origin. Copernicanism must be assumed to be at least a set of propositions dealing with these topics. Such an assumption is, like all similar ones, subject to revision in the course of the investigation.

Finally, even when dealing with direct, explicit evidence, we need another distinction, namely between relatively significant and relatively insignificant. To some extent this can be based on an almost quantitative criterion, which involves a correlation between the length of explicit writings and their significance. This is sufficient to make us discard, as being relatively unimportant, documents like Galilean letters where the name Copernicus is only incidentally mentioned. Notice, however, that the correlation is far from perfect and that sometimes very brief references and statements are significant.[3]

When all these steps are taken and these assumptions made, one arrives at a set of documents, constituting the data base, to be searched, analyzed, and interpreted so as to obtain answers to the questions posed earlier. It would be useful to list and briefly identify all these documents, in order to have an overview of the evidence and as a way of orienting oneself. Moreover, this list might be taken as a documentary definition of the historical entity on the basis of which we are testing methodological theories of appraisal.

My plan for the analysis of these documents is not to give summaries of their content and interweave them into a narrative account of the development of Galileo's Copernicanism. This would be neither manageable nor especially incisive nor wholly relevant. Moreover, on a few occasions there would be no point in repeating work done elsewhere, either by others or by the present author.[4] So, in examining the data base, not only shall I emphasize, but indeed I shall limit myself to, points that strike me as especially relevant, controversial, or novel.

4. THE PERIOD OF TENTATIVE PURSUIT (BEFORE 1609)

Prior to his telescopic discoveries of 1609, Galileo's most explicit description of his attitude toward Copernicanism is in a letter to Mazzoni, dated 30 May 1597. Here we learn that he holds "the opinion of the Pythagoreans and of Copernicus about the motion and location of the earth . . . as much more probable than the other one of Aristotle and Ptolemy" (2: 198).[5] Galileo is silent about his motivation, and his rationale cannot be inferred from the rest of the letter.

Some clues about Galileo's rationale are found in the letter he wrote to Kepler a few months thereafter. Its interpretation, however, is complicated by the fact that this document has usually been translated and interpreted in a very questionable manner. I will therefore first give my proposed translation of the relevant portion of the letter. Referring to Kepler's book *Mysterium Cosmographicum*, a gift copy of which he had just received and for which he was thanking Kepler, Galileo said that he had read only the introduction and planned to read the rest. He then added:

Indeed I shall do it so much more gladly, inasmuch as I came to Copernicus's view many years ago, and from such a position I am devising the causes of many physical effects, causes which are without a doubt inexpressible in terms of the common hypothesis. I have worked out many reasons as well as refutations of arguments to the contrary, which however I have not dared to publish, afraid of the fate suffered by Copernicus himself, our teacher, who, though he earned immortal fame with some, nevertheless became the target of ridicule and scorn with innumerable others (such is the number of fools). I would certainly dare to publish my thoughts if there were more people like you; since there are not, I shall wait on the matter (10: 68).

I take the first sentence of this passage as an expression of Galileo being engaged in a program of physical research which fits well with Copernicanism, but not at all with the Aristotelian–Ptolemaic view. We might say that he is pursuing the physical side of Copernicanism. This must be a reference to the sort of theory of motion which Galileo had been working on and which is recorded in *De Motu* (1: 251–419). I see no reason for reading into this first sentence a statement of commitment, adherence, or acceptance *vis-à-vis* Copernicanism, based on the possession of an allegedly conclusive physical argument. This is not to deny that here we have an implicit reference to Galileo's geokinetic explanation of tides,[6] but we have to stress that he is saying he has many phenomena in mind and not just one, and that he is expressing certainty at most about their geostatic inexplicability and not about their geokinetic explicability. In fact, the rest of the passage, about Galileo's fear to publish, is an explicit comment on his view of the strength of his pro-Copernican arguments; he obviously does not think they are conclusive or even strong enough to convince someone who, unlike Kepler, is not already favorably inclined.

Now that we have a general description of the type and strength of Galileo's rationale, let us see whether we can identify some of these arguments more precisely. I have already noted that one of them is the well-known tidal argument. I believe another can be found in the section dealing with the earth being motionless in his *Treatise on the Sphere, or Cosmography*. This work deserves much more scholarly attention than it has received so far. It is well known, however, that this *Treatise*, which Galileo never published, was not meant to be an original contribution to knowledge, but rather a concise and elementary introduction to spherical astronomy for beginning students; and it is likely that he used it both in his university courses and his private tutoring. Moreover, it cannot be denied that the work is generally conservative, Aristotelian, and Ptolemaic in content. In regard to its form, however, it has some original and interesting elements. For example, I believe there is a large element of truth in Wallace's conclusion that "a careful reading reveals no epistemic commitment on Galileo's part to the Aristotelian-Ptolemaic system" (Wallace 1984b, p. 37). But this is not to say we have a commitment to the Copernican system. In fact, the topic is explicitly mentioned only once, and this brings us to the passage referred to above.

Unlike other sections, the one entitled "That the Earth Is Motionless" begins with an admission that this is a controversial question: "The present question is worthy of consideration since there has been no lack of very great philosophers and mathematicians who, deeming the earth to be a star, have given it motion" (2: 223). But he is quick to add: "Nevertheless, following the opinion of Aristotle and Ptolemy, we shall adduce those reasons why one may believe that it is completely motionless" (*ibid*). Then he goes on to argue why the earth cannot have rectilinear motion, and after that he takes up the question of its possible rotation, concerning which we have the following very important passage:

But, that it may move circularly has more verisimilitude, and therefore some have believed it; they have been moved principally by their considering it almost impossible that the whole universe except the earth should experience a rotation from east to west in the period of 24 hours, and hence they have believed that it is rather the earth which undergoes a rotation from west to east during such a time (*ibid.*).

He does not say whether or why he rejects this argument. Instead, in his usual impersonal style he adds that "having considered this opinion, Ptolemy argues as follows in order to destroy it" (*ibid.*), and then he goes on to summarize the traditional objections from falling bodies, birds, clouds, ships, and whirling.

The argument in the passage quoted above is important because it is obviously an appeal to Galileo's own doctrine of natural and neutral motions, which he had elaborated at least as early as 1590 in his *De Motu* (1: 304–307; cf. 2: 279). This is the theory according to which there are three basic types of motions: natural motion, or motion where the object approaches its natural place; forced motion, or motion where the object recedes from its natural place; and neutral motion, or motion which is neither natural nor forced. For Galileo, examples of neutral motions would be the rotation of a homogeneous sphere at the center of the universe or the rotation of a sphere around its center of gravity even if the sphere is located elsewhere. Finally, for him to start a body moving with neutral motion a force as small as you like is sufficient. Given this doctrine, it presumably follows that the earth would have axial rotation even in an otherwise Ptolemaic universe.

Galileo explicitly recognized this type of argument as one favorable to Copernicanism both in the *Sunspot Letters*, where he argued in support of solar rotation (5: 133–135), and in the *Dialogue* where he gave it at the beginning of Day II in support of terrestrial rotation (7: 146–147). So it is not unlikely that he had made the connection even in the pre-1609 period.

Tycho's letter to Galileo dated 4 May 1600 provides further support for our account, since Galileo never answered it. Here we should add that he also did not answer the letter dated 13 October 1597 which Kepler wrote after receiving his. Kepler was asking to be informed of Galileo's Copernican arguments and wanted him to make certain observations to try to detect stellar parallax. Although there were undoubtedly external factors that contributed to Galileo's lack of cooperation, the internal and methodological reasons can only be his lack of interest in pursuing the astronomical side of Copernicanism, and his dissatisfaction with the strength of his physical arguments.

We may summarize our results for the pre-telescopic period by saying that Galileo's attitude toward Copernicanism was one of partial and qualified pursuit. His judgement was based largely on factors other than empirical accuracy. More positively, it was based largely on the problem-solving success of its associated theories, primarily Galileo's theory of motion; this may suggest to us the ability to solve problems outside the domain of its initial success, but I am not sure this was the specific motivating factor for Galileo. There is no evidence that his judgement concerning Copernicanism was based on predictive novelty. In regard to explanatory coherence, no relevant evidence has been examined yet, but some will be found below, which bears a date of 1614 but refers to this earlier period.

5. THE PERIOD OF FULL-FLEDGED PURSUIT (1609–1616)

There is no question that the telescopic discoveries led Galileo to a significant reappraisal of Copernicanism. Less easy to ascertain is what the change was exactly, how sudden or slow it was, and what the motivating reasons were. The first piece of significant evidence here is *The Starry Messenger*. There are three relevant passages in this work.

One occurs in the author's dedication to the Grand Duke of Tuscany. Since this is often taken as the first published evidence that Galileo accepted Copernicanism,[7] I will provide my own translation of this passage. At this point in the dedication he is calling attention to the satellites of Jupiter and to the fact that he had named them Medicean stars, after the ruling family of Tuscany. The paragraph is as follows:

Behold, then, reserved to your famous name four stars, belonging not to the ordinary and less distinguished multitude of the fixed stars, but to the illustrious order of the wandering stars; like genuine children of Jupiter, they accomplish their orbital revolutions around this most noble star with mutually unequal motions and with marvelous speed; and at the same time, all together in common accord they also complete every twelve years great revolutions around the center of the world, certainly around the sun itself (3: 56).

The final clause of this passage is the expression often taken as evidence of unequivocal commitment to Copernicanism. But that is done by translating the original Latin to read, or by interpreting it to mean:[8] "around the center of the world, that is the sun". However, I believe it is more correct to translate it the way I have rendered it above, which embodies a certain ambiguity. That is, a body that follows Jupiter's orbit is enclosing the sun, and in that sense it is moving around it; and this is certain, with both sides of the controversy agreeing. But the clause could also mean "around the sun as center", and then we would have an unequivocal Copernicanism. At any rate, this would involve directly *only the heliocentric thesis*, and not necessarily the other guiding assumptions in the set. So, although there is no question that Galileo is expressing a favorable attitude toward Copernicanism, and that he is now involved with its astronomical part, we cannot provide a more precise definition of his stance.

This conclusion corresponds very well with the actual content of *The Starry Messenger* in which we find only two places where the status of Copernicanism is discussed explicitly. They both involve rebuttals of traditional anti-Copernican objections. Galileo feels he is now ready to answer the objection that the earth cannot be a planet because it is devoid of light; his telescopic discoveries about the optical properties of the moon enable him to say that the earth is not essentially different from the moon in this respect (3: 75). On the other hand, the ability of Jupiter's satellites to keep up with Jupiter as it revolves in its orbit allows Galileo to refute the objection that the moon clearly orbits the earth and so would be left behind in a Copernican universe (3: 95). Of course, these refutations of objections do not prove Copernicanism; they merely strengthen it. That is why the attitude expressed by Galileo toward it is such that it may be described as a slightly higher degree of pursuit, acceptability, or favorable appraisal than any expressed earlier.

The next significant document for this period is the letter to Giuliano de' Medici, Tuscan ambassador to Prague, dated 1 January 1611. Its main purpose was to decipher the anagram Galileo had sent him in an earlier letter. When properly transposed the anagram stated that Venus shows phases like the moon, a phenomenon he had been able to observe with the help of the telescope. The attitude he displays now is not, as some scholars have alleged,[9] complete acceptance of the Copernican system, but rather acceptance of two specific theses in it. This is as clear as his emphasis on empirical accuracy:

From this marvelous observation we have sensible and certain demonstration of two great questions, which so far have been debated by the greatest minds of the world: one is that planets are all dark (since the same thing happens with Mercury as with Venus); the other is that Venus necessarily revolves around the sun, as do also Mercury and all other planets (11: 11–12).

He goes on to describe the changes in his attitude as one from belief without to one of belief with empirical proof: "this had indeed been believed by the Pythagoreans, Copernicus, Kepler, and myself, but not sensibly proved, as done now for Venus and Mercury" (*ibid.*). I am not sure this transition can be equated with that from pursuit to acceptance. More likely, it is a change from one kind of acceptance to another, the difference being the grounds for the acceptance. In other words, we have a development from a situation in which the acceptability of some guiding assumptions was judged on the basis of factors other than empirical accuracy, to a situation in which it was judged on the basis of empirical accuracy itself. This is not to say, however, that the state of belief or nonempirical acceptance can be extended back to before the telescope. In fact, it is important to notice that, besides distinguishing between these two attitudes, Galileo makes a still finer discrimination between his previous state of belief without empirical proof and his still earlier stage of nonbelief, pursuit, or exploration. This is suggested when he goes on to congratulate Kepler and other Copernicans, but does not include himself:

thus, Mr. Kepler and the other Copernicans will have reason to be proud of their having believed and philosophized correctly, although they have been, and will continue to be, regarded by all bookish philosophers as incompetents and almost fools (*ibid.*).

The next important milestone in our story is a letter to Prince Cesi; its importance lies in what it reveals about simplicity. This is a criterion that deserves discussion, despite the fact that it is not explicitly mentioned in our canonical list, both because simplicity provides the basis for a common interpretation of Galileo's rationale (e.g., Wisan 1984; but cf. Finocchiaro 1985) and because such an interpretation constitutes an obvious alternative to accounts stressing the criteria in our canonical list.

On 20 June 1612 Cesi had written to Galileo expressing his attraction to Copernicanism because of its doing away with epicycles and eccentrics and asking his opinion on whether one was obliged to use these in the case of the terrestrial and lunar orbits, given the periodic changes in distance between the earth and the sun and between the earth and the moon (11: 332–333). Ten days later, Galileo replied that

we must not desire that nature should accommodate herself to what seems better arranged and ordered to us, rather it is appropriate that we should accommodate our intellect to what she has done, certain that this and nothing else is the best (11: 344).

He then goes on to apply this principle by arguing that if by epicycles is meant orbits not encompassing the earth, then we must admit their reality (e.g., the revolutions of Jupiter's satellites around the planet and the orbits of Venus and Mercury around the sun). And if eccentrics are defined relative to the earth, then the orbit of Mars clearly encompasses the earth's orbit, since the telescope reveals that its apparent magnitude is 60 times greater at certain places of its orbit than at others. This explicit theoretical articulation of the limitations of simplicity, together with this particular example of actual usage on his part, are completely in accordance with Galileo's behavior during the pre-telescopic period; at that time, as we have seen, he did not attach enough weight to the greater simplicity of the Copernican system either to accept it or to ground his pursuit on that. This is not to say, of course, that he completely disregarded the criterion of simplicity, but only that he did not attach significant or decisive weight to it.[10]

The *Sunspot Letters* were written in 1612 and published in 1613. As is well known, they contain the strongest endorsement of Copernicanism that Galileo ever published. What is less well known and seldom discussed is the exact form this endorsement takes and the context in which it was presented. It occurs on the last page of the last letter, written 1 December 1612. He finishes with the topic of sunspots about three pages from the end, but goes on to make several very daring and rather precise predictions about the periodic reappearance and re-disappearance of Saturn's rings up to the summer of 1615. He does not, however, reveal the conjecture on which he based these bold predictions, but promises to do so later, after events would confirm or disconfirm him. He concludes:

But if I do not yet doubt their returns, I have reservations about the other particular events, now merely based on probable conjecture; yet whether things fall out just this way or some other, I tell you that this planet also, perhaps no less than horned Venus, agrees admirably with the great Copernican system on which propitious winds now universally are seen to blow to direct us with so bright a guide that little [reason] remains to fear shadows or crosswinds (translated in Drake 1978, p. 198).[11]

Two main claims in the last part of this quotation deserve attention. The first is that *perhaps* Saturn's behavior too confirms Copernicanism. Despite the fact that Galileo never did reveal the theoretical basis of his prediction and the connection with Copernicanism, and despite the fact that, remains a puzzle for scholars,[12] the judgement is obviously based on the criterion of empirical accuracy. The second claim is that now he thinks all evidence is pointing toward Copernicanism and he seems to have little doubt about its correctness. Although Galileo does not explicitly include sunspots in this evidence, the connection is so obvious that it can easily be attributed to him. As he will argue later in Day I of the *Dialogue* (7: 76–80), sunspots contribute to the empirical undermining of the earth-heaven dichotomy, and thus to the strengthening of Copernicanism, by removing the corresponding objection and by supporting the corresponding element of the Copernican system.

In the Postscript that was added to the published version of the *Sunspot Letters* there is another important clue that at about this time Galileo was finding another piece of evidence which he believed could only be explained in terms of the earth's orbital revolution. The phenomenon involved the eclipsing of Jupiter's satellites and the variations in the duration of these eclipses. The details of the argument were never written up by Galileo and they are too technical to report here. Stillman Drake deserves the credit for having tried to identify the relevant documentary evidence and for piecing together the main points (1983, pp. xix, 133–135). However, Galileo's claim in the Postscript is as clear and unambiguous as one could wish:

But a more wonderful cause of the hiding of any of these is that which arises from various eclipses to which they are subject, thanks to the differing directions of the cone of shadow of Jupiter's body – which phenomenon, I confess to you, gave me no little trouble before its cause occurred to me. Such eclipses are sometimes of long and sometimes of short duration, and sometimes are invisible to us. These differences come about from the annual movement of the earth, from differing latitudes of Jupiter, and from the eclipsed planet's being near to or farther from Jupiter, as you shall hear in more detail at the proper time (translated in Drake, 1978, p. 208; cf. 5:248).

The evolution of Galileo's Copernicanism comes to a climax in 1614, as we can see in his letter to Baliani dated March 12. In it we find for the first time a genuine expression of certainty, together with a summary of his reasons, as well as a reasoned rejection of Tycho's theory. The crucial passage is the following:

As regards Copernicus's opinion, I really hold it as certain, and not only because of the observations of Venus, of sunspots, and of the Medicean planets, but because of his other reasons, and because of many other particular reasons of mine, which seem conclusive to me . . . In Tycho's opinion I still find all the very great difficulties which make me abandon Ptolemy, whereas in Copernicus I find nothing which gives me the least scruple, and least of all the objections which Tycho makes to the earth's motion in certain letters of his (12: 34–35).

There is no question that we have here an endorsement of Copernicanism stronger than any we have seen up to this date and than we find anywhere else subsequently. And it is clear that Galileo uses labels that are epistemically loaded: 'certain' to characterize the position and 'conclusive' to describe the supporting arguments. Nevertheless, I do not think we have here a qualitative jump from what came immediately before; it is a change of degree. Moreover, the notions of certainty and conclusiveness do not seem to me to be discrete, but to admit of degrees within them. Finally, Galileo's words must be balanced against the following.

First, this certainty and conclusiveness may be no more literally true than is Galileo's assertion in the same passage that he finds nothing in Copernicus which gives him 'the least scruple'; for as the later *Dialogue* (7: 424–426) clearly explains, Galileo never accepted Copernicus's third motion, according to which the terrestrial axis was supposed to rotate with an annual period in order to compensate for the orbital revolution; and, as the *Dialogue* (7: 404–416) also shows, he never refuted the objection from stellar parallax, but rather accepted the difficulty and suggested a way of testing for it. Second, this is a private letter, and there is no analogous expression in any published work. Third, Galileo was not at this time writing the work on the 'System of the World' which he had promised in *The Starry Messenger*. Although there were many causes for this delay, including external ones, one contributing factor may have been that he did not yet have all the arguments and evidence required to be really sure, as he was claiming in this letter; hence the crucial words here are perhaps merely that – words.

Despite all these provisos, the endorsement neither can nor should be ignored. So let us look at the motivating reasons. Notice that there are three groups of reasons: the observational ones depending on the telescope and involving the phases of Venus, sunspots, and Jupiter's satellites; Copernicus's own reasons; and Galileo's own "other particular reasons". The first group, the chronologically latest, had been accumulating during the several years since his first use of the telescope, and they seem to have caused a qualitative change in Galileo's attitude, from something like pursuit to something like acceptance. Clearly they are intimately related to the criterion of empirical accuracy. The second group of reasons was never, not even in the *Dialogue*, discussed by Galileo in any great detail; in regard to their methodological character, however, we do know, as we saw above, that they were not for Galileo simplicity considerations, at least not in any straighforward sense. I believe that they reduce primarily to the criterion of explanatory coherence or anti-ad-hocness, though this cannot be elaborated here.[13] The important point for us now is that Galileo was not insensitive to this sort of consideration, however insufficient he may have regarded it as a basis of acceptance or even as a sole basis of pursuit. It should also be noted, in view of his qualms about the principle of simplicity simply understood, that this conception of Copernicus's own reasons is also different from the simplicity interpretation. Finally, Galileo also had a third group of reasons, and these he had had for a very long time. Collectively they reduced, as we have seen already, to general problem-solving success and perhaps also to external problem-solving success; that is, to the success of his theory of motion, a theory quite compatible with Copernicanism and quite at odds with the Aristotelian-Ptolemaic system, and to the fruitfulness of Copernicanism in a field – physics – outside the domain of its original success. These reasons included such arguments as the one from tides and the one from neutral motion. Since Galileo claims he has 'many' such arguments, it is interesting to speculate what else he had in mind.

One clue to what they might be is in Galileo's remark about the Tychonic system. He says he has the same basic difficulties with it as with the Ptolemaic system. These can only be problems stemming from his theory of motion, or what today we would call dynamical difficulties. They are sketched and summarized at the beginning of Day II of the *Dialogue* (7: 139–50). We cannot elaborate them here, but two points are relevant to our analysis. These objections apply with equal force to both Tychonic and Ptolemaic versions of the geostatic system, and they are explicitly labeled as probable and not conclusive by Galileo. This helps us to resolve one last question about the present passage.

Galileo asserts that his reasons are "conclusive". The actual expression ("which seem conclusive to me") and punctuation he uses is ambiguous, since the conclusiveness could be referring to either his third group of reasons individually, his third group of reasons as a whole, or all his three groups of reasons collectively. I believe that the last one is meant, as suggested by the just-mentioned passage in the *Dialogue*. But one difficulty now remains. Since the tidal argument is obviously included in the third group, as one of Galileo's physical reasons, and since this argument is not discussed in the same passage at the beginning of Day II but in Day IV of the *Dialogue*, could he not here be attributing conclusiveness to the tidal argument? It is conceivable that he could, but in fact he is not. I have argued elsewhere (1980, pp. 6–24; and 1986, pp. 246–270) that the tidal argument is presented in the *Dialogue* as an inductive, probable, hypothetical, non-necessitating argument, but my interpretation of that text is insufficient here, since it is well known that the *Dialogue* was written after the anti-Copernican decree of 1616 and, hence, Galileo had external motives for such a presentation. So we need to examine the tidal argument as it exists in the version written by him in January 1616, two months before the Decree of the Index. At any rate, this so-called "Discourse on the Tides" is one of the relevant documents in its own right.

A well-kept secret about the "Discourse on the Tides" is that it is not just about tides but also about winds. In other words, we have not one but two arguments for the earth's motion, the first based on the tides, the second on prevailing easterly winds. This immediately suggests that neither one is believed to be absolutely conclusive, for the wind argument would be superfluous if the tidal argument had that quality, and vice-versa.

This suggestion is reinforced by frequent explicit remarks on Galileo's part. For example, after listing several possible causes of the motion of water in general, he introduces the connection between tides and the earth's motion with the following words:

When I examine these and other facts pertaining to this cause of motion considered last, I would be greatly inclined to agree that the cause of tides could reside in some motion of the basins containing sea water; thus, attributing some motion to the terrestial globe, the movements of the sea might originate from it. If this did not account for all particular things we sensibly see in the tides, it would thus be giving a sign of not being an adequate cause of the effect; similarly, if it does account for everything, it will give us an indication of being its proper cause, or at least of being more probable than any other one advanced till now (5: 381).

My conclusion is that there is no doubt the certainty expressed by Galileo in the letter to Baliani was an inductive, practical, not absolute kind of certainty, based on the practical conclusiveness of all the arguments taken together, physical, telescopic, and Copernicus's original ones. Certainly he did not think that any one individual argument or piece of evidence was absolutely conclusive.

6. CONCLUSION

Our critical examination of the evidence is far from complete yet. So far we have not even touched upon any of the documents pertaining to the theological controversy and the Inquisition proceedings. Far from being irrelevant to the present inquiry in general, they constitute a classic test case for a specific methodological thesis, namely whether or not the acceptability of a set of guiding assumptions is judged largely on the basis of its relation to religious beliefs. However, this falls outside the scope of this paper.

In conclusion, I have examined the evolution of Galileo's Copernicanism in order to determine whether he judged its acceptability largely on the basis of empirical accuracy, general problem-solving success, predictive novelty, external problem-solving success, explanatory coherence, or simplicity. The result has been that his various judgements of the acceptability of the Copernican set of guiding assumptions fall primarily into three periods: the pre-telescopic stage, when his cognitive stance may be described as partial pursuit; the full-blown middle period from 1609 to 1616, which constitutes a qualitative change from the preceding one, which may be described as either full-fledged pursuit or partial acceptance and which contains varying and increasing degrees of endorsement or commitment up to a stance which may be labeled tentative or practical acceptance; and a problematic post-1616 stage involving his relationship with the Church, which I have not discussed since it raises issues of a completely different sort that fall outside the scope of the present investigation. During the first period, he judged Copernicanism largely on the basis of its general and external problem-solving success in the physics of motion and its explanatory coherence in the astronomical field; during the second stage, he judged it largely on the basis of these criteria *plus* empirical accuracy; and after 1616, he judged it largely on the basis of these four criteria *plus* its relationship to his religious beliefs. At no time did he judge its acceptability largely on the basis of predictive novelty or of simplicity.

Finally, I should mention one other thesis for which my investigation provides an overwhelming amount of relevant evidence, although admittedly I did not explicitly or deliberately devise my test with it in mind. That is thesis (GA2.5) that during a change in guiding assumptions (i.e., during a scientific revolution), these assumption change abruptly and totally. My evidence from Galileo's Copernicanism shows that this is not so, but rather that his attitude toward it changed slowly, gradually and in a piecemeal fashion. This conclusion is, I believe, so important that one cannot fail to mention it despite the fact that it is primarily an unintended consequence of my test.[14]

NOTES

[*] This paper was completed in 1986–1987, during the author's sabbatical leave and tenure as Barrick Distinguished Scholar at the University of Nevada, Las Vegas; this support is hereby gratefully acknowledged.
¹Galilei (1890–1909). This is called either the Favaro edition, after the editor, or the "National Edition" (*Edizione Nazionale*). Its chief lacunas are its failure to include all of Galileo's notes on motion, concerning which see Drake (1978, especially pp. 76–78) and Wisan (1974, especially pp. 125–127), and Galileo's epistemological notebooks, also called the "Logical Questions", concerning which see Wallace (1984a, especially pp. 28–32).
²Nevertheless, the works most useful for the present purpose are perhaps Drake (1978), Wallace (1984a), and Clavelin (1968/1974).
³For example, Galilei (1890–1909), 1: 326, and 19: 361–362.
⁴Drake (1978), Wallace (1984a), and Clavelin (1968/1974) are generally reliable, though I do not accept everything they say. See also my (1980) and my (1986).
⁵Hereafter, references to the 20 volumes of Galilei (1890–1909) will be made simply by giving the volume and page number in parenthesis in the text. So this reference is to vol. 1, p. 198. All translations are mine, except as noted.
⁶See, for example, Drake (1978), pp. 40–44.
⁷Flora (1953), p. 4, n. 2; and Drake (1957), p. 24, n. 2. In more recent works Drake (1983), pp. 14 and 223, n. 5, attributes to Galileo a weaker Copernican commitment.
⁸ Flora (1953), p. 5; Drake (1957), p. 24; and Drake (1983), p. 14.
⁹Gingerich (1982).
¹⁰See Galilei (1890–1909), 7: 139–150, 349–368, and 416–425; cf. my (1980), pp. 113–114, 128–129, 133–134, 145–150; cf. also Finocchiaro (1985).
¹¹Cf. Galilei (1890–1909), 5: 238; and Drake (1957), p. 144.
¹²For some light on the matter, see Drake (1978), pp. 198, 278.
¹³The argument here would be essentially the one given by Lakatos (1978), pp. 168–192, whose thesis may be accepted when limited to this specific issue and qualified in this manner.
¹⁴ This also corresponds to an interpretation of Galileo elaborated and defended by Drake (1978).

REFERENCES

Clavelin, M. (1968/1974), *The Natural Philosophy of Galileo*, Cambridge: M. I. T. Press, 1974. Translation of *La philosophie naturelle de Galilee*, Paris: Librairie Armand Colin, 1968.
Drake, S. (ed.) (1957), *Discoveries and Opinions of Galileo*, Garden City, New York: Doubleday.
Drake, S. (1978), *Galileo at Work*, Chicago: University of Chicago Press.
Drake, S. (1983), *Telescopes, Tides, and Tactics*, Chicago: University of Chicago Press.
Feyerabend, P. (1981), *Rationalism, Realism and Scientific Method: Philosophical Papers, Vol. I*, Cambridge: Cambridge University Press.

Finocchiaro, M. A. (1980), *Galileo and the Art of Reasoning*, Dordrecht: D. Reidel.

Finocchiaro, M. A. (1985), 'Wisan on Galileo and the Art of Reasoning', *Annals of Science* **42**: 613–616.

Finocchiaro, M. A. (1986), 'The Methodological Background to Galileo's Trial', in William A. Wallace (ed.), *Reinterpreting Galileo*, Washington Catholic University of America Press, pp. 243–272.

Fleck, L. (1979), *Genesis and Development of a Scientific Fact*, Chicago: University of Chicago Press.

Flora, Ferdinando (ed.) (1953), *Galileo Galilei: Opere*, Milan: Ricciardi.

Galilei, Galileo (1890–1909), *Opere*, National Edition in 20 volumes edited by A. Favaro, Florence: Barbera.

Gingerich, O. (1982), 'The Galileo Affair', *Scientific American* **247**, No. 2, August, pp. 132–143.

Kuhn, T. S. (1970), *The Structure of Scientific Revolutions*, 2nd ed., enlarged. Chicago: University of Chicago Press.

Kuhn, T. S. (1977), *The Essential Tension*, Chicago: University of Chicago Press.

Lakatos, I. (1978), *The Methodology of Scientific Research Programmes*, Cambridge: Cambridge University Press.

Laudan, L. (1977), *Progress and Its Problems*, Berkeley: University of California Press.

Laudan, L. (1984), *Science and Values*, Berkeley: University of California Press.

Laudan, L. *et al.* (1986), 'Scientific Change: Philosophical Models and Historical Research', *Synthese* **69**: 141–223.

Wallace, W. A. (1984a), *Galileo and His Sources*, Princeton: Princeton University Press.

Wallace, W. A. (1984b), 'Galileo's Early Arguments for Geocentrism and His Later Rejection of Them', in P. Galluzzi (ed.). *Novita' Celesti e Crisi del Sapere*, Florence: Barbera, pp. 31–40.

Wisan, W. L. (1974), 'The New Science of Motion: A Study of *De Motu Locali*', *Archive for History of Exact Sciences* **13**: 103–306.

Wisan, W. A. (1984), 'On the Art of Reasoning', *Annals of Science* **41**: 483–487.

B. J. T. DOBBS

NEWTON'S REJECTION OF THE MECHANICAL AETHER:
EMPIRICAL DIFFICULTIES AND GUIDING ASSUMPTIONS

1. INTRODUCTION

The case to be examined is Isaac Newton's rejection of a mechanical aether for gravitation in late 1684 or early 1684/85. I have chosen this episode as particularly appropriate for testing theories of scientific change primarily because of its historical importance, the rejection of a mechanical aether being a prerequisite for the establishment of Newtonian gravitational theory.

In addition, certain other aspects of the case help to isolate significant issues. Newton was in the preliminary stages of writing his *Principia* in 1684 and 1685. He was working alone, so he was not subject to the group dynamics of a research team. And perhaps because he worked alone, and had few or no colleagues with which to discuss ideas and evidence, he tended to work out his thoughts on paper. Many of those papers have survived and constitute the foundation of this test of the theory of scientific change. Last, but not least, Newton's historical milieu provided him with an unusually rich range of choices regarding sets of guiding assumptions.

In the event Newton made a dramatic break with the orthodox mechanical philosophy of his day, the philosophy that was generally understood among advanced thinkers at the time to be the most promising method of approaching the study of the natural world. He did not reject the entire system of mechanical thought, but he did reject one of its most basic assumptions: that force could be transferred only by the impact of one material body upon another. For Newton it is not quite correct to say that the mechanical philosophy had provided him with his most powerful set of guiding assumptions before 1684, but certainly he had accepted the concept of a mechanical aether, acting by impact, as the 'cause' of gravitation. That assumption came in for serious questioning in his mind in 1684, and through that historical fact we may examine the anomaly cluster.

What did Newton do when he began to doubt the presence in the natural world of a mechanical gravitational aether? Prior to 1684 he

thought that certain pendulum experiments he performed in the 1660s
or early 1670s provided empirical evidence for the existence of such an
aether. His doubt forced him to question his previous *interpretation* of
those early experiments and to devise a new experimental test in the
form of a greatly refined set of pendulum experiments. Thus, Newton's
initial response was to reinterpret old experiments and create new ones.
The results of the latter forced him to reject a very important guiding
assumption from mechanical philosophy, but whether he immediately
developed a rival set of guiding assumptions to replace the one rejected
must remain an open question. Newton was blessed – or cursed – with a
plethora of alternative sets of guiding assumptions, which may have
given him an extraordinary degree of mental flexibility, but the surviving evidence suggests that, although he may have sought to do so, he
probably did not at first make a definitive choice among them.

2. NEWTON'S ACCEPTANCE AND REJECTION OF A MECHANICAL AETHER FOR GRAVITY

In his student notebook (mid to late 1660s),[1] in his alchemical treatise
"Of Natures obvious laws & processes in vegetation" (early 1670s),[2] in
the "Hypothesis of Light" he sent to the Royal Society in 1675,[3] in his
letter to Boyle of 1678/79,[4] and in correspondence and unpublished
papers of the early 1680s,[5] there exists a continuous record of Newton's
use of speculative aethereal mechanisms as an explanation for gravity.
There were variations in his proposed systems, but there is also a very
strong continuity in that all involved the impact or pressure of fine
particles of matter imperceptible to the senses.

Some scholars have erroneously assigned Newton's undated papers
De aere et aethere and *De gravitatione* to this early period before 1684[6]
and so have obscured the continuity of Newton's use of aethereal
mechanisms by the implication that he engaged in abrupt and rapid
oscillation between fundamentally different explanatory paradigms,
since neither *De aere et aethere* nor *De gravitatione* offers a gravitational
aether and the latter explicitly rejects it. There are many reasons for
doubting that either manuscript was composed before 1684, and both of
them will be treated here as marking the time (1684 and early 1684/85)
when Newton rejected the mechanical aether for gravitation.[7] The
general popularity of aethereal speculation among mechanical philoso-

phers argues against an earlier date. Aethereal speculation was so commonplace, and was indeed such a necessary concomitant of impact physics, that Newton would have needed an urgent reason to abandon it. That reason, it would seem, struck him with its full force only in 1684, when he realized that the precision of planetary motion in effect precluded aethereal gravitational mechanisms. Not until 1684 do Newton's papers reflect the clarity of thought on dynamical principles that enabled him to launch the *Principia*,[8] and only in the course of writing that work did Newton confront the problems that inhered in all his various aethereal gravitational systems up to that time.

The first fruits to be made public from the new dynamical work that Newton undertook at Halley's instigation in 1684, and that eventuated in the *Principia* in 1687, were sent to the Royal Society in November 1684. Entitled *De motu corporum in gyrum*, the first tract concentrated on central-force orbits.[9] In it Newton defined centripetal force for the first time: "that by which a body is impelled or attracted towards some point regarded as its centre".[10] Although he also defined resistance – "that which is the property of a regularly impeding medium", he immediately noted that in his first several propositions "the resistance is nil".[11] Of course that was not the case in the propositions on ballistics, in which the terrestrial atmosphere constituted an impeding medium, but Whiteside has observed that the ballistics problems were probably added as an afterthought, and that regarding resistance Newton originally wrote: "Bodies are hindered neither by the medium nor by any other external causes from yielding perfectly to their innate and to centripetal forces."[12] That general statement, presumably written when the tract consisted only of propositions on bodies in orbit, conclusively demonstrates that by November 1684 Newton knew the motions of celestial objects were not impeded by the medium through which they moved.

Theorem 1 of Newton's first tract on motion was a mathematical demonstration of a general area law for bodies revolving about an immovable center of force, a demonstration that remained substantially unchanged through all the versions of *De motu* and then became Proposition I, Theorem I of Book I of the *Principia*:

The areas which revolving bodies describe by radii drawn to an immovable center of force do lie in the same immovable planes, and are proportional to the times in which they are described.[13]

The area law Kepler had determined empirically but could not really demonstrate mathematically was of course subsumed in Newton's general demonstration.

The problem that Newton encountered here at the earliest stage of his writing is simple and obvious in retrospect but one that proved insurmountable within the context of the mechanical philosophy: the observed area law of Kepler should *not* fit so closely with the exact area law derived mathematically by Newton if the heavens are filled with a retarding medium. Unless the medium is somehow disposed to move with exactly the same variable speed that the planetary body exhibits, the planet should encounter enough resistance from the medium to cause an observed deviation from the mathematical prediction, just as projectiles in the terrestrial atmosphere are observed to deviate from mathematical prediction.

Newton's realization that no form of the hypothetical gravitational aether of mechanical philosophy could be reconciled with actual celestial motions must have been rather a shock to him. From the time of his introduction to mechanical philosophy in the 1660s until the early autumn of 1684 he left an extensive record of his aethereal speculations, as we have seen. Even as he modified his schemata from time to time, Newton had never doubted that the 'cause' of gravity was material. But if the heavens were filled with a material aether, then its presence should produce some notable retardation on the motions of bodies passing through it. Yet Newton said in the first draft of *De motu* that none was in evidence. Newton had had to re-think all of his aethereal mechanisms in order to make that statement, and one result of his re-thinking was already in evidence in his definition of centripetal force. Whereas in earlier documents Newton had offered explanations of apparent attractions in terms of aethereal impulsions in the traditional fashion of mechanical philosophers, in the new definition he equivocated. Bodies are "impelled or attracted" by a centripetal force, he said. No causal mechanism was suggested nor any preference indicated between the two ways of describing the action of the force, a stance soon to be adopted in the *Principia* itself.

Newton probably composed his short but formal essay on air and aether when he first became aware of the problem of celestial retardation, or, rather, aware of the problem generated by the lack of celestial retardation. *De aere et aether* is closely related to Newton's letter to Boyle of 1678/79 and covered much the same set of phenomena, but whereas in his letter to Boyle Newton had proposed aethereal mechanisms for capillarity and the other phenomena, in the first chapter of *De aere et aethere* (on air) he undertook to explain them all by aerial density gradients or, tentatively, by a repulsive force. The second chapter, on the aether, is much abbreviated and consists only of a single paragraph, of which the following is the first part.

> And just as bodies of this Earth by breaking into small particles are converted into air, so these particles can be broken into lesser ones by some violent action and converted into yet more subtle air which, if it is subtle enough to penetrate the pores of glass, crystal and other terrestrial bodies, we may call spirit of air, or the aether.[14]

Reminiscent of passages that characterized the great circulatory systems of Newton's earlier speculations in which air 'relents' into its 'first principle' when it reaches the aetheral regions above the atmosphere, this definition implied that aether differed from air only in the size of its particles and so was a very subtle but still material fluid.

In the remainder of the paragraph Newton marshalled such empirical evidence as he could for the existence of the aether. He cited the Boylean calcination experiments in which metals gained weight when heated in hermetically sealed glasses. "It is clear", Newton said, "that the increase is from a most subtle saline spirit" that came through the pores of the glass. He cited pendulum experiments *in vacuo* in which the motion of the bob was damped almost as quickly as in air. He cited the case of iron filings arranged in curved lines by a loadstone: "I believe everyone who sees [that] . . . will acknowledge that these magnetic effluvia are of this [aethereal] kind." He began an argument that electrical attraction was "caused in the same way by a most tenuous matter of this kind. . . ."[15] And there he broke off, leaving the manuscript incomplete.

In that brief tract on air and aether one may see the efforts of a man who has suddenly had a fundamental explanatory device called into question. It is as if he had asked himself whether he could find evidence of any sort that would sustain his twenty-year-old belief in aethereal mechanisms. He was willing to, and in this tract did, assign many short-range functions of his earlier aethers to the air – all except the chemical, magnetic, and electrical ones that still seemed to him to offer evidence for the existence of the aether. Short-range phenomena had always been functionally separate from gravity in Newton's aethereal systems even when they were quite closely assimilated to the gravitational aether in the early and middle 1670s. But it was the gravitational aether that had been called into question in 1684, the cosmic gravitational aether that seemed to be precluded by the lack of retardation in celestial motions. And there was only one item in the evidentiary list of the second chapter of *De aere et aethere* that was directly related to the issue of celestial retardations.

That one item was the damping of pendulum motion even when the retarding air had been evacuated. Newton had understood the *in vacuo* pendulum experiments to indicate the presence of aether within the glass even after the air was removed. By definition, aether had the ability to penetrate the pores of glass and all other terrestrial bodies. Therefore, the pendulum motion "ought not to cease unless, when the air is exhausted, there remains in the glass something much more subtle which damps the motion of the bob".[16] For Newton that experiment had indicated the presence of a gravitational aether that had acted in conjunction with the air to slow the motion of the bob when air was present and had acted almost as effectively to damp the motion when air was absent. But the experiment had only been designed to separate the effects of air and aether; with the very presence of the postulated aether in question Newton needed further evidence. Was there any way he could experimentally justify the hypothesis that a gravitational aether remained within the evacuated container? Was there a property of the aether that could be detected by its varying reactions under experimentally varied conditions?

The answer was yes. Since, by the hypothesis, aether penetrated not only the pores of the glass container but also every other terrestrial body, it should interact with the internal parts of the bodies it permeated. By holding constant the retardation due to the presence of air (presumed to act only on the surface of the pendulum bob) and varying the quantity of matter within the interior of the bob (where the aether alone was presumed to interact with particles of matter), one should expect an increase in aethereal retardation with an increasing quantity of matter. If, that is, the gravitational aether really existed.

Newton reported just such a set of experiments in the *Principia*, in which he had first used an empty box as a bob, then filled the same box with lead or another metal.[17] The total resistance experienced by the empty box was to the total resistance experienced by the full box as 77 to 78; therefore the increased quantity of matter made very little if any difference, certainly not enough to justify the existence of the gravitational aether. Westfall has made the suggestion that Newton performed these experiments not long after he had abruptly broken off the composition of *De aere et aethere*, and the logic of Newton's development makes that suggestion highly plausible.[18] Although Westfall's dating of these events to 1679 is made doubtful by Newton's continued use of vortex terminology until 1684, these new and more refined pendulum experiments were exactly the sort of test Newton might have devised to shore up a wavering faith in an aethereal mechanism for gravity. In the event the new experiments showed him that the resistance on the internal parts of the box was nil, or, if not nil, negligible, and that the gravitational aether as he had always envisioned it probably did not exist. If he wrote *De aere et aethere* and performed the new pendulum experiments shortly before he wrote the first draft of *De motu* in 1684, as seems most likely, one has there a full and sufficient explanation for Newton's equivocal use of "impelled or attracted" in the first definition of centripetal force.

It seems fair to say that Newton's developement to this point substantiates (GA2.1), that when a set of guiding assumptions runs into empirical difficulties scientists believe this reflects adversely on their own skills rather than on inadequacies in the guiding assumptions. If the chronology of events sketched above is correct, Newton's initial reaction was first to question his own interpretation of experimental data, then to devise and perform a new set of experiments that probed more directly and profoundly for the existence of a gravitational aether.

Newton did not, however, depart fully from aethereal systems as a result of the new pendulum experiments. In what seems to be his first written attempt to come to grips with the difficult information on non-resistance, he described the resistance of "pure aether" as "either non-existent or extremely small". But one must note that it was the *resistance* of the aether and not the aether itself that Newton described as non-existent. The passage was in a scholium in the augmented tract on motion, composed about December 1684. The new tract had the general title *De motu sphæricorum Corporum in fluidis* and was divided into two sections: "De motu corporum in mediis non resistentibus" and "De motu Corporum in mediis resistentibus".[19] The former section contained the scholium of special interest for our line of analysis.

Thus far I have explained the motions of bodies in non-resisting mediums, in order that I might determine the motions of the celestial bodies in the aether. For I think that the resistance of pure aether is either non-existent or extremely small. Quicksilver resists strongly, water far less, and air still less. These mediums resist according to their density. . . . Therefore the solid matter of air may be made less, and the resistance of the medium will be diminished nearly in the same proportion until it reaches the tenuousness of aether. . . . All those sounder astronomers think that comets descend below the orb of Saturn, who know how to compute their distances from the parallax of the Earth's orbit, more or less; these therefore are indifferently carried through all parts of our heaven with an immense velocity, and yet they do not lose their tails nor the vapour surrounding their heads, which the resistance of the aether would impede and tear away. Planets will persevere in their motion for thousands of years, so far are they from experiencing any resistance.[20]

The medium in which celestial bodies moved was a non-resisting one; nevertheless it was the aether still. The argument on diminishing resistances was designed not to argue the aether away completely but to highlight its extreme 'tenuousness.' Furthermore, the aether was still a medium, and, since Newton discussed it under the general rubric of the motion of spherical bodies in fluids, it was to him presumably a fluid medium, comparable in some ways to the other fluids he mentioned. Did it still cause gravity? One might think that Newton's non-resisting aether had become much too tenuous to provide a mechanical cause for gravity by impact, but if the heavens were completely filled with it, as Descartes had taught him to believe, perhaps that was not how Newton saw it. If the cosmic aether was for Newton plenistic and corporeal still, as the terms 'fluid' and 'tenuous' imply, probably he did still think in terms of a material substrate for gravity.

It seems to be the case then that Newton, at the very end of 1684, still thought the heavens to be filled with an aethereal medium, albeit a 'non-resisting' one, despite the empirical difficulties into which the mechanical gravitational aether had fallen. Such an interpretation is certainly congruent with the surface meaning of his words in the scholium to the first section of *De motu sphæricorum Corporum in fluidis*, and if indeed this interpretation is correct it serves to substantiate (GA2.6): that scientists often introduce hypotheses which are not testable in order to save the guiding assumptions. By what conceivable means could one test for the existence of a non-resisting medium? The hypothetical aethers of mechanical philosophy had always been assumed *not* to interact with the human sensory apparatus but had been supposed to interact with the interior parts of permeable bodies. But Newton had ruled out that possible mode of detection and had substituted the untestable hypothesis of a non-resisting aether. But his postulate of a non-resisting medium in the celestial regions did have the advantage of preserving (temporarily) that most fundamental postulate of the mechanical philosophy: the assumption that force could be transferred from one body to another only by contact. The non-resisting medium, for all of its flaws, did still preserve corporeal contact between celestial bodies.

Let us pause to summarize Newton's situation at the close of 1684. Traditionally, mechanical philosophers had held in the case of gravity that the required impacts were supplied by the action of particles of an imperceptible aether, and so had Newton. Newton had thought he had empirical evidence for the existence of such an aether, but when the lack of celestial retardations offered counter-evidence he then argued for an extremely tenuous version of the aether, so tenuous as to be non-resisting. All he said in the augmented *De motu* was that gravity was "one kind of centripetal force". "Motion in the heavens, therefore, is ruled by the laws demonstrated."[21] The fact that those statements on gravity came in the same scholium as, and directly following, the long paragraph on the non-resistance of the aether is suggestive but far from conclusive. At most one may infer that gravity was still intimately connected in Newton's mind with the aethereal medium, as it had always been. Persistent empirical difficulties for the mechanical gravitational aether had indeed arisen, but although Newton had modified the system he had not yet abandoned it. Had nothing further happened, the Newtonian case would serve to justify (GA2.2), that scientists are prepared to leave the difficulties unresolved for years. But something further did happen very quickly: Newton came to recognize that no matter how finely divided and "tenuous" his aethereal medium might be, if it were still a corporeal fluid, it would offer just as much resistance as a less subtle fluid would, or, perhaps, even more. So in the Newtonian case the difficulties were not left unresolved for years.

Newton clarified his thinking by rejecting the plenistic and corporeal nature of the medium "utterly and completely" in a formal essay of considerable length, left untitled, but with the incipit, *De Gravitatione et aequipondio fluidorum et solidorum in fluidis scientiam*:

as water offers less resistance to the motion of solid bodies through it than quicksilver does, and air much less than water, and aetherial spaces even less than air-filled ones, should we set aside altogether the force of resistance to the passage of bodies, we must also reject the corporeal nature [of the medium] utterly and completely. In the same way, if the subtle matter were deprived of all resistance to the motion of globules, I should no longer believe it to be subtle matter but a scattered vacuum. And so if there were any aerial or aetherial space of such a kind that it yielded without any resistance to the motions of comets or any other projectiles I should believe that it was utterly void. For it is impossible that a corporeal fluid should not impede the motion of bodies passing through it, assuming that . . . it is not disposed to move at the same speed as the body. . . .[22]

Newton introduced the manuscript by saying, "It is proper to treat the science of gravity and of the equilibrium of fluid and solid bodies in fluids by two methods." One of those methods was to demonstrate propositions "strictly and geometrically"; the other was a "freer method of discussion, disposed in scholia, [so that it] may not be confused with the former which is treated in Lemmas, propositions, and corollaries".[23] Since the two sciences mentioned – of gravity and of the equilibrium of fluid and solid bodies in fluids – bear a strong resemblance to Books I and II of the published *Principia* and since the published *Principia* was constructed 'strictly and geometrically' except for a number of scholia utilizing a 'freer method of discussion', the introductory paragraph of *De gravitatione* could almost serve for the published work. Probably Newton did intend to be writing an introduction to the *Principia* as he envisioned it late in 1684 or early in 1684/85 when he wrote *De gravitatione*. Although many ideas from the first part of *De gravitatione* appear in later papers, subsequent mathematical and experimental discoveries produced alterations in his planned discourse, and he abandoned the essay in the midst of an attempt to establish mathematically the properties of a 'non-gravitating fluid'.[24]

With his propositions on a non-gravitating fluid Newton was searching for a method of treating an incorporeal aether (or one that was almost so) 'strictly and geometrically'. However, as he must soon have realized, as long as the aether did not retard the bodies passing through it, there was really no necessity for bringing it under mathematical law; he could ignore the presence of the aether in mathematizing the motions of the bodies in it since most of the aether was in fact empty space, as he had already said in the essay, and thus did not require to be treated as a fluid of any sort.

But lest any doubt remain, it should be observed from what was said earlier that there are empty spaces in the natural world. For if the aether were a corporeal fluid entirely without vacuous pores, however subtle its parts are made by division, it would be as dense as any other fluid, and it would yield to the motion of bodies through it with no less sluggishness; indeed with a much greater, if the projectile should be porous, because then the aether would enter its internal pores, and encounter and resist not only the whole of its external surface but also the surfaces of all the internal parts. Since the resistance of the aether is on the contrary so small . . . there is all the more reason for thinking that by far the largest part of the aetherial space is void, scattered between the aetherial particles.[25]

The lack of retardation forced Newton to conclude that the aether consisted mostly of 'vacuous pores'. This passage from De gravitatione seems quite clearly to depend upon the result of the more sophisticated pendulum experiments that had shown the resistance of the aether to be null or very small, for when Newton described the hypothetical interaction of any corporeal aether with the surfaces of all the internal parts of bodies, he described the rationale for those experiments. Applied to the heavens, the results of those pendulum experiments demonstrated that "by far the largest part of the aetherial space is void", and in the next version of De motu, that Whiteside argues probably dates from the winter or early spring of 1684/85, the relevant section is entitled "De motu corporum in spatijs non resistentibus".[26] The shift from the non-resisting media of the augumented De motu of December 1684 to the non-resisting spaces of the revised De motu of early 1684/85 is a significant one, for Newton had rejected the mechanical gravitational aether once and for all. From the time of the writing of De gravitatione in late 1684 or early 1684/85 Newton's universe consisted mostly of space that was empty of body.

Let us now assess (GA2.3), (GA2.4) and (GA2.5) in terms of Newton's revolutionary rejection of the mechanical gravitational aether. Thesis (GA2.3) asserts that scientists often refuse to change guiding assumptions that have run into empirical difficulties. Certainly Newton hesitated for some months before rejecting impact physics but he did not in the end refuse to do so. (GA2.5) suggests that empirical difficulties become grounds for rejecting guiding assumptions only if the problems persistently resist solution, and that thesis does seem to apply to the Newtonian case quite well. Newton had first attempted to resolve the difficulty experimentally. Then, finding terrestrial experimental evidence for the non-resistance of the aether, he had attempted to reconceptualize the celestial aether as so 'tenuous' as to account for its non-resistance to the motion of celestial objects. The breaking point finally came when he recognized an internal contradiction: no matter how tenuous he made it, "a corporeal fluid entirely without vacuous pores" would retard bodies moving through it. Persistent difficulties had presumably made the difference. But, if Newton had been able somehow to adapt the hypothesis of a dense corporeal aether, so it could accommodate the new evidence of celestial non-resistance, probably he would have done so, for he certainly seems to have made attempts to do so, and that fact lends some support to (GA2.4): that scientists ignore difficulties as long as guiding assumptions anticipate novel phenomena successfully. Only after the persistent difficulties with the dense aether convinced him that he had to do so did Newton say "that by far the largest part of the aetherial space is void. . . ."

3. ALTERNATIVE SETS OF GUIDING ASSUMPTIONS

Having concluded that the largest part of the cosmic aether was empty space and could not serve as a mechanical cause for gravity as he had once thought, Newton published in the *Principia* the mathematical principles of gravity only – "not defining in this treatise the species or physical qualities of forces, but investigating the quantities and mathematical proportions of them".

I here use the word *attraction* in general for any endeavor whatever, made by bodies to approach to each other, whether that endeavor arise from the action of the bodies themselves, as tending to each other or agitating each other by spirits emitted; or whether it arises from the action of the air, or of any medium whatever, whether corporeal or incorporeal, in any manner impelling bodies placed therein towards each other. In the same general sense I use the word impulse. . . .[27]

Whether a medium "corporeal or incorporeal", whether bodily contact of the aether or the air or "spirits emitted", might be a cause of gravity he would not then say. But in the post-*Principia* period Newton struggled on with each of the options still open to him. Some of the options – the corporeal medium, involving bodily contact with air particles or the extra-fine corpuscles of aether – had been excluded by the new evidence and were never to return. But there remained as possibilities the 'incorporeal medium' and the 'spirits emitted', the latter of which Newton apparently conceived as quasi-material inhabitants of the grey area between the complete incorporeality of God and the full solidity of body. Newton's search for a cause for gravity was not over, and, although limitations of space preclude consideration of that search here, one may at least observe that this list of options from the *Principia* serves to substantiate (GA2.7), that only *after* the rejection of one set of guiding assumptions do scientists seriously attempt to develop rival sets. When Newton first rejected impact physics, he did not have an immediate replacement, only a number of possibilities for future exploration.[28]

The question arises, nevertheless, as to why Newton was willing to give up Cartesian vortex explanations for celestial motions when many of his brightest contemporaries were not.[29] It is not impossible that the answer to that question lies in another set of guiding assumptions held by Newton but not, or at least not so strongly, by his continental critics: the assumptions of voluntarist theology.[30] A voluntarist was convinced that God's will was his primary attribute and that the world was totally contingent and dependent on God's will. God's will being absolute, He could do anything that did not involve a contradiction, and man's puny mind was not the measure of what God could will to do. So although mechanically-acting aethereal vortices and impact physics might well appear to be the only reasonable and rational explanation of celestial motions to human beings (as it did, for example, to Leibniz and Huygens), Newton's voluntarist stance insisted upon God's unlimited freedom to order the heavens by any means whatsoever, even non-mechanical ones.[31] The 'reasonableness' and 'rationality' of mechanical principles thus did not necessarily carry so much force and conviction for Newton as for others.

4. CONCLUSION

The case of Isaac Newton's rejection of a mechanical aether for gravitation substantiates most elements of the anomaly cluster. Newton seems first to have doubted his own interpretative skills, then to have devised a new experimental test. The new experiments gave negative evidence on the existence of a gravitational aether, but he still hesitated to reject the principles of impact physics and devised an untestable hypothesis to save the guiding assumptions. But he did not leave the difficulties unresolved for years; on the contrary, when he discovered an internal contradiction in his own new but untestable hypothesis, the balance of evidence seemed to him to have shifted, and he did then reject the dense Cartesian aether as a mechanical explanation of gravity. It is possible, even probable, that certain guiding assumptions from systems of thought outside of mechanical philosophy may have enabled Newton to make his dramatic break with orthodox mechanism, but at least at first he was unable to decide upon the most appropriate alternative.

NOTES

[1] McGuire and Tamny (1983).

[2] Newton (*c.* 1672).

[3] Newton (1959–1978), I, 362–382 (Newton to Oldenberg, 7 Dec. 1675).

[4] Ibid, II, 228–296 (Newton to Boyle, 28 Feb. 1678/79).

[5] Ibid., II, 319–334 (Newton to Burnet, 24 Dec. 1680; Burnet to Newton, 13 Jan. 1680/81; Newton to Burnet, ? Jan. 1680/81); Newton MS, University Library, Cambridge, Portsmouth Collection Add. MS 3964.14, f.613r, as dated and cited in Whiteside (1970), pp. 14 and 18, n.42.

[6] Both papers were originally published in Newton (1962B). The Halls argued for composition dates of 1673–1675 for *De aere et aethere* and 1664–1668 for *De gravitatione*; subsequent scholars have tended to follow the Halls' dating with relatively minor variations, no one suggesting a date as late as 1684.

[7] I will explore these reasons at greater length in my forthcoming book *The Janus Faces of Genius: The Role of Alchemy in Newton's Thought.*

[8] Whiteside (1970); Newton (1967–1980), VI.

[9] Newton (1967–1980), VI, 18–19, 30–75.

[10] Ibid., VI, 30–31.

[11] Ibid., VI, 32–33.

[12] Ibid, VI, 32, n.7.

[13] Newton (1962A), I, 40–41; Newton (1972), I, 88–90; Westfall (1980), pp. 411–414; Herivel (1965), pp. 258–259, 278; Newton (1967–1980), VI, 35–37, 539–542.

[14] Newton (1962B), pp. 214–228, quotation from pp. 220 and 227 (Halls' translation).

[15] Ibid., pp. 220 and 227–228, (Halls' translation).

[16] Ibid. (Hall's translation).

[17] Newton (1962A), I, 325–26; Newton (1972), I, 461–463.

[18] Westfall (1971), pp. 375–377 and 410.

[19] Newton (1967–1980), VI, 74–81; Newton (1962B), pp. 243–245, 247–270, 271–292.

[20] Newton (1962B), pp. 261 and 285–286 (Halls' translation).

[21] *Ibid.*, pp. 261 and 286 (Halls' translation).

[22] *Ibid.*, pp. 90–156, quotation from pp. 112–113 and 146–147 (Halls' translation).

[23] *Ibid.*, pp. 90 and 121 (Halls' translation).

[24] *Ibid.*, pp. 117–121 and 152–156.

[25] *Ibid.*, pp. 113 and 147 (Halls' translation).

[26] Newton (1967–1980), VI, 122.

[27]Newton (1962A), I, 192; Newton (1972), I, 298.

[28]Newton's exploration of the explanatory possibilities of the "incorporeal medium" and the "spirits emitted" constitute a part of my forthcoming *The Janus Faces of Genius*.

[29]See Baigrie (1988).

[30]McGuire (1968), esp. pp. 187–208.

[31]In Newton (c. 1672, f.4v) one finds a succinct restatement by Newton of voluntarist theology in the context of arguments for non-mechanical causal principles in 'vegetation'.

REFERENCES

Baigrie, Brian S. (1988), 'The Problem of Vortex Motion 1687–1713: Its Significance for Lakatos' Account of the Coordination of Scientific Research', this volume, pp. 85–102.

Herivel, John (1965), *The Background to Newton's Principia. A Study of Newton's Dynamical Research in the Years 1664–84* Oxford: Clarendon Press.

McGuire, J. E. (1968), 'Force, Active Principles, and Newton's Invisible Realm', *Ambix* 15: 154–208.

McGuire, J. E., and Tamny, Martin (1983), *Certain Philosophical Questions: Newton's Trinity Notebook*, Cambridge: Cambridge University Press.

Newton, Isaac. (*c.* 1672), 'Of Natures obvious laws & processes in vegetation', Dibner Library of the History of Science and Technology of the Smithsonian Institution Libraries, Washington, D.C., Dibner Collection MSS 1031 B.

Newton, Isaac. (1959–1978), *The Correspondence of Isaac Newton*, ed. by H. W. Turnbull, J. P. Scott, A. R. Hall, and Laura Tilling (7 vols.), Cambridge: Cambridge University Press.

Newton, Isaac. (1962A), *Sir Isaac Newton's Mathematical Principles of Natural Philosophy and His System of the World* (1729, trans. Andrew Motte), ed. by Florian Cajori (reprint of the 1934 edn. (2 vols.)), Berkeley/Los Angeles: University of California Press.

Newton, Isaac. (1962B), *Unpublished Scientific Papers of Isaac Newton: A Selection from the Portsmouth Collection in the University Library, Cambridge*, ed. and trans. by A. Rupert Hall and Marie Boas Hall, Cambridge: Cambridge University Press.

Newton, Isaac. (1967–1980), *The Mathematical Papers of Isaac Newton*, ed. by Derek T. Whiteside with the assistance in publication of M. A. Hoskin (8 vols.), Cambridge: Cambridge University Press.

Newton, Isaac. (1972), *Isaac Newton's Philosophiae naturalis principia mathematica. The Third Edition (1726) with Variant Readings*, assembled and ed. by Alexandre Koyré and I. Bernard Cohen with the assistance of Anne Whitman (2 vols.), Cambridge, MA: Harvard University Press.

Westfall, Richard S. (1971), *Force in Newton's Physics. The Science of Dynamics in the Seventeenth Century*, London: Macdonald; New York: American Elsevier

Westfall, Richard S. (1980), *Never at Rest. A Biography of Isaac Newton*, Cambridge: Cambridge University Press.

Whiteside, Derek T. (1970), 'Before the *Principia*: The Maturing of Newton's Thoughts on Dynamical Astronomy', *J. History Astron.* 1: 5–19.

BRIAN S. BAIGRIE*

THE VORTEX THEORY OF MOTION, 1687–1713: EMPIRICAL DIFFICULTIES AND GUIDING ASSUMPTIONS

1. INTRODUCTION

It is now widely recognised by philosophers of science that the existence of empirical difficulties need not entail the rejection of the set of guiding assumptions which generated them, i.e., when confronted by difficulties, scientists are prepared to leave them unresolved for years (GA2.2) and they often refuse to change their guiding assumptions (GA2.3). For the most part, this phenomenon has been interpreted in a Kuhnian light, as signifying that scientists are often *dogmatic* about their theoretical commitments. This interpretation, in turn, has occasioned a great deal of theoretical activity on the part of philosophers, since it is not immediately apparent how dogmatic behavior on the part of scientists can be reconciled with our supposition that science is the apex of rationality. Furthermore, if scientists are prepared to set aside empirical difficulties, it is not clear how their guiding assumptions can so frustrate them that scientists are willing to abandon their theoretical commitments.

This paper advances an historical framework for testing two theses concerning the emergence of serious difficulties for sets of guiding assumptions. These theses are: (GA2.4) and (GA2.5). Both are based upon the Kuhnian notion that empirical difficulties are 'mere puzzles',[1] and that any inflation in their importance is a function of the shortcomings of the guiding assumptions, i.e., if the guiding assumptions cease to anticipate novel phenomena, or if numerous attempts to disarm the difficulties fail, this is regarded as sufficient warrant for modifying the guiding assumptions or, when all else fails, rejecting them altogether. The operative principle of this paper, in contrast, is that the emergence of serious difficulties for sets of guiding assumptions challenges our understanding of just how empirical difficulties are *transformed* from mere puzzles into vexing problems during the protracted rivalry between rival sets of guiding assumptions. Whereas (GA2.4) and (GA2.5) advance putative criteria governing scientific rationality that justify an inflation of interest in some empirical difficulties during periods of crisis, the historical elements that inform my test case have been selected on

the supposition that empirical difficulties are articulated in the same way
as are guiding assumptions, and that any inflation of interest in them
mirrors the fact that their significance for the rivalry between sets of
guiding assumptions has been disclosed by this process.

An examination of some of the events of the period 1687–1713 – i.e.,
the period bounded by the publication of the first two editions of
Newton's *Principia* – presents an interesting set of conditions for examining the emergence of serious difficulties. In 1687 Isaac Newton
advanced what proved to be a vexing, if not fatal, objection against the
reigning vortex conception of Descartes; namely, that the mechanism of
vortical motion was incompatible with Kepler's laws of planetary motion (hereafter, the problem of Keplerian motion). What distinguishes
this historical episode is that: (1) although vorticists and Newtonians
were divided on a number of issues,[2] the problem of Keplerian motion
dominated their exchanges to such an extent that my conclusions are not
biased if these other issues are set aside; (2) the problem of Keplerian
motion was formulated in a rigorous way which was acceptable to both
groups, thereby dissolving the nagging philosophical worry concerning
the relativity of standards; and, most importantly, (3) this problem
inspired three distinct responses from scientists, that it was (a) fatal to
the vortex theory, (b) an idle curiosity which would be routinely disarmed, and (c) an important problem which could be resolved – and
thereby strengthen the vortex theory – if its guiding assumptions were
modified. What my analysis discloses is that all three responses were
rational in the years immediately following the publication of Newton's
Principia, whereas the application of (GA2.4) or (GA2.5) is consistent
with one or another of (a) or (b) but not (c). Furthermore, my test case
indicates that the conceptual depths of this difficulty were articulated in
the quarter century following the publication of the *Principia* in a way
which disclosed its debating potential, compelling proponents of vortical
explanations to respond to the problem of Keplerian motion as dictated
by (c). While admitting that (GA2.4) and (GA2.5) are compatible with
the response of scientists to some empirical difficulties, the problem of
Keplerian motion constitutes a notable exception. The debating potential of some empirical difficulties, I will conclude, is articulated in a way
which not only sheds welcome light on (GA2.2) and (GA2.3), but is
inconsistent with (GA2.4) and (GA2.5).

2. THE PROBLEM OF KEPLERIAN MOTION

The Copernican hypothesis generated a new problem for astronomy: if the earth is one planet among many, planetary motions could no longer be explained by imputing a natural, circular motion to celestial bodies. Descartes solved this problem by postulating fluid vortices to replace the solid spheres of Eudoxus and Aristotle. These fluid vortices provided a mechanism for planetary motion, as had the solid spheres, but Descartes substituted a pressure or purely mechanical impulsion for the animistic explanation characteristic of ancient astronomy.

Descartes' theory was wonderfully successful in accounting for the *gross* phenomena of a heliocentric system; among others, the vortex theory explained the appearance and disappearance of comets, the motions of the planets in approximately circular orbits, and the movement of all the planets in the same plane and in the same direction around the sun. The vortex theory even explained such phenomena as sun spots. The track record of the vortex conception was especially startling given that it was conceived in isolation from practical astronomy, notably the *technical* details of planetary motion extracted by Kepler from the mass of astronomical data compiled by Tycho Brahe. Nevertheless, Descartes' physical theory was so appealing to Giovanni Cassini, Director of the renowned Paris Observatory and perhaps the most eminent astronomer of this period, that when he discovered the clash between Kepler's law of elliptical orbits and Descartes' retention of circular orbits, Cassini abandoned Kepler's law (in lieu of ovals of his own design) rather than renounce the vortex model. Although Descartes was not alert to the finer points of astronomy, this testifies that the very conception of a vortex was not a problem either for the likes of Cassini, or for such prominent thinkers as Huygens, Leibniz, and the Bernoullis, all of whom preferred vortical explanations to atomism. The vortex conception was not even a problem for the young Newton,[3] who was still exploring the potential of vortices as astronomical devices as late as 1675.

Newton's *Principia* of 1687 was presented to the Royal Society as "a mathematical demonstration of the Copernican hypothesis as proposed by Kepler" (Birch, 1756–1777, pp. 479–480). In his review of the *Principia* Edmond Halley wrote that Newton's first eleven propositions were "found to agree with the *Phenomena* of the Celestial Motions, as discovered by the great Sagacity and Diligence of *Kepler* . . .".[4] Halley's enthusiasm embodied a challenge to the vorticists to match Newton's achievement by unravelling the exact mathematical relations of Kepler's laws, a challenge explicitly formulated by Newton in Book II of the *Principia*, where it is objected that the vortex theory could neither account for the elliptical orbit described in accordance with the area law nor for Kepler's third law of periodic times.

The Cartesian camp was disadvantaged, to some extent, because it had remained indifferent to Kepler's work, despite the fact that some of its more prominent members, such as Cassini, Huygens, and Leibniz, were familiar with the modifications imposed upon the Copernican hypothesis by Kepler. Cassini was exposed by Riccioli to the new astronomy of Kepler early in his career (see Taton, 1971). Leibniz learned about Kepler's first two laws of planetary motion during an exchange of letters with Newton about mathematical questions (Turnbull and Scott, 1959–1977, Volume 1, pp. 36, 144). A year prior to the publication of the *Principia* Huygens declared his support for the vortex theory: "Planets swim in matter. For, if they did not, what would prevent the planets from fleeing, what would move them? Kepler wants, but wrongly, that it should be the sun" (cited by Koyré, 1965, p. 117). This passage testifies that Huygens was aware of Kepler's ideas: recognising their inconsistency with the vortex theory, Huygens evidently viewed this clash as insignificant. In this respect, his response was in step with Leibniz and, indeed, the entire vorticular camp, which was aware of Kepler's ideas but indifferent to them.

Although this reception suggests a dogmatic preference for their guiding assumptions, there is a straightforward rationale for the response of vorticists to Keplerian astronomy: they *did not recognise* the problems implicit for their guiding assumptions in Kepler's laws simply because they had *no pressing reason* to view these empirical generalizations as more than idle details. In the first place, Kepler's laws of planetary motion were without any foundation in physical theory: Kepler set out to discover the *causes* of planetary motion, but his most plausible candidate was the attractive vortex model which was compatible with the Copernican system of uniform, circular motion.[5] Secondly, each of his laws were problematic in important ways. The area law could not be employed because of vexing difficulties involved in computing areas, prompting the astronomical community to employ alternatives suggested by Boulliau, Seth Ward and Mercator in order to ascertain the orbital velocity of the planets. The third or harmonic law could be easily confirmed, but could not be used to predict the positions or the orbits of the planets. Newton would remark in 1687 that "this proportion, first observed by *Kepler*, is now received by all astronomers . . .". (1934, p. 404). Although he would employ this law in conjunction with the law of centrifugal force to produce the inverse-square law for circular planetary orbits, Newton viewed this derivation as an elementary mathematical achievement because the centrifugal rule had already been published by Huygens in 1673 (cf. Cohen, 1980, pp. 229–230). And we have already seen that Cassini had devised a capable alternative to the elliptical law. Although Kepler's ideas were endorsed by some astronomers (see Russell, 1964), no one prior to 1687 discerned in Kepler's ideas the framework for a dynamical account of celestial phenomena. The vorticists, therefore, were at liberty to set Kepler's laws aside a riddles that would eventually yield to systematic investigation. Although the vortex theory as proposed by Descartes clashed with Keplerian astronomy, its proponents had no *compelling* reason to suppose that the vortex model was the locus of this difficulty.[6]

While the vorticists were *justified* in regarding Kepler's laws as oddities, the relations detected by Kepler – especially the relationship between time and distance expressed in his third law of periodic times – suggested a functional relationship between the sun as the center of the planetary orbits and the motions of the planets. Further to this, in Book I of the *Principia* Newton demonstrated that Kepler's law of areas is an implication of an inertial physics. Since Newton recognised further that Kepler's area law and the elliptical orbit rise and fall together – i.e., that the areas must be computed before the orbit can be determined – Newton advanced an argument to the effect that any inertial physics which fails to address Kepler's laws of planetary motion is inadequate.[7] In view of Newton's arguments and the fact that Descartes' vortex theory postulated inertial motion, it was therefore rational for Newton to submit Kepler's laws of planetary motion as empirical difficulties for the vortex theory.

The first question for our evaluation of (GA2.4) and (GA2.5) concerns the severity of this challenge. If we are to test these theses, we must isolate what they predict about the reaction of vorticists to Newton's argument. The only options available to us are (i) that Newton's argument *obliged* vorticists to follow in his footsteps and repudiate the vortex theory, or (ii) that this empirical difficulty was insignificant and appropriately set aside. Since Newton had advanced a *new* difficulty

which had not been examined in a systematic way, the application of
(GA2.5) is consistent with (ii). Evaluating (GA2.4) is somewhat more
troublesome, since neither camp, to my knowledge, endorsed the conse-
quentialist theory of justification which is implicit in this thesis. Indeed,
Newtonians and vorticists alike regarded the way in which hypotheses
are generated as determining the excellence of scientific theories and
not, as (GA2.4) implies, the extent to which their empirical predictions
were confirmed by subsequent investigation. The point is that even if
the vortex theory failed to produce new knowledge, it is doubtful that
this would have been interpreted as a strike against it. Moreover,
granted that the vortex theory was remarkably successful in explaining
celestial phenomena and that it inspired some fruitful investigations,
such as Huygens' study of the spheroidal shape of the earth (see Aiton
1972, p. 261), the application of (GA2.4) would appear to support
setting aside Newton's challenge as well.

Having extracted this prediction for both theses, the second question
for evaluating them is whether these predictions jibe with the relevant
facts of the matter. Before we consider this, however, it is worth
highlighting a few features of Newton's argument that influenced its
reception. The first is that Newton did not simply claim that Keplerian
motion is an implication of any inertial physics and grant vorticists the
liberty to ascertain when and how this difficulty would be tackled.
Rather, Newton fashioned a remonstration of the vortex theory, one
which attempted to demonstrate the *implausibility* of the mechanism of
a vortex in the light of Keplerian motion. Newton, in effect, attempted
to draw a line between some empirical difficulties that merely test our
ingenuity (such as explaining why the planets orbit the sun in the same
direction and in the same plane), and others which actually undermine
theoretical commitments. To this end, Newton argued that a solid
(planet) could not remain continually in the same orbit unless it were
the same density as the fluid, in which case it would rotate exactly as the
fluid (*Principia*, Book II, prop. 53). And this would imply a circular
orbit which, of course, is inconsistent with Kepler's law of elliptical
orbits. He also noted that if the planets were carried around the sun by a
vortex, the periods of revolution would vary as the square of the
distance, whereas Kepler's third law requires the periodic layers to be
proportional to $x^{3/2}$. Reconciling these powers requires suppositions
relating density and velocity to distance which Newton rejected (see
Book II, scholium to prop. 52). "Let philosophers then see", Newton

proclaimed, "how that phenomenon of the 3/2th power can be accounted for by vortices" (Newton, 1934, p. 394).

The second feature of Newton's argument which should be noted is that the entire second book of the *Principia*, which deals with "the circular motion of fluids", can be read as a *development* of vorticular themes. Aside from its anti-Cartesian potential (which we should be careful not to emphasize too much), Newton was well-versed in the vortex theory and was the first scientist to comprehend the full dynamical significance of Kepler's ideas. In view of subsequent events, it is somewhat ironic that Newton was perhaps the only scientist in a position to turn Cartesian science away from its relatively unsophisticated treatment of gross celestial phenomena to the finer details of celestial dynamics. This point has been overlooked by philosophers who still portray Newton as the incarnation of Descartes' demon, rather than the individual who single-handedly elevated the vortex theory to a new level of sophistication. Reading Newton's argument as essentially anti-vorticular presupposes that empirical difficulties function solely as *potential blows* to sets of guiding assumptions (in the event that the skill of the scientific community cannot disarm them). This presupposition drives the Kuhnian view that difficulties are regarded by scientists as *mere* puzzles (since it would be irrational for scientists to regard potential blows in any other way). The upshot is that (GA2.4) and (GA2.5) in principle rule out the possibility that scientists regard empirical difficulties as affording *potential advances* for their theoretical commitments. But if we study the development of scientific ideas, we see that this is precisely the way some vorticular theorists viewed Newton's challenge.

3. THE ARTICULATION OF THIS PROBLEM

When the *Principia* was published, it did not engender a great deal of interest, except in Britain. Reviews were scarce and, for the most part, openly critical. Only the *Acta eruditorum* and *Bibliothéque universelle* noted Newton's criticism of vortices, but neither review rendered a verdict on its validity. No Cartesian abandoned the vortex theory during the twenty-four year period examined in this study, a fact which seems to support (GA2.4) and (GA2.5). However, few vorticists could have been in a position to do so. Rouse Ball (1893, p. 67) estimates that less than 300 copies of the *Principia* were printed and, even among those who managed to acquire a copy, such as Leibniz and Huygens, the consensus was that it was a wonderful exercise in mechanics, completely lacking a foundation in physical principles.[8] No doubt this explains, in part, why many vorticists simply ignored Newton's criticism. A second qualification is that the *Principia* was an exacting book even for competent mathematicians, let alone for the average vorticular theorist whose primary scientific activity consisted in constructing mechanical models to explain gross celestial phenomena. It is instructive that the only vorticists who responded to Newton's challenge were those like Leibniz and Huygens who not only received copies of the *Principia* but were mathematically proficient as well.

Now, it may very well be the case that (some) scientists respond to empirical difficulties in a way which accords with (GA2.2) and (GA2.3), but this apparent dogmatism can be interpreted as a purely contingent matter and need not be regarded as germane to the rationality of scientific practice. It is true that most vorticists simply ignored Newton's challenge, but surely their response was affected by the fact that they were not in a position to deal with this empirical difficulty. One does not want to assert that ignoring Newton's challenge was rational if the only warrant was a lack of intellectual ability or unfamiliarity with the latest developments in celestial theory. After all, many vorticists simply were not alert to the dynamical potential of Kepler's laws; it need not have been the case that they were *blind* to the potential of these laws as well. Moreover, Newton's initial challenge was arresting since it pointed to a genuine lacuna in the vortex theory, one which required no hindsight for recognition. Although Descartes had promised a mathematical analysis of celestial phenomena, his *Principles of Philosophy* (1644) was virtually barren of even the most elementary mathematical models. This lacuna was significant because any analysis of vortical motion carried out with reference to Keplerian astronomy involved reconciling the mathematical relations involved in Kepler's laws. Finally, Newton's analysis of the dynamical conditions of vortical motion was far from decisive. While it represented a genuine (if unwelcome) development of Cartesian ideas, Newton's presentation of fluid motion did not fully rule out the prospect that some tinkering with the vortex theory would disarm his allegations. In particular, Newton's analysis made a number of arbitrary assumptions; it assumed, for instance, that friction within a fluid is proportional to the relative velocity of the parts of the fluid, a thesis which had no empirical warrant whatsoever. As Westfall has suggested (1980, pp.

456–458), defenders of the vortex theory could just as well advance the equally baseless hypothesis that friction is proportional to the four-thirds power of the relative velocity and thereby arrive at Kepler's third law.

Although Newton's challenge was not fatal to the vortex theory, what these considerations suggest is that it would have been irrational for a vorticist to ignore Newton's challenge. Even if we were to allow that the vortex theory was successful and, further to this, that the problem of Keplerian motion was a *new* difficulty, it is far from obvious that vorticular theorists would respond in the way specified by (GA2.2) and (GA2.3). While admitting that dogmatism is sometimes unavoidable, there was nothing to prevent vorticists from regarding this problem as a serious one which could be disarmed by modifying the vortex theory in appropriate ways. And moreover, given the mathematical machinery developed by Newton's "demonstration of the Copernican hypothesis as proposed by Kepler", there was arguably everything to be gained. We have seen, however, that this critical response is ruled out by (GA2.4) and (GA2.5). The question, therefore, is whether this prohibition is consistent with actual scientific practice. On this note, I now turn to two vorticular theorists, Leibniz and Huygens; their response to the *Principia*, I submit, discredits (GA2.4) and (GA2.5).

After reading the eighteen page review of the *Principia* printed in the *Acta Eruditorum* in 1688, Leibniz immediately responded to Newton's challenge with his "Essay Upon the Causes of the Celestial Motions", which appeared in the February 1689 issue of this journal. Leibniz's paper is arresting for two reasons: (1) he had been aware of Kepler's laws of planetary motion for more than two decades, but he had not attempted to relate them to the vortex model; and (2) Leibniz's paper set a precedent for vorticular science by fully accepting the validity of Kepler's laws; indeed, Kepler is characterized by Leibniz as "that incomparable man" whom "the fates have watched over that he might be the first among mortals to publish the laws of the heavens, the truth of things, and the principles of the gods" (cited by Cohen, 1975, p. 12). These considerations testify that Leibniz *learned* something from the *Acta* review of the *Principia*: that the challenge presented by Newton to the vortex theory was indeed a *valid* (if not sound) one. What he learned, in effect, is that if a physics based upon the impulsion of fluids was to provide a sound basis for astronomy, it would have to accommodate Kepler's laws of planetary motion. On Leibniz's part, this discovery *entailed* abandoning Descartes' simple vortex model, and replacing it with a detailed mathematical analysis of fluid motion.

As against Descartes' presentation of the vortex, Leibniz postulated an aether of concentric layers moving in harmonic circular motion which transport the planets that remain at rest in them. While the vortex rotates in circular layers, Leibniz argued that the planets move in ellipses. All the while, the planets are governed by the fundamental Keplerian law – the 'Harmonic Law' – which asserts their velocity as being inversely proportional to the distance or the radius of the circular layer, a conjecture which is compatible with Kepler's area law because the relationship between velocity and distance, on Leibniz's view, is not applicable to the elliptical planetary orbits. In order to reproduce the elliptical orbit, Leibniz argued that the planet itself follows two motions: a trans-radial motion conforming to the circulation of the aether, and a paracentric motion from layer to layer whereby it approaches and recedes from the sun. Leibniz's celestial dynamics, therefore, explained planetary motion in terms of the action of three forces: the action of the fluid vortex which produced the trans-radial motion, a centrifugal force occasioned by the orbital motion, and a gravitation toward the sun.

In a second, unpublished version of his manuscript, and in an article published in the *Acta Eruditorum* in 1690 titled "The Cause of Gravity", Leibniz presented his views on gravitational attraction. His basic idea is that the *causa gravitatis* is an outward impulse, analogous to the rays of light, occasioned by the circulations of the aethereal matter in great circles, and diminishing in the inverse-square proportion of the radius. Leibniz's explanation of gravity demanded that the speeds of the layers of the vortex vary inversely as the square of the distances. Accordingly, this hypothesis did not accord with harmonic circulation, which required the speeds to vary inversely as the distances. The consequence is that the visible centre of a vortex – say, the earth – would have two different speeds at the same time.[9]

A distinguishing feature of his treatment of vortical motion is that the inverse square proportion is deduced from Kepler's first law, rather than the third law, the course taken by Newton, Hooke, Wren, and Halley. Koyré contends that this confirms the originality of Leibniz's conception (1965, p. 133). But even if we acknowledge the novelty of Leibniz's account of planetary motions, it does not completely remove the sting of David Gregory's intimation (1972, pp. 172–182) that Leibniz was simply imitating Newton's recipe. Consider this exerpt from a letter of Leibniz to Huygens (late 1690):

For harmonic circulation alone has the property that a body circulating in that manner keeps precisely the force of its direction or preceding impression exactly as if it were moved in the void by its impetuosity alone joined with its weight (cited by Westfall 1971, p. 306).

This passage testifies that Leibniz conceived the circulating aether as transporting a planet exactly as it would move in a vacuum under the influence of gravity. What his statement indicates, then, is his desire *to duplicate the function of gravity* with the mechanism of a circulating aether. But on Leibniz's view, gravity merely varies the planet's distances from the sun; it does not serve the Newtonian function of diverting it from its rectilinear path. We can debate the pros and cons of Leibniz's proposal, but even if his theory manages to produce a consistent causal framework for planetary motions, it testifies to the impact exerted by the *Principia* upon Leibniz's cosmological speculations. In particular, the fact that Leibniz's celestial mechanics was *designed* to reproduce the central theses of Newton's theory that were compatible with the mechanism of vortical motion testifies that sets of guiding assumptions are not insulated from criticism in the manner specified by (GA2.4) and (GA2.5).

The case of Huygens is similar, although the modifications that he imposed upon the vortex theory were in many ways even more arresting than Leibniz's. Huygens recognised that a *cause* acts on the planets to divert them from their rectilinear paths, and he was willing to accept Newton's claim that this cause is the same as that which operates in free fall. Like Leibniz, Huygens was therefore willing to accept a gravitation toward the sun and that the effect of gravity diminishes in the reciprocal proportion to the squares of the distances, largely because this supposition "produces precisely the effect of the elliptical orbit that Kepler had guessed and proved by observation . . ." (Huygens, 1888–1950, vol. 21, p. 472; see Koyré, 1965, p. 122). Nevertheless, Huygens wrote in 1690 that

I do not agree with a Principle . . . according to which all the small parts that we can imagine in two or several different bodies mutually attract each other. . . . That is something I would not be able to admit because I believe that I see clearly that the cause of such an attraction is not explainable by any of the principles of mechanics, or of the rules of motion. Nor am I convinced of the necessity of the mutual attraction of whole bodies . . . (*Ibid.*, Volume 21, p. 471; cf. Koyré 1965, p. 118).

While admitting two of Newton's theses, Huygens would not endorse Newton's central notion of the *mutual attraction* of bodies because this was incompatible with a mechanical explanation for gravity.

While granting that "vortices [have been] destroyed by Newton", Huygens elected "to rectify the idea of vortices" but substituted vortices "very distant the one from the other, and not, like those of M. Des Cartes, touching each other" (*Ibid.*, Volume 21, pp. 437–439; cf. Koyré 1965, p. 117) in order to avoid the 'absurdity' of mutual attraction. To this end, he was even willing to grant the existence of a vacuum. He could not accept a perfect vacuum because his own wave theory of light required a mechanical basis for the transmission of light, but as an atomist he was willing to separate the different vortices by vast reaches of empty space.

It will be countered by proponents of (GA2.4) that the views of Leibniz and Huygens did not represent any 'theoretical growth'. Since these modifications of the vortex theory were guided by the successes of Newton's guiding assumptions, they could be taken as evidence that the vortex theory had entered a 'degenerative phase', and therefore that it was no longer anticipating novel phenomena successfully. However, in order to interpret the work of Leibniz and Huygens as confirming evidence for (GA2.4), we would have to overlook the fact that their efforts were the first in a series of studies which analysed vortical motion in terms of the finer Keplerian details of planetary motion (see Aiton 1972 for a thorough discussion of the many versions of the vortex theory developed in the first half of the 18th century). What their work supports, rather, is the claim that while Newton's challenge was not decisive, it was pointed enough that the guiding assumptions of the vortex model had to be modified. As Aiton remarks (1972, p. 7), the gist of Newton's challenge was that Descartes had not presented a robust theory of fluid motion and that the greater part of the task remained to be completed by advocates of mechanical explanations. And in the period 1687–1713 it was not yet apparent that the vortex theory would not be able to surmount this challenge.

The hurdle to accepting this interpretation is the notion that empirical difficulties are regarded by scientists as insignificant unless, that is, their guiding assumptions fail the community in other respects. Although this approach is appealing, it clashes with the fact that the vorticular guiding assumptions *evolved* at the hands of Leibniz and Huygens in response to Newton's criticism, a conceptual process first detected by Laudan.[10] What philosophers have failed to notice, though, is that the evolution of guiding assumptions is produced by, and corresponds to, a parallel evolution of the empirical difficulties. As difficulties are articulated, their debating potential is disclosed in a way which requires that (some of) the guiding assumptions of one or another rival camp is modified and, in some cases, repudiated altogether. The application of (GA2.4) and (GA.2.5) compels us to view empirical difficulties as insignificant or (in extraordinary circumstances) fatal, but the facts surrounding the demise of the vortex theory indicate that difficulties are explored in a way which transforms (some of) them into serious problems.

The articulation of empirical difficulties is reflected not only in the modified vortex theories advanced by Leibniz and Huygens, but also in the fact that Leibniz's theory inspired further exploration of its conceptual depths. For example, in *The Elements of Physical and Geometrical Astronomy* of 1702, D. Gregory advanced a number of criticisms of Leibniz's harmonic vortex. He remarked that the different planets move in accordance with Kepler's third law, which is inconsistent with the harmonic circulation; i.e., the periodic times of the planets are in the sesquiplicate (i.e., 3/2) ratio of the distances from the center, and not as the squares of the radii, as required by a fluid circulating harmonically (1972, p. 179). Gregory also noted the inconsistency in the thesis that the comets obey the area law although they are not carried by the harmonic circulation (1972, pp. 177–178). In effect, Gregory pointed out that the problem of Keplerian motion was more general than anticipated by Leibniz's harmonic vortex. Newton had been interested in comets since he recognised in 1680 that comets give evidence of the universality of gravitation, for if comets are not subject to the same accelerations as are the planets at given distances from the sun – i.e., if they are not 'a kind of planet' – gravitation cannot be universal. However, Gregory added a new wrinkle to Newton's initial critique by advancing the motions of comets specifically as Keplerian phenomena which must be reconciled with vortical motion as well. In the second edition to the *Principia* published in 1713, Newton would exploit Gregory's argument while shoring up his theory of cometary motion by incorporating a mass of new data on comets.

And now a final historical note, one which testifies to the impact exerted by the articulation of the problem of Keplerian motion. In 1697, a decade after the publication of the *Principia*, Samuel Clarke published a Latin translation of Jacques Rohault's Cartesian treatise, *Traité de physique*. Although he included a series of notes on Newton, Clarke overlooked Newton's criticism of vortices. These notes are greatly expanded in the 1702 edition. Only in the edition of 1710 are the Cartesian vortices described as fictions.[11] No doubt, Clarke's assessment reflects the growing support for Newton's cosmology in Britain and the associated repudiation of the vortex theory. However, at a more fundamental level, it also corresponds to the articulation of the challenge originally presented by Newton in 1687, and the failure of the vortex theory to disarm it. Clarke's presentation of Rohault's Cartesian views alongside Newton's cosmology, and the dynamics of this debate as mirrored in the series of modifications rendered by Clarke, testify that the coordination of scientific research around a cluster of important problem is the product of the interaction between alternative sets of guiding assumptions.

4. CONCLUSION

There is a great deal of evidence which can be construed as supporting the claim that scientists are dogmatic about their guiding assumptions, that "anomalies are listed but shoved aside in the hope that they will turn, in due course, into corroborations of the programme" (Lakatos, 1970, p. 137). After all, most vorticular theorists overlooked the problem of Keplerian motion, an indifference which persisted in certain quarters until well into the 18th century. Villemot, for example, composed a version of the vortex theory in 1707 which completely ignored Kepler's laws (see Aiton, 1972). A qualification is that Villemot had not read Newton's *Principia*, which suggests that he simply was not aware that a problem existed. Nevertheless, in view of the indifference in some Cartesian treatises to Newton's argument, it is conceivable that the vorticists were dogmatic about their guiding assumptions, either because their program was successful or because the problem of Keplerian motion had not been subjected to systematic analysis. In either event, it is arguable that the problem generated by Kepler's laws, and the response of vorticists to it, was determined by the vorticular guiding assumptions, and not by any critical interaction between the core doctrines of the vorticular and the Newtonian programs. As their guiding assumptions began to bend beneath the weight of criticism, vorticists were forced to respond to the tension produced by Kepler's laws. If we restrict our attention to the vorticular response to Newton's challenge, or at least to the response characteristic of defenders of such explanations, the bottom line is that we simply cannot completely rule out (GA2.4) and (GA2.5).

However, this counter-argument *presupposes* that scientists are *insulated in their research activities*; i.e., that guiding assumptions blind scientists to the potential of alternative points of view, thereby compelling scientists to regard challenges to their guiding assumptions as insignificant. And this contention runs afoul of the fact that vorticular theorists such as Huygens and Leibniz accepted Newton's baton and attempted to use it to fortify their theoretical commitments. If we set aside our belief that science is essentially an adversarial activity, and simply examine the imprint imposed by Newton, Leibniz, Huygens, and Gregory upon the problem of Keplerian motion, it is evident that each contributed in a *positive* way to its articulation and the discovery of its anti-vorticular potential. It is evident, in other words, that the importance of this problem was determined by the interaction between two sets of guiding assumptions and the role this problem played in adjudicating between them. Although (GA2.4) and (GA2.5) assert that the importance attached to empirical difficulties is a function of their relationship to a single set of guiding assumptions, these considerations testify that importance is (at least) a three-place relation between two contending sets of guiding assumptions and some problem which makes a difference in their scientific standing.

Ultimately, of course, the problem of Keplerian motion contributed to the demise of the vortex theory. But if we resist the temptation to be wise after the fact, and recognise that the unfolding of the conceptual dimensions of this problem may just as well have fortified the vortex conception by elevating it to a new level of sophistication, the entire process appears as a joint effort to ascertain the viability of the vortex theory as a dynamical framework for Kepler's laws.

NOTES

*I wish to thank Larry Laudan and the anonymous referee for a number of suggestions which greatly improved this paper.

[1]See Kuhn's (1962) which claims that "the researcher is a solver of puzzles not a tester of paradigms" (p. 144), and that "no part of the aim of normal science is to call forth new sorts of phenomena; indeed those that will not fit the box are often not seen at all" (p. 24). On Kuhn's view, guiding assumptions constrain scientists from regarding empirical difficulties as anything more than challenges to their skill. Any inflation in their importance is therefore to be regarded as an 'extraordinary' event (the ingenuity of the scientific community is frustrated by them), and not as the by-product of articulating their conceptual depths. The point here is that Kuhn's treatment of empirical difficulties is decidedly *kinematic*: problems are not shown to be important as the guiding assumptions are articulated, but it is simply the interest of scientists that is inflated by their persistence.

[2]The significance of the problem of Keplerian motion in relation to some of the other problems debated by Cartesians and Newtonians is discussed in my (1987). For a thorough comparison which is sympathetic to the Cartesian theory, see Aiton (1972).

[3]The reference is to Newton's "An Hypothesis Explaining the Properties of Light Discoursed of in My Severall Papers", in Turnbull and Scott (1959–1977), Volume 1, pp. 362–386.

[4]Isaac Newton (1958), p. 405. Since Newton failed to give Kepler credit for the laws that bear his name, no doubt Cohen (1972), pp. 148–50, is right that Halley's remark was meant to be critical.

[5]In his *Astronomia nova* of 1609, Kepler developed the view (based upon his reading of Jean Taisner and William Gilbert) that magnetic lines – acting in the planes of the planetary orbits and diminishing inversely with distance – emanated from the sun and propelled the planets. This hypothesis accorded with the observation that the linear velocities of the planets decrease in relation to their distance from the sun, a fact which was formalized by Kepler in his inverse-distance law: viz. that the orbital velocity of the planets is inversely proportional to their distance from the sun. Although he compared his *species immateriata* with rays of light, Kepler's hypothesis did not accord with his own demonstration that the intensity of light diminishes reciprocally as the square of the distance. Further to this, the analogy between a sun and a magnet fails in many ways; a magnet, for instance, will not cause another magnetic body to rotate around it, nor will it attract a given planet for one-half of its orbit and repel it for the other (see Wilson, 1970, pp. 106–107).

[6]Compare my account with Aiton (1972) who argues that "it was not until the advent of the theory of perturbations in the second half of the eighteenth century that physical theory could be usefully applied to the practical problems of astronomy. As a result, in the period between Descartes and Newton, the fact that the vortex theory made no attempt to explains Kepler's laws . . . was not an obstacle to its acceptance as a general physical explanation of the planetary motions (p. 6)." If Aiton is right, it is not clear why Leibniz, Huygens and other vorticists were willing to modify their guiding assumptions in response to Newton's challenge even prior to the advent of the theory of perturbations. Adding another layer of confusion, Aiton contends that Kepler's laws "were [always] important details which the vortex theory should attempt to explain" (p. 260), a claim which implies that the vorticists who set aside Kepler's laws were acting irrationally. Impressed by the

role these laws ultimately played in the rivalry between Newtonians and vorticists, Aiton supposes that the difficulties generated by these laws were always important and, therefore, that the postulation of Newton's cosmology did not transform them from mere puzzles into pressing difficulties. On the supposition that the problem of Keplerian motion was always important but that it nevertheless did not constitute an obstacle to the acceptance of the vortex theory, Aiton is unable to provide a rational account of the repudiation of this theory; indeed, he claims that it was abandoned "not because it had been disproved . . . but because scientists lost faith in it" (1972, p. 7).

[7] For a discussion of the relationship between Kepler's area law and the law of elliptical orbits, see Wilson (1972). Cohen (1980, chapter 5) discusses the treatment these laws received at the hands of Newton.

[8] One reviewer wrote: "In order to make an *opus* as perfect as possible, M. Newton has only to give us a Physics as exact as his Mechanics. He will give it when he substitutes true motions for those that he has supposed" (*Journal des Sçavans*, 2 August 1688, cited by Koyré 1965, p. 115).

[9] Recognising this inconsistency, in a letter composed for Huygens but withheld, Leibniz postulated two independent vortices, one consisting of very tenuous matter which caused the gravitational solicitation, and another vortex consisting of less tenuous matter carrying the planets about the sun in harmonic circulation.

[10] See Laudan (1977), p. 100: "the core assumptions of any given research tradition are continuosly undergoing conceptual scrutiny. Some of these assumptions will, at any given time, by found to be strong, and unproblematic. Others will be regarded as less clear, less well founded. As new arguments emerge which buttress, or cast doubt on, different elements of the research tradition, the relative degree of entrenchment of the different components will shift. During the evolution of any active research tradition, scientists learn more about the conceptual dependence and autonomy of its various elements; when it can be shown that certain elements, previously regarded as essential to the whole enterprise, can be jettisoned without compromising the problem-solving success of the tradition itself, these elements cease to be a part of the 'unrejectable core' of the research tradition." I quote Laudan at length because my test case confirms his anti-Kuhnian thesis that guiding assumptions evolve in response to criticism.

[11] This work was later translated into English by Samuel's brother, John Clarke, in 1723 under the title *Rohault's System of Natural Philosophy, Illustrated with Dr. Samuel Clarke's Notes taken Mostly out of Sir Isaac Newton's Philosophy*. It was reprinted in 1723 and again in 1729. Cohen describes this book as "one of the chief textbooks in general use" and claims that "many generations of students . . . learned what Newtonianism they did along with the Cartesianism of this 'ambivalent' book which presented one kind of physics in the text and another in the footnotes" (1966, p. 94).

REFERENCES

Aiton, Eric J. (1972), *The Vortex Theory of Planetary Motions*, London: Macdonald.
Baigrie, Brian S. (1987), 'Kepler's Laws of Planetary Motion, Before and After Newton's *Principia*: An Essay on the Transformation of Scientific Problems', *Studies in History and Philosophy of Science* 18: 177–208.
Ball, Rouse (1893), *An Essay on Newton's "Principia"*, London: Macmillan.

Birch, Thomas (1756–1977), *The History of the Royal Society of London for Improving of Natural Knowledge*, vol. 4, London: A. Millar.

Cohen, I. Bernard (1966), *Franklin and Newton: An Inquiry into Speculative Newtonian Experimental Science*, Cambridge, Mass: Harvard University Press.

Cohen, I. Bernard (1972), *Introduction to Newton's 'Principia'*, Cambridge, Mass: Harvard University Press.

Cohen, I. Bernard (1975), 'Kepler's Century, Prelude to Newton's', *Vistas in Astronomy* **18**: 3–36.

Cohen, I. Bernard (1980), *The Newtonian Revolution*, Cambridge: Cambridge University Press.

Descartes, R. (1983), *The Principles of Philosophy*, Trans. V. R. Miller and Reese P. Miller, Dordrecht: D. Reidel.

Gregory, David (1726), *Elements of Physical and Geometrical Astronomy*, London: Midwinter, reprinted (1972) in London: Johnson Reprint Corporation.

Huygens, Christiaan (1888–1950), *Oeuvres complètes de Christiaan Huygens*, vols. 21 and 22, The Hague: Martinus Nijhoff.

Koyré, A. (1965), *Newtonian Studies*, Chicago: University of Chicago Press.

Kuhn, T. (1962), *The Structure of Scientific Revolutions,* Chicago: University of Chicago Press.

Lakatos, I. (1970), 'Falsification and the Methodology of Scientific Research Programmes', *Criticism and the Growth of Knowledge*, eds. I. Lakatos and A. Musgrave, Cambridge: Cambridge University Press, pp. 91–196.

Lakatos, I. (1981), 'History of Science and Its Rational Reconstructions', *Scientific Revolutions*, ed. I. Hacking, Oxford: Oxford University Press, pp. 105–127.

Laudan, L. (1977), *Progress and Its Problems*, Berkeley: University of California Press.

Leibniz, G. W. (1689), 'Tentamen de motuum coelestium causis', *Acta Eruditorum*, February, 38–47. Reprinted in Leibniz (1849–1863), Volume 6, pp. 144–161.

Leibniz, G. W. (1849–1863), *Leibnizens Mathematische Schriften*, vol. 6, ed. C. I. Gerhardt, Berlin: Verlag von Asher.

Newton, I. (1934), *Sir Isaac Newton's Mathematical Principles of Natural Philosophy and his System of the World*, trans. A. Motte, revised by Florian Cajori, Berkeley: University of California Press.

Newton, Isaac (1958), *Isaac Newton's Papers and Letters on Natural Philosophy and Related Documents*, eds. I. Bernard Cohen and R. E. Schofield, Cambridge, Mass: Harvard University Press.

Russell, J. L. (1964), 'Kepler's Laws of Planetary Motion: 1609–1666', *British History of Science* **2**: 1–24.

Taton, René (1971), 'Cassini', *Dictionary of Scientific Biography* vol. 3, ed. C. C. Gillispie, New York: Charles Scribner's Sons.

Turnbull, H. W., Scott, J. F., Hall, A. R. and Tilling, Laura (eds.) (1959–1977), *The Correspondence of Isaac Newton*, Cambridge: Cambridge University Press.

Westfall, Richard S. (1971), *Force in Newton's Physics: The Science of Dynamics in the Seventeenth Century*, London: Macdonald.

Wilson, Curtis A. (1970), 'From Kepler's Laws, So-Called, to Universal Gravitation: Empirical Factor', *Archive for History of Exact Sciences* **6**: 89–170.

Wilson, Curtis A. (1972), 'How Did Kepler Discover His First Two Laws?', *Scientific American* **226**: 92–106.

2. CHEMISTRY FROM THE 18TH TO THE 20TH CENTURIES

C. E. PERRIN

THE CHEMICAL REVOLUTION: SHIFTS IN GUIDING ASSUMPTIONS

1. INTRODUCTION

Science is a seemingly inexhaustible source of novelty. Paradoxically, scientists have been portrayed in recent decades as rather conservative – resisting innovation in their discipline sometimes to the point of dogmatism. Yet if a discovery or a novel theory is ever to become part of accredited scientific knowledge, it must be adopted by members of a scientific community. During those profound transformations known as scientific revolutions, one set of guiding assumptions dominating a discipline is replaced by another. Needless to say, the changes that cut most deeply tend to be the most vigorously opposed by the community. Consequently, a scientific revolution may often be detected not only by the depth of the conceptual shift it entails, but by the drama of the accompanying debate.

How is such a debate resolved? The question has received insufficient attention both in historical studies of famous cases and in theoretical models of scientific change. My paper will pursue this issue for one of the classic episodes, the chemical revolution of the late eighteenth century. I shall analyse the pattern of response of scientists confronted by novel claims rather than the nature of those claims – the form rather than the causes or justifications of the revolutionary shift. Among the theses proposed for testing at this conference, I shall deal generally with those describing a change in guiding assumptions (cluster GA4), but the principal question to be examined is *whether or not guiding assumptions change abruptly and totally*, (GA4.3). At the same time the episode is an interesting test of the popular claim of a generational factor in theory change, sometimes known as Planck's hypothesis (GA4.5).[1] In the analysis that follows, I shall sketch what the chemical revolution was about, examine the availability of Lavoisier's views, and review the key events of the debate they engendered, identifying its principal phases. Finally, I shall represent communal reception of the new assumptions as a composite spectrum of individual responses.

105

2. THE CHEMICAL REVOLUTION

The chemical revolution has been characterized as both the overthrow of phlogiston theory and the birth of modern chemistry. The first expression claims too little and the second too much.[2] Let us, briefly, peel back the successive layers that reveal the complexity of the transition. On the surface the episode did involve the overthrow of G. E. Stahl's phlogiston theory of combustion and its replacement by A. L. Lavoisier's oxygen theory. The conflict between oxygen and phologiston was its most palpable and hotly disputed issue. But when Lavoisier first proclaimed a revolution in chemistry in 1772, he had in mind not the overthrow of phlogiston but the establishment of the chemical role of air.[3] On this second level Lavoisier's contributions have often been conflated with British studies of air, under the heading of pneumatic chemistry. However, there is an important distinction. Where the pneumaticists investigated the formation of different kinds of air and their effect on processes such as respiration and putrefaction, Lavoisier initially understood air to be a unique element and sought to define its role in chemical composition and change. Lavoisier's success in this area, combined with discoveries of gases by the pneumaticists, added a new dimension to chemical theory and practice.

Pursuit of the chemical role of air led to the overthrow of traditional notions about which substances are more simple and which more compounded. Lavoisier's discoveries revealed in succession the composition of atmospheric air, of the mineral acids, of water, and of organic substances. On a third level of complexity, then, a novel view of chemical composition emerged, one grounded in a new set of simple substances (defined pragmatically as the actual end products of analysis) and expressed in a novel system of chemical nomenclature (in which the name reveals the elemental composition of the compound).

On a fourth level Lavoisier's innovations had an inadvertent consequence – the undermining of quality-bearing principles as legitimate chemical entities. The success of 18th-century chemistry owed much to two guiding assumptions, the respective doctrines of elective chemical affinities and of essential chemical principles (such as phlogiston). These doctrines had both taxonomic and explanatory applications and could be used to make predictions. The ability of chemists to manipulate materials according to these precepts convinced them of their validity. Lavoisier's innovations left the affinity theory intact (except that affinities had to be associated with different sets of entities). But by arguing the non-existence of phlogiston (the very cornerstone of the doctrine of essential principles), he challenged that entire system. Lavoisier, himself, continued to use the language of principles, calling oxygen the principle of acidity and caloric the principle of heat and expansibility. However, he justified his usage by pointing not to the common property exemplified in the principle, but to the concrete results of analysis and synthesis. What gradually replaced the doctrine of chemical principles was a combinatorial view of composition in which the properties of the compound bore no necessary relation to the properties of the constituents.

Although the elimination of essential qualities from chemistry was perhaps the most subtle consequence of Lavoisier's innovations, it depended upon another departure of equal significance. His success derived from methods unlike the qualitative and intuitive approach that had traditionally guided chemical investigation. Lavoisier's work was characterized by his sytematic use of quantification, of measuring instruments and what has been called his balance-sheet approach. In that way he was able to establish a new standard of clarity and precision in chemical experimentation. These same methods, in Lavoisier's opinion, brought chemistry and physics closer together than they had ever been.

In summary, Lavoisier did much more than introduce a new theory of combustion or eliminate 'protean' phlogiston. He developed an entire new system of chemistry. My purpose in reviewing its conceptual and methodological dimensions is not only to make explicit what the chemical revolution was about, but to emphasize its complex and multi-layered nature. Such a system did not spring fully developed from Lavoisier's brow, but was laboriously conceived and articulated over a period of more than twenty years. Lavoisier published no full and sytematic exposition of his views until 1789. But by then many of his discoveries and ideas had been disseminated; much of the task of persuading his colleagues had been accomplished.

3. THE AVAILABILITY OF LAVOISIER'S VIEWS

One of the oversimplications vitiating traditional historical accounts of the chemical revolution (as well as philosophical and sociological analyses based upon them) is the assumption that two conceptual packages, labeled phlogiston and oxygen, were simply available for scientists to choose between. Such accounts are heavily colored by hindsight. The situation during the generation when Lavoisier was developing and promoting his theories was far more subtle and fluid than such naive models imply. Even in the period before he had any impact upon chemistry, there was hardly unanimity among its practitioners on matters of abstract theory. Although near universal assent was accorded the pragmatic use of affinity and essential principles, on the *cause* of affinity or the *nature* of phlogiston, the consensus quickly fragmented.

Nor was Lavoisier's system of chemistry so readily available as is often supposed. It embraced distinct components conceived and announced at different times. Moreover, even after his discoveries and interpretations were deemed sufficiently developed for public presentation, there was the non-trivial problem of bringing them to the active attention of chemists.

To illustrate the availability of Lavoisier's views as a function of time, I have charted the pattern of his publications relating to issues of the chemical revolution (Figure 1).[4] Lavoisier's work appeared in three successive waves of increasing mass, each wave following a period of intense research activity. In each case there was *local* awareness of his results prior to formal publication, for he read preliminary versions to the Academy of Sciences. Access to Lavoisier's views depended, then, upon one's situation in the scientific community. Colleagues in the Academy typically learned of his ideas well before publication, but even they were often informed only after months, or years, of private reflection or discussion with collaborators. For scientists outside the capital, or outside the French community, access was further hampered by limited circulation of the Academy's expensive volumes and their publication in French.

Let us take a closer look at what the successive waves of publication reveal about the content and aim of Lavoisier's researches. The first wave consists of a book, his *Opuscules physiques et chimiques*, an article on the calcination of tin, and another on the principle absorbed by metals during calcination. These works advance Lavoisier's thesis that combustibles and metals gain weight in burning by absorbing a portion of atmospheric air. They contain only minor dissent from Stahl's theory and leave the door open to compromise. Lavoisier's formal posture toward Stahlian theory, in that early phase, was one of accommodation. The articles of the second wave contain further reflections and discoveries on combustion, the constitution

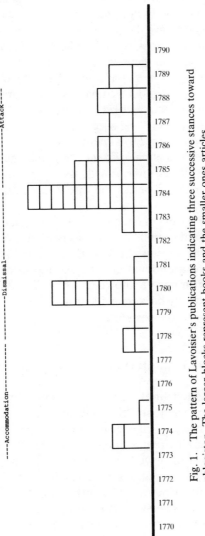

Fig. 1. The pattern of Lavoisier's publications indicating three successive stances toward phlogiston. The larger blocks represent books and the smaller ones articles.

of acids, and the formation of the vapor state. Most momentous was a paper on combustion in general announcing Lavoisier's intention to account for those phenomena without using the concept of phlogiston; if he should succeed in doing so, Stahl's theory would be "shaken to its foundations". With the second wave the gauntlet was thrown down, but Lavoisier did not pursue an attack on Stahlian chemistry, opting instead to develop the positive side of his own theories. The massive third wave includes his famous "Reflections on Phlogiston" and other direct assaults on that venerable concept, as well as detailed papers on calorimetry, the composition of water, fermentation and other major topics. The same period witnessed the appearance in rapid succession of three books: a collaborative work on the new chemical nomenclature (1787), another rebutting Richard Kirwan's defense of phlogiston (1788), and finally Lavoisier's own *Elementary Treatise of Chemistry*. With the appearance of these works there was no longer any doubt of the threat Lavoisier's work posed for conventional theory.

In short, the new system of chemistry was unveiled not all at once, but sequentially. Even after publication it was neither uniformly nor simply available outside the capital. Lavoisier's early publications posed no explicit challenge to traditional guiding assumptions, though his early doubts about phlogiston were known to a few. The threat to traditional chemistry only gradually became evident to the chemical community at large. We turn now to the response.

4. THE FRENCH DEBATE – AN OVERVIEW

To keep the discussion brief and avoid complicating factors, such as language barriers or national bias, my analysis will be restricted to response from the French-language community.[5] Figure 2 lists representative events and turning points in the French debate. During the initial stage when Lavoisier displayed an accomodating posture toward traditional assumptions, his relations with established chemists might be described as cordial. Criticism he hazarded about Stahl's theory, at the Academy's Easter 1773 session, was sternly criticized by A. Baumé – who complained (without mentioning names) about "certain physicists" who wished to substitute fixed air for phlogiston. But in the Academy's official report on the *Opuscules* (which tactfully avoided provocative

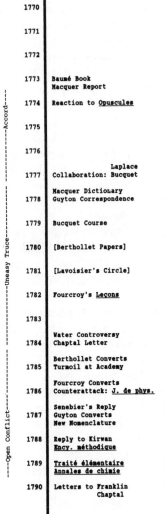

1770	
1771	
1772	
1773	Baumé Book Macquer Report
1774	Reaction to Opuscules
1775	
1776	
1777	Laplace Collaboration: Bucquet
1778	Macquer Dictionary Guyton Correspondence
1779	Bucquet Course
1780	[Berthollet Papers]
1781	[Lavoisier's Circle]
1782	Fourcroy's Leçons
1783	
1784	Water Controversy Chaptal Letter
1785	Berthollet Converts Turmoil at Academy
1786	Fourcroy Converts Counterattack: J. de phys.
1787	Senebier's Reply Guyton Converts New Nomenclature
1788	Reply to Kirwan Ency. méthodique
1789	Traité élémentaire Annales de chimie
1790	Letters to Franklin Chaptal

----Open Conflict----------------------Uneasy Truce------------------------Accord-----

Fig. 2. The French debate over Lavoisier's views indicating three phases of response.

claims), France's most eminent chemist, P. J. Macquer, spoke very favorably of Lavoisier's experimental methods and the reasonableness of his explanation of the weight gain of metals during calcination. Lavoisier's book was not seen as revolutionary. Rather it was received as one among a rash of publications on the theme of fixed air, collectively relating interesting discoveries and not a few extravagant claims. In fact, a principal criticism of Lavoisier's work was that it was not new – that his results had been anticipated by Jean Rey, Robert Boyle, or others. The lucidity of his style was admired and distribution of his book won him memberships in several scientific academies.

In the late fall of 1777, after Lavoisier read his paper formally dismissing phlogiston, the reception turned cooler. An exchange of correspondence early in 1778 between Macquer and his Dijon correspondent, L. B. Guyton de Morveau, shows that both men were displeased with Lavoisier's new departure – though relieved to find his argument less strong than they had feared. Meanwhile Lavoisier had been working closely with the chemist J. B. M. Bucquet. By the time Bucquet gave his last complete chemistry course in 1778–1779, he was offering phlogistic and Lavoisian explanations side-by-side, clearly favoring the latter. During the late 1770s and the early 1780s Lavoisier tacitly avoided confrontation over phlogiston with Academic chemists. But in the same period he began to invite selected colleagues to an apartment near the Louvre, following Academy meetings, for informal discussions of scientific topics. Among the regulars at these sessions (or at Monday dinners in his home at the Arsenal) were the mathematicians Laplace, G. Monge, A. Vandermonde and young aspirants to the Academy, including C. L. Berthollet and A. F. Fourcroy. In that friendly atmosphere Lavoisier's circle engaged in free discussion of pneumatic chemistry, heat researches, and other topics of current interest. Lavoisier used the forum as a sounding board for new ideas and results before presenting them to the full Academy. While some established chemists preferred to deprecate the new discoveries, the. more moderate ones showed degrees of interest. Berthollet, in a series of papers in the late 1770s and early 1780s, challenged Lavoisier's interpretations of phenomena, but in a way that avoided both personal rancour and endless debate over untestable assumptions. The stance adopted by Fourcroy in his 1782 textbook illustrates the response of a well-informed and flexible Parisian chemist during the the middle phase of reaction. He gave intelligent explanations of phenomena according to traditional theory and to Lavoisier's hypotheses, always alert to possible compromise. But on the devisive issue of the *existence* of phlogiston he was cautious, finding that the evidence did not yet allow a definitive choice. It was, on the whole, a time of subdued debate and uneasy truce.

From the outset of the third phase Lavoisier adopted a more aggressive strategy, openly attacking the concept of phlogiston in his addresses to the Academy. Several factors had strengthened his hand, most notably the discovery of the composition of water. That discovery provided a rebuttal to the phlogistic claim that inflammable air, evolved during the solution of metals in acid, arose from phlogiston in the metal; decomposition of water now offered an alternative source. The new strategy, backed by new evidence and arguments, persuaded a trickle of the fence-sitters to align themselves publicly with Lavoisier's position. In the summer of 1784 J. A. C. Chaptal wrote to Lavoisier from Montpellier announcing himself as a 'zealous disciple' of Lavoisier's view on acids and citing the analysis of water as an 'experimentum crucis' in the debate. In the spring of 1785 Berthollet openly adopted Lavoisier's view in a paper he read to the Academy on dephlogisticated marine acid (chlorine). Fourcroy's espousal of Lavoisier's chemistry in the 1786 edition of his textbook was a further influential endorsement. Finally, the conversion of an ardent defender of phlogiston, Guyton de Morveau, during his visit to Paris in the early months of 1787 completed the core of Lavoisier's lieutenants in the struggle to transform chemistry.

During the same period opposition to the new chemistry began to coalesce. Lavoisier's "Reflections on Phlogiston" (read in the summer of 1785) provoked such turmoil that members of the audience found it difficult to hear the speaker. It is only from this time that we can identify an open sense of crisis in the French community. Apprehension among its members over Lavoisier's attack found a voice in the *Journal de physique*, now under the editorship of J. C. Delamétherie. With each escalation of the debate Delamétherie's review articles and editorial comments became more stridently critical of the new chemistry; he actively sought contributions opposing it. Early in 1787 the Genevan naturalist, J. Senebier, published in the *Journal* a reply to Lavoisier's "Reflections". But it was publication of the new chemical nomenclature by Lavoisier and associates that provoked the greatest consternation. Moderate critics objected that it was a premature attempt at systematization; more extreme foes saw it as a blatant attempt to impose use of the new concepts. Despite mounting opposition between 1785 and 1789, the trickle of converts became a stream. Although Lavoisier would probably not yet have won a headcount, the growing number of converts, their conviction, and their influence in the scientific community convinced him that victory was at hand.

5. A SPECTRUM OF RESPONSES

To gauge the response of an entire scientific community to a new set of guiding assumptions is no straightforward matter. There was, of course, no referéndum or poll of 18th-century chemists to which we can turn. Our characterization of the communal reception of the new chemistry must be based upon a composite of individual responses. The task leaves us with a dual problem of methodology. (i) to delineate the community in question and (ii) to characterize attitudes of community members toward the new chemistry.

One approach to the first problem is to work with a predefined sample of chemical articles or chemists.[6] However, the 18th century was still a time when disciplinary lines were not sharply drawn. More to the point, many of Lavoisier's key collaborators had previously worked in fields remote from chemistry; to limit my analysis to response from a community of chemists would omit much influential activity and opinion. My composite, therefore, includes any contributor to science who played a documented part in the debate over the new chemistry (however brief or minor). To prepare a list of participants that is as complete as possible, I have surveyed the major French scientific periodicals and monographs. In addition, I have searched manuscript sources, especially letters and lecture notes, for opinions scientists confided to colleagues and information teachers passed on to students. The list is not exhaustive, but it is extensive and includes the major and many minor actors.

It is doubly difficult to characterize the stance of participants toward the new chemistry, that is, to document and interpret attitudes. As a starting point, the secondary literature is sprinkled with citations of supporters of the Stahlian or, alternatively, the Lavoisian theory (culled from remarks published at different times).[7] But it was quite a different thing to be a supporter of phlogiston in 1775, say, as compared with 1785 or 1795; the state of the debate and the prevailing opinion in the community shifted dramatically during that period. Such a simple approach neglects the *time dependence of the choice*. In my experience, a little further digging has often revealed that an individual whom tradition has cast as a 'phlogistian' (with the negative overtones the term carries in Whiggish histories) subsequently shifted his allegiance. Moreover, my own and other detailed studies of individual responses to the new chemistry show that conversion was typically a *gradual change of attitude* toward its claims, rather than a sudden shift to a new way of seeing the world.[8]

To represent the communal reception, I have prepared a spectrum of responses from sixty-nine French participants, in rough chronological order of their adopting the new theory (Figure 3).[9] Attaching a specific date may appear antithetical to my understanding of the conversion process. Let me therefore emphasize that these dates refer not to quasi-mystical moments of insight, but to documentation that allows us to say that a scientist has – by that time – endorsed or employed key components of the Lavoisian system. There remains an ambiguity of *how much* of the new chemistry an individual need embrace in order to be deemed a convert, for it encompassed diverse components. Among the most important, in approximate order of their unveiling, were (i) the absorption of air in combustion and calcination, (ii) the analysis of atmospheric air into two distinct gases, (iii) the oxygen theory of acids, (iv) the caloric theory of heat and the vapor state, (v) the composition of water, (vi) the rejection of phlogiston, and (vii) the new nomenclature. The same order represents, more or less, an ascending degree of difficulty of adoption.

Between those who campaigned wholeheartedly for acceptance of the Lavoisian system and those who denied even its most modest claims, there were many shades of response. Components (i) and (ii) were fairly easily reconciled with modified versions of traditional theory; they were granted by such prominent defenders of phlogiston as Macquer and Berthollet in the 1770s. Concession on (iii) and (iv) is more significant, often implying a measure of sympathy with Lavoisier's approach. For an individual to yield on (v) is a stronger indicator of reorientation, for Lavoisier's view on the composition of water was highly controversial. Even that concession, however, need not signal abandonment of all hope of salvaging phlogiston. Finally, public renunciation of phlogiston and exclusive use of the new nomenclature are the strongest signs of conversion.

A few examples will help to clarify the criteria of inclusion in the

Fig. 3. A spectrum of response.

NAME	BORN	EXPOSURE	ADOPTION	AGE IN 1785
Trudaine de Montigny	1733		(1773)	d 1777
Bayen, P.	1725		(1774)	60
*Laplace, P. S.	1749		1777	36
Bucquet, J. B. M.	1746	1773	1778/9	d 1780
Chaptal, J. A. C. (Montpell.)	1756	1777-9	1784	29
*Monge, G.	1746		1784	39
*Meusnier de la Place, J. B.	1754		1784	31
Clouet, J. F. (Mézières)	1751		(1784)	34
Berthollet, C. L.	1748	c.1775	1785	37
*Vandermonde, A.	1735		1785	50
*Brisson, M. J.	1723		(1785)	62
Gengembre, P.	?		1785	?
Fourcroy, A. F.	1755	c.1778	1785/6	30
Fougeroux de Bondaroy	1732		(1785)	53
Picot de la Peyrouse (Toul.)	1744		(1785)	41
Reboul, H. (Toul.)	1763		1785	22
*Hassenfratz, J. H.	1755		1786	30
Adet, P. A.	1763		1786	22
Brogniart, Alex.	1770		1786	15
*Cousin, J. A. J.	1739		1786	46
*Schurer, J. L. (Stras.)	1734		(1786)	51
Guyton de Morveau (Dijon)	1737	1774	1787	48
Chaussier, F. (Dijon)	1746		(1787)	39
Riche, C. A. (Montpell.)	1762		1787	23
Puymaurin, fils (Toul.)	?		1787	?
*Ehrmann, F. L. (Stras.)	1741		(1787)	44
Nicolas, P. F. (Nancy)	1743		(1787)	42
Tessié du Clouseau (Angers)	?		1787	?
Bonjour, F. J.	1754		1787	31
*Lefèvre-Gineau, L.	1751		1788	34
*Saussure, H. B. de (Gen.)	1740		1788	45
*Lagrange, J. L.	1736		(1788)	49
*Caritat de Condorcet	1743		(1788)	42
*Dionis du Séjour, A. P.	1734		(1788)	51
Macquart, L. C. H.	1745		1788	40
La Rochefoucault d'Enville	1743		(1788)	42
Poulletier de la Salle	1719		(1788)	66
Dietrich, P. F. de	1748		(1788)	37
Seguin, A.	1767		1789	18
*Schurer, F. L. (Stras.)	1764		1789	21

Fig. 3. *continued*

NAME	BORN	EXPOSURE	ADOPTION	AGE IN 1785
Van Mons, J. B. (Bruss.)	1765		1789	20
Gosse, H. A. (Geneva)	1753		(1789)	32
*Pictet, M. A (Geneva)	1752		1789	33
Proust, L. J. (Spain)	1754		(1789)	31
Cadet de Vaux, A.	1743		1789	42
-----------------------	--		----	--
Bouillon-Lagrange, E. J. B.	1764		1790	21
Vauquelin, N. L.	1763		1790	22
Pelletier, B.	1761		1790	24
*La Croix, S. F. (Besançon)	1765		1790	20
*Arbogast, L. F. A. (Stras.)	1759		1790	26
Haussmann, J. M. (Colmar)	1749		1790	36
Bouvier, A. M. J.	1746		1790	39
Silvestre, A. F. baron de	1762		1790	23
Chappe, C.	1763		1790	22
Senebier, J. (Geneva)	1742		1791	43
Hallé, J. N.	1754		1791	31
*Beraud, J. J. (Marseille)	1753		1791	32
Lefebre d' Hellancourt, A. M.	?		1791	?
Gillet-Laumont, F. P. N.	1747		1791	38
Leblond, J. B., (Guyana)	1747		1791	38
Thouret, M. A.	1749		1791	36
Dizé, M. J. J.	1764		(1791)	21
*Libes, A. (Béziers)	1752		1792	33
Curaudau, F. R. (Vendôme)	1765		1792	20
Bartholdi, C. G. (Colmar)	1762		1792	23
Saussure, N. de (Geneva)	1767		1792	18
Deyeux, N.	1745		1793	40
Parmentier, A.	1737		1793	48
Darcet, J.	1725		1790s	60

table. The most straightforward category consists of individuals who collaborated with Lavoisier in developing the new chemistry or campaigned on its behalf.[10] Their conviction is evident and the public nature of their involvement aids in dating their conversion. They include members of Lavoisier's immediate circle in Paris (such as Berthollet, Fourcroy, Laplace, Monge, Hassenfratz and Adet) and proselytes outside the capital (Guyton de Morveau, Chaptal, Reboul, Tessié du Clouseau, F. L. Schurer, Pictet, Van Mons, among others). More ambiguous is the case of individuals who did not campaign actively for the new view, but spoke favorably of or implicitly adopted some of its components. For example, both Fougeroux and Picot in 1785 endorsed Lavoisier's demonstration of the compound nature of water, but it is not clear that their support at that time extended to exclusion of phlogiston. Also problematic are supporters who did not write on chemical topics (the physicists and mathematicians), but apparently lent moral support. Lagrange is included, not only because a colleague (J. B. Delambre) later recalled his infatuation with Lavoisian chemistry, but because Fourcroy in 1788 included Lagrange in a short list of supporters. On the other hand, C. A. Coulomb, described by biographers as a friend and supporter of Lavoisier, is omitted simply because I have not yet found contemporary evidence of when (or how) that support was manifested.

With these qualifications in mind let us review what Figure 3 can tell us about the pattern of communal response. During the first phase of reception there was really no one who could be labeled a convert to Lavoisier's assumptions, several of which had not yet been articulated or divulged. Trudaine de Montigny collaborated with him on experiments and encouraged his investigation of the chemical role of air, but such support need not imply a fundamental break with tradition. Trudaine, in any case, kept his views to himself. P. Bayen was emboldened by Lavoisier's discoveries to voice doubts he already had concerning phlogiston. At the outset of the second phase, Lavoisier worked in close concert with Laplace and Bucquet, who largely shared his views. But it was only from the beginning of the third phase – the open campaign against phlogiston – that several chemists came out in his defense.

The years 1784–1789 were the key period of transition. For that interval I have identified forty-one individuals who adopted the Lavoisian position on key issues. Their numbers include core members of the antiphlogistic party as well as some more obscure participants. Fifteen members of the group (more then a third) can be identified as mathematicians or physicists, confirming Lavoisier's claim of substantial support from those disciplines. Four are naturalists (Fougeroux, Picot, Reboul and Dietrich); three are perhaps best described as amateurs and promoters of the sciences (La Rochefoucauld, Poulletier and Cadet de Vaux). The nineteen remaining had a primary orientation toward chemistry, although nearly all were trained in medicine or pharmacy.

Participation in the debate differed for members of the various disciplines. Among the mathematicians or engineers, a few (Monge,

Vandermonde, Meusnier and Hassenfratz) were drawn into major chemical investigations, largely through Lavoisier's personal intervention. However, most of them were cast in more casual roles, offering supportive remarks in reports to the Academy, book reviews or conversation. A few young men who opened new courses of experimental physics during the period (Lefèvre-Gineau, F. L. Schurer and Pictet) were enthusiasts of the 'pneumatic doctrine', as it was called, and played a prominent part in its local promotion. Among the amateurs and the naturalists (with the notable exception of the young Reboul) support was largely behind the scenes. In the end it was the *chemists* among the converts who were the key agents in assimilating the new system. Their number includes many of France's most promising chemists and, importantly, its most popular teachers of chemistry. They played a predominant role in training the next generation of French scientists.

The struggle for the new chemistry in France was mostly acted out by 1790. Although division remained in the community, it was clear by then, to participants in both camps, that the tide was turning; a fourth phase of consolidation began. Those who had shifted during the earlier transitional period were enthusiasts who took a risk by identifying with an unpopular cause. The mix of converts after 1790 is a little different, including beginners who adopted the new chemistry without learning the old, fence-sitters who had been waiting to see how the wind changed, and skeptics whose objections had been worn down.

Clearly, not all constituencies with an interest in the debate are proportionally represented here. There must have been several hundred students each year who attended lectures by the antiphlogistians (Chaptal, alone, claimed to have audiences of up to four or five hundred). Presumably, many accepted what they heard, but only the active few appear in my sample. Among established chemists, on the other hand, there were many who neither challenged nor endorsed the new views: some explicitly adopted a neutral or wait-and-see attitude, others simply avoided comment on theoretical matters.[11] Of these silent observers, some quietly adopted the new system at a later date; others simply disappeared from chemistry by reason of death, retirement, emigration or discontent.

Where, then, are the *opponents* of the new chemistry (of whom we have heard little)? The simple answer is that they are represented on the same list. Nearly all converts initially *resisted* Lavoisier's interpretations. Even his chief allies among the chemists (Bucquet, Chaptal,

Fig. 4. Continued opposition

Name	Born	Continued Resistance	Age in 1785
Marivetz, baron E. C. de	1731	1788	54
Machy, J. F. de	1728	1788	57
Romé de Lisle, J. B. L.	1736	1788	49
Para de Phanjas, abbé	1724	1788	61
Opoix, C. (Provins)	1745	1789	40
Thouvenel, P.	1747	1790	38
Delamétherie, J. C.	1743	c.1795	42
Lamarck, J. B. de	1744	1796	41
Baumé, A.	1728	1798	57
Monnet, A. G.	1734	1798	51
Sage, B. G.	1740	1818	45

Berthollet, Fourcroy, Guyton) went through a conversion lasting several years from the time of initial engagement with Lavoisier's ideas until a public display of their support. There remains a further interesting group of opponents, those who (as far as I can determine) took their opposition to their deathbed. Eleven recalcitrants are listed in Figure 4, along with the date of a major diatribe against the new chemistry. The response of this group may be contrasted with that of the converts not only by the persistence of their opposition but by its animosity. In contrast to the reasoned opposition of a Berthollet, a Guyton or a Senebier, the late attacks from this group were acrimonious and directed as much against the antiphlogistians as against their theories.

What characteristics distinguish this group from those who converted or at least bowed to the inevitability of the new doctrine? In the first place there is an age differential. The average age of the recalcitrants in 1785 (the year of welling crisis) was 48.6. On the other hand, the average age (in the same year) of all converts for whom I have birthdates was 35.4. For an even more striking contrast, consider the nineteen chemists who actively supported the new theory between 1785 and 1789: their average age in 1785 was 30.7 – nearly a generation younger than the recalcitrants. Typically, the recalcitrants were conservative, their opposition extending not only to Lavoisier but to modern chemistry in general. Finally, most of them were marginal to chemistry or became so in the 1790s; typically, they complained of neglect or unfair treatment from the antiphlogistians.[12]

Like a study of the Darwinian shift in Britain, conducted by Hull *et al.* (1978), these results indicate an age differential between converts and opponents. But Hull's group concluded that the age factor has been exaggerated. Is Planck's principle, then, borne out for the case of the chemical revolution? Perhaps a clearer answer is possible if we distinguish strong and weak forms. Planck's strong statement implies that adoption of new scientific truth requires a new generation of scientists. The present study shows that several senior chemists gave Lavoisier's work a fair hearing from the start. A few eventually became strong supporters, while others gradually accustomed themselves to and employed the new ideas. On the other hand, a weaker version of Planck's hypothesis receives better support, for there was a sharp age differential between the early *enthusiasts* and the stubborn resistors. The correlation of age with receptivity, however, must be qualified, for it is not simply a matter of years. Social factors, for one, impinge: as Hufbauer (1982) has suggested, fringe figures and beginners in the field had less to lose than established chemists by staking their reputations on a new and unproven theory. Moreover, the quality of relations with the theory's promoters seems to have been important in the stubborn resistance displayed by the recalcitrants.

6. CONCLUSION

This study disconfirms the thesis that guiding assumptions change abruptly and totally (GA4.3). The conclusion is supported for the case of the chemical revolution by several different analyses: (1) The new guiding assumptions included several components of differing degrees of compatability with traditional assumptions. They could be (and, in fact, were) adopted sequentially and with different degrees of commitment. (2) Detailed studies of individual conversions (not reported here because of length restrictions) show they followed a gradual and stepwise pattern, with successive adaptation of new components into a modified traditional framework. (3) Debate within the community broke down into four phases: an early period in which Lavoisier's assumptions were not widely known and his discoveries were considered benign; a second phase when his intentions became clear, solutions were sought to problems raised by his innovations, theories proliferated, but controversy was avoided; a third phase of crisis as a few scientists rallied to

Lavoisier's side, resistance grew, and the community fragmented (GA4.2); finally, a cooling period (in the 1790s) during which resistance dwindled as many active scientists quietly adopted the new assumptions, though a handful continued stubborn opposition (GA4.5). (4) A profile of communal response to the new chemistry shows converts came over to the new view not all at once but over a period of several years – with active young chemists serving as the primary agents of dissemination, and senior chemists showing a greater degree of reluctance or skepticism.

The entire process took more than twenty years, though most active French chemists shifted within a decade of open crisis. Communication broke down to some extent, but that was due as much to personal factors (age, personality, background) and social factors (links to promoters, peer pressure and status concerns), as to incommensurability of assumptions. Beginners in chemistry and members of other disciplines – with little commitment to traditional chemistry – were able to adopt the new theory with less self-recrimination. Nevertheless, flexible senior chemists were able to understand and adopt the new views (after a period of adjustment), if usually with less fervor than their younger counterparts. The popular perception by scientists – such as Planck, Darwin and Lavoisier – that their views found a more ready reception among the young, is not unfounded. But Planck was overly pessimistic. The case of the chemical revolution shows that scientists can and do convert to novel and controversial theories, though the process is a gradual and complex one.

NOTES

[1]Planck phrased it this way: "new scientific truth does not triumph by convincing its opponents and making them see the light, but rather because its opponents eventually die, and a new generation grows up that is familiar with it" (Hull *et al.*, 1978).

[2]Crosland (1980) has recently reviewed problems of the chemical revolution.

[3]For new evidence concerning the origins and aim of Lavoisier's intended revolution in chemistry, see Perrin (1986b).

[4]Information on Lavoisier's publications is gleaned from Duveen and Klickstein (1954); I simply weeded out items conveying no information about the revolutionary ideas and

plotted the rest. In Figure 1, the larger blocks represent books and the smaller ones articles.

[5]For a fuller discussion of the campaign for the new chemistry in France and the resistance it generated, see Perrin (1981).

[6]H. G. McCann (1978) attempted an exhaustive survey of original chemical articles, published in France and Britain between 1760 and 1795, as the basis of his quantitative study of the transition from phlogiston to oxygen; K. Hufbauer (1982), more fruitfully, used prosopographic techniques to profile the German chemical community and their collective response to Lavoisier.

[7]See, for example, Partington (1962, pp. 488–494).

[8]I have developed this theme in an unpublished paper 'Gestalt Switch or Attitude Change? Cognitive Response to Lavoisier's Chemistry', read at Virginia Tech on 11 April 1985. Examples of my detailed analyses of the conversions of J. Black and Chaptal may be found in Perrin (1982) and Perrin (1986a).

[9]The sources of information represented in Figure 3 are too numerous to list here. An individual's location is given in parentheses; where none appears, he was in Paris at the time of adoption. An asterisk indicates a disciplinary orientation toward mathematics or physics. The column *Exposure* gives approximate dates, for a few key converts, of early encounter with Lavoisier's claims. Where my sources are equivocal on the extent or timing of the conversion, I have made a qualified inclusion by placing the year in parentheses.

[10]Unqualified acceptance of all of Lavoisier's claims would be too rigid a criterion of conversion (that even Lavoisier himself might not satisfy). If Berthollet, for example, retained some doubts about the acid theory, or if Laplace did not accept the material theory of heat, such differences did not impair their enthusiastic support for the Lavoisian system as a whole. Such differences were regarded as unresolved topics within the Lavoisian framework, rather than challenges to it.

[11]In his survey of some 858 French and British chemical articles, McCann (1978) found only a handful of neutral ones. He has taken this result as confirmation of Kuhn's analysis of conversion as an all-or-nothing gestalt switch. It should be pointed out, however, that research articles may not be the most likely place to find expressions of neutrality or hesitancy. While some chemists with strong convictions expressed them in print during the transitional period, others who experienced doubt or uncertainty were perhaps more likely to *avoid* theory. (Some 43% of McCann's articles were 'no information'; moreover, others were pegged as 'phlogiston' or 'oxygen' on implicit grounds.) My approach, through individuals rather than articles, has revealed many cases of partial adoption, attempts at compromise, and expressions of formal neutrality.

[12]After the revolution, when the Academy was reformed in 1795 as the First Class of the Institute (with Lavoisier's former associates firmly in control), Sage and Baumé did not regain their regular places, but only associate memberships. Sage had also lost his positions at the Mint and at the School of Mines. In 1798, Baumé suffered the humiliation of unsuccessfully soliciting votes (from former junior associates) for a vacant place in the Academy. It is no coincidence that Baumé's major attack on the antiphlogistians was written that same year, or that Sage continued for the rest of his life to blame his misfortune upon former colleagues such as Fourcroy and Chaptal.

REFERENCES

Crosland, M. (1980), 'Chemistry and the Chemical Revolution', in G. S. Rousseau and R. Porter (eds.), *The Ferment of Knowledge*, Cambridge: Cambridge Univ. Press, pp. 389–416.

Duveen, D. I. and Klickstein, H. S. (1954), *A Bibliography of the Works of Antoine Laurent Lavoisier, 1743–1794*, London: Dawson and Weil.

Guerlac, H. (1975), *Antoine Laurent Lavoisier, Chemist and Revolutionary*, New York: Scribner's.

Hufbauer, K. (1982), *The Formation of the German Chemical Community (1720–1795)*, Berkeley: Univ. of California Press.

Hull, D. L., Tessner, P. D. and Diamond, A. M. (1978), 'Planck's Principle', *Science* **202**, 717–723.

Kuhn, T. S. (1962), *The Structure of Scientific Revolutions*, Chicago: Univ. of Chicago Press, 2nd edn., enlarged, 1970.

McCann, H. G. (1978), *Chemistry Transformed: The Paradigmatic Shift from Phlogiston to Oxygen*, New Jersey: Ablex, Norwood.

Partington, J. R. (1962), *A History of Chemistry*, vol. 3, London: Macmillan.

Perrin, C. E. (1981), 'The Triumph of the Antiphlogistians', in H. Woolf (ed.), *The Analytic Spirit: Essays in the History of Science in Honor of Henry Guerlac*, Ithaca: Cornell Univ. Press, pp. 40–63.

Perrin, C. E. (1982), 'A Reluctant Catalyst: Joseph Black and the Edinburgh Reception of Lavoisier's Chemistry', *Ambix* **29**, 141–176.

Perrin, C. E. (1986a), 'Of Theory Shifts and Industrial Innovations: the Relations of J. A. C. Chaptal and A. L. Lavoisier', *Annals of Science* **43**, 511–542.

Perrin, C. E. (1986b), 'Lavoisier's Thoughts on Calcination and Combustion, 1772–1773', *Isis* **77**, 647–666.

SEYMOUR H. MAUSKOPF

MOLECULAR GEOMETRY IN 19TH-CENTURY FRANCE: SHIFTS IN GUIDING ASSUMPTIONS

1. INTRODUCTION

The histories of crystallography and mineralogy have been comparatively neglected by historians and philosophers of science. Recently,
however, this state of affairs has begun to change with the appearance of
a number of books and smaller monographs by historians of science.
There is general consensus that the work of the French scientist
René-Just Haüy (1743–1822) marked a watershed in the development of
these fields. Haüy devised the first comprehensive quantitative model of
crystal structure;[1] subsequently he turned to mineral classification and
developed a system whose basic unit, the species, was dependent upon
his theory of crystal structure. With these achievements, crystallography
and perhaps mineralogy are considered to have reached scientific ma-
turity.

But if Haüy's work brought crystallography and mineralogy to ma-
turity, the ascendancy of his work was short lived in the views of a
number of the historians who have addressed these subjects, barely
outlasting Haüy himself. In the Introduction to his book, *Origins of the
Science of Crystals*, John G. Burke characterized the fate of Haüy's
work thus:

> The science of crystals was the science of Haüy. But when the deficiencies in Haüy's
> theory became apparent, when scientists produced new facts contradicting certain impor-
> tant foundations of the science as it had been erected by Haüy, there was a fresh approach
> which incorporated only the undeniably valid portions of the previous one. A new paradigm
> was established, and thereafter crystallographers sought to extend its usefulness.[2]

Later in his book, Burke was more specific and even more emphatic:

> The discovery of isomorphism and polymorphism and the recognition of a definite
> relationship between optical properties and crystal form were among the principal causes
> for the abandonment of Haüy's crystallographic method and the mineralogical system he
> had based upon it.[3]

Norma E. Emerton has recently echoed this interpretation:

> Those who followed Haüy found that his emphasis on cleavage and the crystal nucleus, the

125

geometrical form of the integrant molecule, and the concrete embodiment of the primitive form were an embarrassment now that they no longer acknowledged the older chemical and corpuscular interpretations of form in which these ideas were rooted. So they retained his mathematical treatment while abandoning "certain inconvenient and even unphilosophical views embraced by the method of Haüy"; these "unphilosophical views" were the solid polyhedral nucleus and rows of angular particles that represented the last stage in the development of the form concept, of which a crystallographer wrote: "The whole theory of molecules and decrements is to be regarded as little else than a series of symbolic characters. . . . We ought to divest our notions of molecules and decrements of . . . absolute reality."[4]

Burke's and Emerton's evaluation of what happened after the death of Haüy in crystallography (at least) supports (GA4.3) and (GA4.4). Rapid and comprehensive change of guiding assumptions is posited as having overtaken the community of crystallographers and mineralogists. Another very recent study, *Die Entdeckung des Isomorphismus* by Hans-Werner Schütt,[5] arrives at a somewhat different evaluation of the fate of Haüy's crystallography and mineralogy. Schütt attempts to use the historical episode of the ascendancy of Haüy's crystallography and mineralogy and the challenge to it by Eilhard Mitscherlich's discovery of isomorphism as a case study of Thomas S. Kuhn's theory of scientific change.[6] Schütt sees Haüy's crystallography and mineralogy, which he leaves essentially undifferentiated, as constituting a dominant paradigm for the period 1800–1820; indeed, he tries to identify those elements of Haüy's theory which would correspond to the categories of Kuhn's disciplinary matrix. He sees Mitscherlich's discovery as having undercut Haüy's concept of mineral species (and thereby his mineralogical system), but he also perceives a Haüyian tradition persisting in France for decades after the impact of Mitscherlich's discovery. I am in general agreement with Schütt's interpretation; this is understandable in light of the fact that Schütt's proposition about the French tradition is based largely upon my own work.[7] However, I do not think that Schütt has systematically assessed which elements of Haüy's crystallography and mineralogy persist, which are modified and which are abandoned.

What I propose to do in this paper is to examine more systematically than has been done heretofore, the theory changes which took place in crystallography and mineralogy in the period after the death of Haüy. I shall concentrate on France because I know the French scientific literature best. What I shall specifically assert is: if much of the relevant European scientific community may have abandoned Haüyian crystallography and mineralogy *toute court* after his death, by no means all did.

In France, a complex development took place in which guiding assumptions (including theories and models) changed neither abruptly nor totally (contra GA4.3). It is, indeed, possible to demonstrate the perseverance of a French tradition rooted in Haüy's crystal structure theory to the end of the nineteenth century.

This raises an issue not directly addressed in any of the theses about scientific change under consideration: the role of such sociocultural perspectives as national traditions. For the case under study here, I believe that sensitivity to national traditions within the community of nineteenth-century crystallographers and mineralogists[8] is important in assessing the nature of the scientific change which occurred after the death of Haüy. Historians like Burke and Emerton who have concluded that a sudden and comprehensive switching of allegiance took place at that time have done so, I believe, with insufficient sensitivity to different national traditions.

2. THE CRYSTAL STRUCTURE THEORY AND MINERALOGY OF HAÜY

Haüy set forth his theory of crystal structure virtually full-blown in his first book, *Essai d'une théorie sur la structure des crystaux*; only minor modifications were made to the theory between 1784 and 1801, the year of publication of his monumental *Traité de minéralogie*.[9] As the title of the latter work indicates, Haüy had turned his attention to mineralogy in this interval and had developed a system of classification whose rationale, at the species level, was dependent on his matter theory as encapsulated in his theory of crystal structure.

What follows is a schematization of the guiding assumptions in Haüy's crystallography and mineralogy. Although the assumptions of the two sciences are closely related, I separate them for the purpose of clarity.

Matter Theory (Crystallography)

I. Two Stage Molecular Theory of Crystal Structure
    A. Determinately-shaped polyhedral integrant molecules (*molécules intégrantes*) composed of determinately-shaped elementary molecules (*molécules constituantes*)
        1. Only one possible polyhedral shape for each kind of integrant molecule.
    B. Crystals ≅ assemblage of integrant molecules. Geometry of

macrocrystal determined by geometry of integrant molecule; form of integrant molecule inferred from structure of macrocrystal.

II. Crystal-Structure Model: Crystals envisioned as built up of integrant molecules stacked like bricks in a pyramid.
  A. Crystal structure analyzed into nuclear or primitive form and secondary form.
    1. Primitive form constant for and common to all crystals of the same species.
    2. Secondary forms (actual external forms) derived from primitive forms through laws of molecular decrement: each layer of integrant molecules laid down on the faces of the primitive form envisioned as recessed by a small integer number of molecules as compared with layer below it. (see Figure 1.)

As Burke summarizes Haüy's view: "The experimental part of Haüy's theory, then, was the establishment of the shapes of the integrant molecule and the primitive form and the application of the correct laws of decrement in order to develop the secondary form. Under no circumstance could the assumption of decrement by rows of fractional molecules be admitted."[10]

III. Empirical Support for Theory
  A. Law of constancy of interfacial angles: "the respective inclination of the same faces were constant and invariable in each crystalline species"[11]
  B. Revelation of primitive forms of many crystals through cleavage.

IV. Issues Not (or Rarely) Addressed
  A. Models of the shapes of constituent molecules and of their arrangement to form integrant molecules
  B. Explanation for why different secondary forms develop in same species.

Mineral Taxonomy: Guiding Assumption Derived From Crystal Structure Theory

V. Doctrine of Fixed Mineral Species
  A. Defined as constant in chemical composition and crystal form (primitive form, constancy of interfacial angles)
  B. Mineralogical 'individual' = integrant molecule from whose fixed

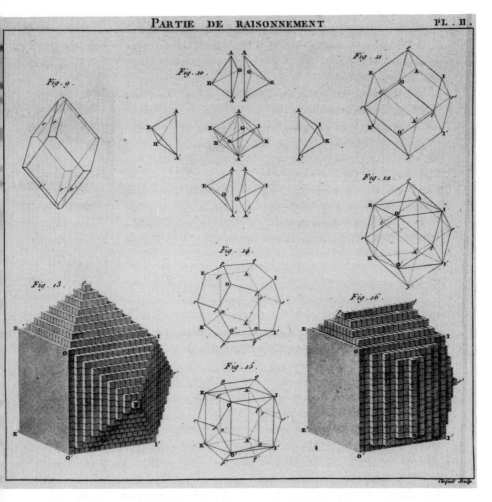

Fig. 1. Haüy's models of molecular crystal structure. From *Traité de minéralogie* (1801), Atlas.

composition and form derive the chemical and crystallographical constancies of the species

C. In cases of apparent discrepancies regarding V. A. crystallography to take precedence over chemistry. Crystallography held to be better able to:

    1. Determine essential identity or difference between minerals

2. Delineate essential form-giving components of minerals from mechanically intermixed foreign substances.

Contemporary Alternatives
VI. Matter Theory
   A. Dynamical non-molecular: Germany; J. J. Bernhardi, C. S. Weiss, F. Mohs
   B. Stacking of spherical/spheroidal particles: England; J. Dalton, W. H. Wollaston, J. F. Daniell.

VII. Mineralogy
   A. Denial of fixed mineral species: J. F. d'Aubuisson, C. L. Berthollet
   B. Denial of precedence of crystallography in species determination: J. J. Berzelius (over chemistry), Wernerians (over external characteristics).

3. DISCUSSION

Although one of the alternative models of crystal structure, that utilizing spherical/spheroidal particles, went back to the time of Hooke and Huyghens, it is generally agreed that Haüy's crystal structure theory was the first comprehensive and quantitative one. Its principal merit lay in its ability to relate crystals of the same chemical composition but widely varying external forms to each other geometrically. All historians who have examined crystallography in the late eighteenth and early 19th century agree that Haüy's theory was dominant during his lifetime, although the two above-mentioned alternatives were also being developed.

The case of Haüy's mineralogy is more complex because there already pre-existed several approaches to mineral classification by the time Haüy turned to this subject. The most prevalent were those based on chemical criteria and exemplified by the systems of the Swedish mineralogists A. F. Cronstedt and T. Bergman.[12] During the last quarter of the 18th century, A. G. Werner at the Bergakademie in Freiberg had developed his own approach designed for practical purposes and based primarily upon a wide range of external characteristics (including crystal form). At the same time, in France a more theoretical mineralogical approach was developed by Haüy's immediate predecessor (and competitor), J.-B Romé de l'Isle and then by Haüy himself. It was based upon chemical and crystallographical characteristics; it postulated the existence of distinct, fixed mineral species; and it identified the putative individuals corresponding to the organic individuals as the integrant molecule.

According to the recent study by Schütt, Haüy's mineralogy achieved the same European-wide ascendancy as did his crystallography after the publication of his *Traité de minéralogie* in 1801 despite the pre-existence of other approaches to mineral classification and the institutional force of Werner's mining school at Freiberg. In Schütt's analysis of the mineralogical 'community' in the first two decades of the 19th century, he could identify only a handful of writers on mineralogy who rejected Haüy's system and doctrine of mineral species.[13]

Towards the end of his life, Haüy published two massive summations of the achievements of his scientific career: a revised edition of his *Traité de minéralogie*[14] and a *Traité de cristallographie*.[15] By the time these works appeared, the ascendancy of Haüy's mineral species concept – and in the view of Burke and Emerton, his crystallography – were seriously threatened. The threat to his concept of mineral species came from Mitscherlich's work on isomorphism and, to a subsidiary degree, dimorphism. It will be helpful to define these two concepts as Mitscherlich defined them. First, isomorphism:

. . . the law of the relation between chemical composition and crystalline form could thus be enunciated: the same number of atoms combined in the same manner produced the same crystalline form and [the same crystalline form] is independent of the chemical nature of the atoms and it is only determined by the number and relative position of the atoms.[16]

Concerning dimorphism, Mitscherlich wrote:

If the relative position of the atoms which have produced a crystal is changed by some circumstance, the primitive form will no longer remain the same.[17]

The concepts of isomorphism and dimorphism accounted for and explained in a manner different from Haüy's several types of discrepancies to his mineral species concept. Isomorphism accounted for those cases where minerals or mineral substances of variable or even different composition appeared to crystallize in the same form. The Haüyian strategy to deal with such cases took one of two forms: either to deny that the crystal forms were exactly the same or to postulate that there was an 'essential' form-giving component of constant composition which composed the true integrant molecules and that the apparent compositional variability was due to mechanically intermixed inclusions of substances foreign to the integrant molecules. On the eve of Mitscherlich's discovery of isomorphism, Haüy's own disciple F.-S. Beudant had studied artificially prepared mixed crystallizations of sulphate salts of various metals in which he had found the 'essential' component to constitute only 2 percent of the mass in some cases.[18]

Dimorphism accounted for the discrepancy in which minerals of the same apparent chemical composition crystallized in different and geometrically incompatible forms.[19] Prior to Mitscherlich, the principal mode of dealing with such cases was either to try to reconcile the geometrical disparity or to search for some hitherto undetected chemical difference.

Mitscherlich came to study what he termed isomorphism in the context of chemical stoichiometry (i.e., determination of quantitative chemical formulas); his perception – 'discovery' – was that those substances which crystallized in the same form and formed mixed crystallizations also had analogous chemical formulas. In place of the attempt to differentiate between 'essential' and mechanically intermixed components, Mitscherlich approached the issue through atomic stoichiometry, as his law of isomorphism makes clear. Similarly, Mitscherlich's atomistic explanation of dimorphism obviated the need to search for chemical differences between such substances.

There is no doubt that Mitscherlich's work on isomorphism and polymorphism seriously undercut Haüy's mineralogy, notably his concept of fixed mineral species and his espousal of the primacy of crystallography over chemistry in their determination. To mention illustrative examples, two of Haüy's earlier supporters, Berzelius and Beudant, came to abjure Haüy's dogma on these points. Evidently the consensus over Haüy's mineralogy which Schütt demonstrated prevailed in the decades before Haüy's death had been destroyed by the mid 1820s.

It is also true that the European-wide ascendancy which Haüy's crystal structure theory had enjoyed was also fracturing by this time. But to suggest that Haüy's crystallography (and mineralogy) were completely discredited is, I believe, in error; those historians who have made this assertion have been insufficiently sensitive to an issue which I think is very important in this case and undoubtedly many others where the evaluation of scientific change is at stake: the existence (and persistence) of different national traditions.

Specifically, what I mean to assert is that while the strong characterization of the overthrow of Haüy's system may apply to some countries, notably Germany and England, it does not describe what transpired in France during the rest of the nineteenth century. The remainder of this paper will give an overview of the French developments. Before moving on to this, I would point out that systematic and comparative study of the development of crystallography and mineralogy along national lines has never been carried out; even Schütt gives only the merest sketch of such development after 1825, aside from France. It is my suspicion that the demise of Haüy's crystallography outside of France was due as much to the pre-existence of other explanatory traditions as it was to the demonstration of deficiencies in Haüy's system.

4. THE FRENCH TRADITION

The case of France indeed illustrates how a clearly identifiable tradition traceable to Haüy's crystal structure theory persisted throughout the 19th century, surviving Mitscherlich's discoveries and even incorporating them into itself. This was made possible by the assimilation of important modifications, notably those proposed by A.-M. Ampère. I shall schematize the modification and the subsequent development of French crystallography and mineralogy. I shall then conclude with a commentary on this schema.

Modification to Molecular Concept Proposed by Ampère (1814)
Ampère was most interested in an issue rarely addressed by Haüy and his followers: compound molecular formation (above, IV.A.) He devised a theory of chemical combination and associated models of elementary and compound molecules.

I. Theory of chemical combination.
   A.   Chemical combination envisioned as the mutual penetration of molecular polyhedra (of the reactants) to form new compound polyhedral molecules.

II. Model of polyhedral elementary molecule.
   A.   All molecules ('particules' in Ampère's terminology) considered to be polyhedra composed of point atoms ('molecules') with Dalto-

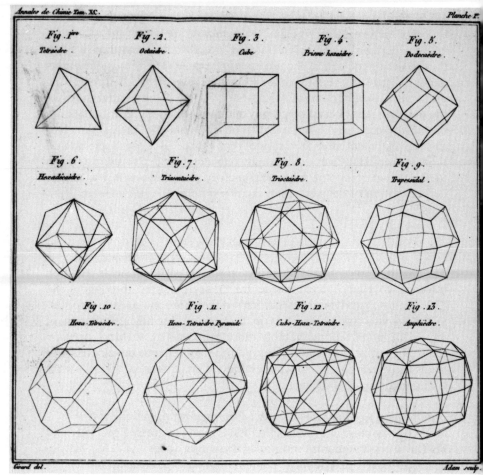

Fig. 2. Ampére's models of molecular polyhedra. From *Annales de Chimie,* Vol. 90 (1814).

nian gravimetric attributes, located at the apices of the solid angles of the polyhedra. (See Figure 2.)

  1. Simplest molecular polyhedra were of the form of five of Haüy's primitive forms of crystals. These posited especially for elementary gaseous molecules.

III. Two stage molecular crystal structure theory. Symmetry of crystal reflected form of crystalline molecules. The crystal structure theory was only implicit in Ampère's formulation since he was more interested in the chemistry of gases than in crystals.

Modified Matter Theory (of M.-A. Gaudin, A.-E. Baudrimont, A. Laurent, G. Delafosse, A. Bravais and later)
I. Two Stage Polyhedral Molecular Theories of Crystal Structure.
   A. Determinately shaped polyhedral compound molecules composed of atoms located at apices of solid angles of polyhedra (except Gaudin).
      1. Concern with chemical formation of compound molecules.
         a. Symmetry considerations (Gaudin, Baudrimont).
         b. Isomorphic substitution – radical, type theories (Baudrimont, Laurent).
         c. Use of crystallographical analogies for molecular formation (Laurent: fundamental and derived radicals).
   B. Crystals = assemblages of compound molecules. Geometry of macrocrystals determined by (and reflects) form of molecules.
      1. Laurent: reconsideration of isomorphism, polymorphism, isomerism.
      2. Delafosse: refinement to Haüy's crystal structure theory. Distinction between integrant molecule of Haüy and physical crystalline molecule particularly for treating hemihedrism. (See Figure 3.
         a. Influence of Laurent and Delafosse on L. Pasteur's discovery of enantiomorphism.
      3. Bravais: (from Delafosse) crystal lattice theory. Fourteen space lattices; molecular polyhedra located at nodes and determine lattice symmetry.

Mineral Taxonomy: Changes in Guiding Assumptions (not to be discussed in Commentary)
I. Doctrine of Mineral Species
   A. Fixed species no longer accepted in full generality by those outside of Haüyian crystallographical tradition;[20] where existence of fixed species is still accepted (Delafosse) now defined in terms of nature, number and arrangement of atoms in polyhedral molecule.[21]
   B. Not all accepted the mineralogical 'individual'; those who did (e.g., Delafosse, de Lapparent) saw it as being in the polyhedral physical molecule composed of chemical atoms.

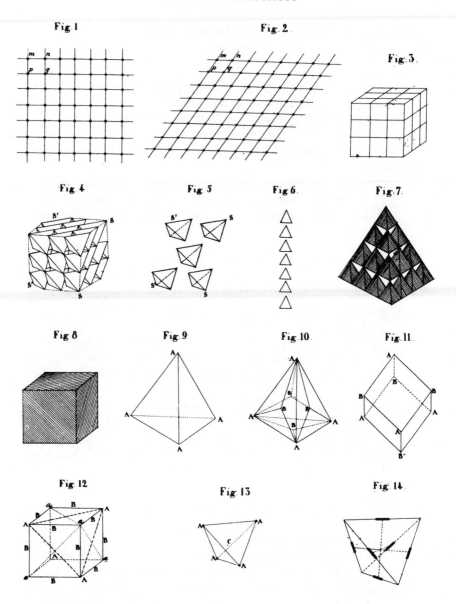

Fig. 3.   Delafosse's model of molecular structure of boracite (Figures 4, 5, and 8). From *'Recherches sur la cristallisation'* (1843).

C. No longer is crystallography considered to take precedence over chemistry; rather both together (plus physical characteristics) taken to define species.

5. COMMENTARY

Although Ampère's ideas on compound molecular formation were published in a lengthy article in the *Annales de chimie* in 1814 (where he also stated – or restated after Avogadro – the gas hypothesis), they apparently went unnoticed by Haüy, his associates or other scientists for many years. Ampère himself lectured on them in the 1820s, but the only scientist to take explicit notice before 1830 was the young chemist, J.-B. Dumas, and his interest was exclusively in the gas hypothesis.

Haüy's own crystal structure theory also appears to have lain dormant during the decade after his death. However, in the early 1830s, three young French scientists began anew to consider questions involving molecular form and crystal structure in a general Haüyian mode but also influenced by Ampère: M.-A. Gaudin (1804–1880), A.-E. Baudrimont (1806–1880) and A. Laurent (1807–1853). It is difficult to see many commonalities between the three, other than their interest in molecule form and crystal structure and the fact that they all were considered rather maverick scientists by their contemporaries. Baudrimont and Laurent were chemists, but each had reached that discipline by a different route; Gaudin is hard to denominate by any specific scientific discipline.

Gaudin was most clearly and directly influenced by Ampère; in expressing his debt to the latter, he also clearly delineated his relation to a Haüyian tradition:

Haüy, by his admirable work, founded mineralogy. It remained to show that cleavages and decrements are powerless, in certain cases, to reach the primitive molecule and above all the intimate arrangement of its atoms: this is the problem so happily opened up by the genius of Ampère and it is through his profound lessons at the Collège de France that I have made resolution to continue his work.[22]

The filiation of the Haüyian tradition to Baudrimont is more obscure since he was trained in pharmacy and medicine. Laurent's initiation to Haüy's theory undoubtedly occurred during his schooling at the École des Mines. Both Baudrimont and Laurent acknowledged the influence of Ampère's models at various times.

Space limitation precludes the presentation of their work in any detail, but a number of general comments on it are in order. All three subscribed to a two-stage molecular model of crystal structure, but they moved beyond Haüy (à la Ampère) to consider the chemistry of molecular formation. For Baudrimont and Laurent this became their focal consideration; they gave relatively little attention to the details of crystal structure theory *per se*. However, formal criteria and analogies from crystal structure theory were most important in their theories of molecular chemistry: as, for instance, in Baudrimont's and Laurent's utilization of the principle of isomorphism in their theories of substitution in organic chemistry. That they were able to incorporate isomorphism into their consideration of molecular formation – and even make it central to it – was due to the fact that they conceived the chemical molecule in Ampèrian terms as polyhedral arrangements of chemical atoms.[23]

Although they were comparatively unconcerned with working out models of crystal structure in detail, all three scientists believed that crystalline symmetry reflected molecular form. In the 1840s Laurent attempted to implement this belief by correlating changes of crystalline form with changes of chemical composition in organic substitution series (the molecule models of which were themselves based on analogy with Haüyian crystal structure theory). This led him to reconsider the relationship between isomorphism, dimorphism and isomerism. As I have shown,[24] this research program was of critical influence on Louis Pasteur in the work leading up to his discovery of enantiomorphism in 1848.

Also of great importance to Pasteur – and a major advance in crystallography *per se* – was the refinement made to Haüy's crystal structure theory by Gabriel Delafosse (1796–1878). In both his career and his scientific work, Delafosse represents the most direct continuation of the Haüyian tradition. He had been Haüy's student and had assisted the master in the preparation for publication of both the *Traité de cristallographie* and the second edition of the *Traité de minéralogie*. In 1843, Delafosse published a long paper on the molecular crystal structure of hemihedral crystals (crystals of incomplete symmetry), a problem which in his opinion could not be explained by Haüy's crystal structure theory. He therefore refined the theory by making a distinction between Haüy's integrant molecules and the true physical crystalline molecules. The former became the spatial unit of the reticulate arrangement of the latter in the crystal – the unit of the crystal lattice.

Regarding the physical molecules themselves, Delafosse was interested in working out models of molecular shape whereby their reticulate arrangement would naturally account for the crystals' hemihedral characteristics. For Delafosse, hemihedry was a reflection of the structural characteristics of the physical molecule.

Delafosse was Pasteur's teacher of mineralogy and I have shown how he, along with Laurent, defined the French crystallographical tradition for Pasteur's research.[25] Delafosse's refinement to Haüy's crystal structure theory also led directly to the mathematician Auguste Bravais' formulation of a crystal lattice theory at the end of the 1840s.

What is significant for this paper is that both Delafosse and Bravais were working in a tradition of Haüyian crystallography as modified by Ampère. They both subscribed to a two-stage polyhedral molecular theory of crystal structure and they both interpreted these polyhedra explicitly in Ampèrian terms:

The principles which have guided us in the arrangement of atoms in crystalline molecules are those of Ampère but suitably modified and extended. We have agreed with him that atoms of the same type are generally located so that their centers of gravity always occupy identical apices of the polyhedra which they form in space. But, in taking this idea for our point of departure, we have combined it with another idea no less essential, namely: that the molecular form ought always agree with that of its [macroscopic] body and consequently it ought to be one of the forms of [that body's] crystalline system.[26]

To summarize, we acknowledge that the molecular polyhedron is symmetrical and that the elements of its symmetry are passed on to the corresponding [crystalline] assemblage in determining its structure.

In truth, it could be objected that the difficulty in explaining the state of symmetry of the assemblage is simply moved back and that it remains to be shown why the molecular polyhedron is symmetrical. The atomic theory furnishes a ready response to that last requirement. . . . By considerations of a completely different kind, Ampère had, in 1814, already arrived at the conclusion that the molecular polyhedron ought to be formed of atoms placed symmetrically about its center of gravity; it is easy to account for the internal equilibrium of the molecule on that assumption.[27]

I have found that the consciousness of a French molecular tradition stemming from Haüy persisted to the end of the 19th century and beyond. It must be said that the molecular aspect of Bravais crystal lattice theory lay dormant for a quarter of a century until François-Ernest Mallard began to utilize it in his development of a molecular theory of crystal structure based on investigations of a variety of optical and physical properties of crystals. By the time of his death (1894), Mallard was deemed by his French colleagues to have effected a révolution nationale. . . . by which French science, returning to the luminous concepts of Haüy, has suceeded in giving a completely rational basis [to crystallography] without ever straying from the fundamental concrete consideration of molecular polyhedra.[28] (See Figure 4.)

In the preface to his *Traité de cristallographie*, Mallard himself wrote:

Since the admirable research of Haüy, crystallography has undergone profound modifications. More a naturalist than geometer, Haüy had, with singular sagacity, built the science opened up by his genius upon a physical hypothesis which might seem hazardous. The German school, guided by Weiss, Ross and Naumann renounced completely the hypothesis concerning the form of the molecules and only saw in crystallographical laws, laws of geometry. . . .

But crystals are not [just] geometrical entities; the form which characterizes then is but a figured expression of the most intimate properties of the matter of which they are composed. Thus, it is an arbitrary restraint and sterilization of the science which is limited to the study of crystalline polyhedra to interdict the search for the physical significance of the geometrical language in which we describe these polyhedra by an exaggerated mistrust of hypothesis. One cannot even see very clearly what objective is being pursued by the scientist who is restricted to studying crystalline form in and of itself, in the words of the German philosophers.

This manner of comprehending the science [of crystallography] has never been that of the crystallographers of our country.[29]

6. CONCLUDING COMMENTS

In this case study of 19th century French crystallography, I have tried to show that a tradition derived from the crystal structure theory of René Just Haüy survived long after Haüy's theory had apparently been discarded in other countries. I have suggested that this was due, in part at least, to the fact that modifications in the guiding assumptions of Haüyian crystallography (specifically Ampère's modification of the model of the compound molecule) were accepted in France. This enabled the Haüyian tradition to persevere and even incorporate into it Mitscherlich's formulations of the principles of isomorphism and polymorphism.

It has been implied several times that the fate of Haüyian crystallography and mineralogy was much different in other countries than in France. Until detailed study of the history of these sciences in the 19th century is carried out, it will not be possible to analyze the details of the demise of Haüy's theory in other countries. However, I can suggest two reasons why the fate of Haüyian crystallography might have been different.

porte le centre de gravité d'un polyèdre moléculaire, auquel on a donné ici la forme d'une pyramide octaédrique.

On peut encore, avec Mallard (1), donner de ce mode de représentation un autre énoncé. Si l'on joint entre eux les centres des noyaux parallélépipédiques, on forme un nouveau système réticulaire, qui n'est autre que l'ancien, transporté parallèlement à lui-même. Alors les centres de gravité des molécules se trouvent situés aux centres des noyaux du nouvel assemblage. Le cristal est donc partagé en noyaux ou *cellules* parallélépipédiques, chacune de ces cellules ayant, dans son intérieur, une *molécule cristalline*, dont le centre de gravité

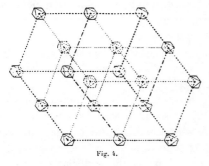

Fig. 4.

coïncide avec le centre de la cellule. Cette dernière constitue par excellence la *particule* individuelle du corps cristallisé.

Molécules complexes. — Le principe qui vient d'être établi comporte une restriction. En effet, la démonstration suppose que *tous les centres de gravité* des molécules sont des points homologues, c'est-à-dire que l'identité des propriétés physiques, suivant les directions parallèles, s'étend même

aux intervalles intermoléculaires. Mais il peut se faire qu'il n'en soit pas ainsi. Dans ce cas, sur une même rangée, les molécules homologues, et conséquemment de même orientation, se retrouveraient de deux en deux, de trois en trois ou, en général, de n en n. Celles-là seules formeraient le réseau cristallin caractéristique du corps. Appelons M et M′ les deux molécules homologues sur une même rangée, et $M_\alpha, M_\beta, M_\gamma$, etc., celles qui les séparent et dont chacune a son orientation particulière. Le réseau formé par les centres de gravité de toutes les molécules telles que M_z sera identique avec le réseau des centres MM′. Il en est de même pour la combinaison M_β, pour celle qui a comme point de départ M_γ, etc. Chacun de ces réseaux, réduit aux lignes qui joignent les centres, est le réseau MM′ transporté. parallèlement à lui-même, d'une certaine quantité. Mais tous diffèrent les uns des autres par l'orientation des molécules qui en occupent les nœuds, et ainsi il y a autant de ces réseaux qu'il y a de molécules comprises entre M et M′.

Dès lors, on peut imaginer deux variétés parmi les corps cristallisés : l'une, où toutes les molécules sont également orientées; l'autre, où les orientations sont variables, mais *alternent régulièrement*. Cette considération nous sera plus tard très utile, pour expliquer rationnellement, avec Mallard, certains faits

(1) *Revue scientifique*, 30 juillet 1887.

Fig. 4. De Lapparent's model of a complex molecule. From *Cours de minéralogie*, 3rd edn. (1899).

(1) First, even if we accept Hans-Werner Schütt's assessment of the European ascendance of Haüy's crystallography and mineralogy during the first two decades of the nineteenth century, it is important to realize that this dominance was superimposed, as it were, upon other indigenous theories of material form and structure (above, VI.A. & B.). During these very decades, scientists in Germany and England developed theories of crystal structure based upon these alternative approaches. Hence, it may well have been the case that the theoretical situation regarding crystal structure theory was always complex in these countries, with parallel sets of theories, models and guiding assumptions coexisting in unstable equilibrium for several decades.

(2) Ampère's modification of the molecular model which, in my analysis, enabled French crystallography to survive Mitscherlich's assaults, appear to have had no discernible influence outside of France. I have suggested that even in France there was a twenty-year lag between Ampère's first publication on his molecular geometry (1814) and its first demonstrable impact upon crystallographical thought. However, during those decades the dominance of Haüyian crystallography had remained sufficiently strong there to facilitate the incorporation of Ampère's modification into it even after the protracted lag. It was through Ampère's modification that the chemical atomic theory was melded with Haüyian crystallography.

The above-mentioned points suggest a general proposition about the analysis of theory change in science: it may not be sufficient to consider science in abstraction from its sociocultural context. Taking into consideration factors such as national tradition may yield a much different and more complex assessment of the fate of a particular set of guiding assumptions than when they are ignored. This is certainly suggested by the case study of crystallography. I would reiterate that this case would benefit from further examination from the standpoint of comparative national development.

In such examination, attention must be paid to the concrete dynamics of national tradition to understand why certain approaches, sets of guiding assumptions, etc. are sustained (and/or modified) in one country and not in others. Factors such as pedagogy, professionalism and patronage and nationalism itself need to be taken into account. I have indicated, a few examples of these factors in my commentary on the post-Ампèrian period of French crystallography, but the French case itself would be illuminated by further exploration of them. This is especially true because until well into the second half of the 19th century, disciplinary coherence in crystallography was quite weak; the group of scientists whom I have discussed is characterized by heterogeneity of background and of professional activity. If Haüy had founded a tradition, he had not founded a school (with the exception of Delafosse). By the same token, however, this disciplinary weakness probably contributed to the considerable permeability between crystallography and other disciplines which also characterizes this period in France.

NOTES

[1]René-Just Haüy, *Essai d'une théorie sur la structure des crystaux* (Paris, 1784).

[2]John G. Burke, *Origins of the Science of Crystals* (Berkeley, 1966), p. 9.

[3]*Ibid.*, p. 107.

[4]Norma E. Emerton, *The Scientific Reinterpretation of Form* (Ithaca, 1984), pp. 284–285. The quotations are taken from an anonymous review of Henry Brooks's *A Familiar Introduction to Crystallography* in *Annals of Philosophy*, 1823.

[5]Hildesheim, 1984.

[6]*The Structure of Scientific Revolutions*, 2d. edn. (Chicago, 1970).

[7]*Crystals and Compounds* (Philadelphia, 1976).

[8]It is by no means easy to define disciplinary boundaries here.

[9]1st edn., 5 vols. (Paris, 1801).

[10]Burke, *Origins*, p. 102.

[11]*Ibid.*, p. 69, paraphrasing J.-B. Romé de l'Isle, the first scientist to enunciate the law in 1783.

[12]David Oldroyd, *From Paracelsus to Haüy: The Development of Mineralogy in Its Relation to Chemistry* (Ph.D. dissertation, University of New South Wales, 1974); 'Mineralogy and the "Chemical Revolution"', *Centaurus* (1975), **19**: 54–71.

[13]Schütt, *Die Entdeckung*, pp. 59–60. The only opponents he lists are J.-F. d'Aubuisson and E.-C. Kramp in France, T. Thomson in England and C. S. Weiss in Germany. In another connection, he also cites the chemist C.-L. Berthollet, pp. 98–104. Schütt's finding is a significant one which merits close scrutiny.

[14]2d. edn., 5 vols. (Paris, 1822).

[15]3 vols. (Paris, 1822).

[16]Quoted in Mauskopf, *Crystals and Compounds*, p. 31. Mitscherlich later modified the second part of this law to take into account the chemical nature of the constituents.
[17]*Ibid*. To the best of my knowledge, Mitscherlich never articulated a theory of crystal structure. But implicit in his formulation of isomorphism and polymorphism was, at the least, a negation of Haüy's theory of compound molecular structure: no longer could the compound or integrante molecule be envisioned as formed by the juxtaposition of determinately shaped constituent molecules in only one arrangement.
[18]Haüy did allow that in the case of so-called limiting form associated with the cubic form there could be identity of form with difference in composition.
[19]The most celebrated case was calcite and aragonite.
[20]A. Dufrenoy, *Traité de minéralogie*, 4 vols. (Paris, 1855–1856), I, p. 202.
[21]G. Delafosse, *Nouveau cours de minéralogie*, 2 vols. (Paris, 1858); A. de Lapparent, *Cours de minéralogie*, 3d. edn. (Paris, 1899), p. 7.
[22]Quoted in Mauskopf, *Crystals and Compounds*, p. 39.
[23]*Ibid*., Chapter 4.
[24]*Ibid*., pp. 74–75.
[25]*Ibid*., pp. 69–70.
[26]Delafosse, *Nouveau cours*, I, p. 513.
[27]A. Bravais, *Études cristallographiques* (Paris, 1866), pp. 204–205.
[28]A. de Lapparent, 'Notice nécrologique', *Annales des Mines*, 9m ser. (1895) 7:277. In his necrology, G. Wyrouboff wrote of how Mallard "a transformé – je dirais presque revolutionné – la cristallographie". "François-Ernest Mallard", *Bulletin de la Sociétie Française de Minéralogie* (1894), **17**: 243.
[29]E. Mallard, *Traité de cristallographie*, 2 vols. (Paris, 1879, 1884), I, p. i.

A. J. ROCKE

KEKULÉ'S BENZENE THEORY AND THE APPRAISAL
OF SCIENTIFIC THEORIES

In three papers published in 1865 and 1866, August Kekulé, professor of chemistry at the University of Ghent, proposed a theory of the structure of benzene that provided the basis for the first satisfactory understanding of aromatic compounds, a very large and important class of organic substances. It would be difficult to overestimate the impact of these papers. Within a decade most chemists accepted the theory as empirically verified and heuristically invaluable, and within a generation one observer could assert not only that Kekulé's theory was the "most brilliant piece of scientific prediction to be found in the whole of organic chemistry", but also that "three-fourths of modern organic chemistry is, directly or indirectly, the product of this theory".[1]

Partly because of its high contemporary visibility and the circumstance that consensus was reached so quickly, the development and reception of the benzene theory provides an ideal test case for examining models of theory appraisal in science. Kekulé's concepts regarding benzene structure pretty much expanded into a theoretical void, requiring no displacement of entrenched ideas or overthrow of pre-existing guiding assumptions. The benzene theory was essentially simply a detailed application of Kekulé's own more general theory of chemical structure – so in essence two theories were being tested when the benzene theory was appraised by Kekulé and others. Like the benzene theory, the structure theory had no real rivals; opponents were forced to retreat into positivism. Consequently, the sparse theoretical landscape here allows an unusually sharp focus on the process of theory testing.

An additional advantage to using this case to examine models of theory appraisal is that chemistry in general, and organic chemistry in particular, has not often been used by philosophers of science as a source for historical examples and so represents fruitful virgin territory for this sort of endeavor. There are, of course, innumerable references in the literature to two famous (inorganic) chemical episodes – the phlogistonist/oxygenist debate and the controversies over atomic theory in the 19th century – but most philosophers use the history of physics preferentially, and many do so nearly exclusively. Although reasons for

145

such a preference are understandable, the result is a regrettable narrowness of vision. The widening of focus which I am attempting here will in many cases add further confirmation to theses based mainly on an examination of the history of physics; in other cases some doubt will be cast on the generality of conclusions that have been reached.

The empirical richness of late-19th-century aromatic chemistry (literally tens of thousands of new benzene derivatives were discovered in the last third of the century) suggests that this field is ideal for examining perhaps the most central aspect of theory appraisal, namely the role of experiment in confirming or falsifying theoretical ideas. I wish to inquire whether this case study provides support or counterexamples for the following theses: (a) the belief (T2.4) that a theory is judged solely on the verification of its new predictions; (b) the suggestion (T2.6) that usually only a very small number of experiments suffice to test a new theory; (c) the tenet (T2.7) that the *rate of problem-solving* is more important than its actual perceived truth value for the favorable reception of a new approach in a scientific community; (d) the claim (T2.8) that the reception of a theory must be viewed relative to its rivals rather than in absolute terms; and (e) the attempted resurrection (T2.9 and T2.10) of crucial experiments, if not in the absolute sense claimed for them by Bacon, at least in the more relative sense of *perception* of criticality by the relevant peer community.

*   *   *

It is important to define precisely what "Kekulé's benzene theory" might be taken to signify, since there can be at least four different historically justifiable interpretations of this phrase. In the least specific sense, Kekulé simply asserted that a characteristic six-carbon nucleus (benzene, C_6H_6) is common to all aromatic compounds and that derivatives are formed by functional groups substituting for hydrogen of the nucleus or effecting transformations in the 'side-chains' attached to the nucleus. I will call this the SCN (six-carbon nucleus) version of the theory. A second more detailed benzene theory specifies a particular form for this aromatic nucleus, namely that it consists of six trivalent CH groups, somehow connected together in a cyclic fashion according to the (then becoming widely accepted) structure-theoretical tenets of carbon tetravalence and self-linking; I will refer to this assertion as the 'cyclohexamethine' ($[CH]_6$) or CHM theory (see Figure 1). Kekulé went even further than this, though, even in his first paper, suggesting that the

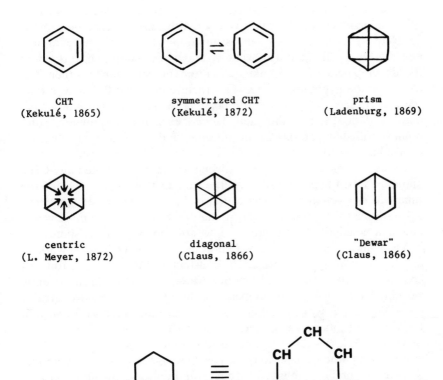

Fig. 1. Cyclohexamethine (CHM) theories

formula for benzene was in fact one particular species of the genus cyclohexamethine – namely cyclohexatriene (CHT), the familiar Kekulé hexagon with alternating single and double bonds. Finally, in 1872 Kekulé proposed a modification of the cyclohexatriene theory, his 'oscillation hypothesis', which would fully symmetrize the CHT formula, and which in most respects is operationally equivalent to the resonance hybrid benzene structures in modern chemistry textbooks.[2]

If one assumes that Kekulé's celebrated benzene dream anecdote is accurate in its essentials – an assumption which, as I have argued elsewhere,[3] is consistent with what we know from published work about the development of his thinking on aromatic compounds – then it is clear that the hypothesis of a CHT structure occurred to him at a time (circa early 1862) when it was not possible to argue for its validity from existing data. In fact, the then-claimed existence of aromatic compounds with five carbons in the nucleus, and of isomers of benzene, contradicted even the minimalist SCN theory, and the claimed existence of isomers of mono-substituted benzene would have falsified any CHT and most CHM hypotheses.[4] It is little wonder that Kekulé decided the time was not yet ripe for publication of his idea. However, empirical aromatic chemistry began rapidly to mature during the next three years, and nearly all of the purported lower homologues and isomers of benzene and of mono-substituted benzene were shown by such workers as Rudolph Fittig and Friedrich Beilstein to have been erroneous conclusions from faulty experiments. Moreover, Beilstein, apparently working in the absence of any benzene theory at all, began to perceive empirically what was eventually to be an essential support for several CHM variants, namely that there are always three possible isomers – no more and no less – for every di-substituted benzene.[5]

It was only when conclusions such as these appeared to be on the way to being securely established that Kekulé published his ideas on aromatic compounds, in January 1865. Even so, his first essay on this subject was remarkably cautious; its nearly exclusive focus was to show how an SCN theory could account for many hitherto unexplained cases of structural isomerism. He did assert the CHT formula, but gave it little emphasis and did not try to justify it empirically. Although by this time four instances of tri-isomerism of di-derivatives had been recognized, in this paper Kekulé referred to just one such case as being securely established and suggested only in a hesitant and obscure fashion the expectation of such tri-isomerisms as a retrodiction from the CHT theory. There was no new experimental work described here.

But now Kekulé was fully engaged with the problem and he turned his enormous energy to examining the implications of what increasingly appeared to him to be an empirically supportable and heuristically valuable theory. On 10 April 1865 he wrote to his friend Adolf Baeyer concerning his present work:

Naturally, everything is aromatic. Facts to prove or disprove my recent aromatic theory . . . A great deal is in the works; the plans are unlimited, for the aromatic theory is an inexhaustible treasure-trove. Now when 'German youths' need dissertation topics, they will find plenty of them here.[6]

His emerging confidence was revealed in a summary article published in early 1866 and in the fascicle of his textbook covering the aromatic field, which appeared that summer. He described a large number of new experiments from his laboratory that supported the theory and now firmly asserted all of the known tri-isomerisms as confirmed retrodictions from the CHT theory.

Still, Kekulé had no illusions of having established CHT in a definitive fashion. He referred to his work as merely "the outlines of a theory", and still quite "incomplete"; until verified, it should be regarded only as a "more or less elegant hypothesis" to which "a certain probability can perhaps no longer be denied" and which further work would either confirm or falsify.[7] In fact, Kekulé was correct in this assessment. Although his case for SCN was fairly solid (if, that is, one accepted the principles of structure theory upon which Kekulé was reasoning), the arguments for CHM were only reasonable and by no means compelling. The apparent non-existence of more than one isomer of each mono-substituted benzene excluded a variety of possible benzene structures, including the only two to have been proposed before 1865,[8] but two others besides CHM would still be possible,[9] and anyway this was only *negative* evidence in a still young field. The cases where a third isomer of a di-derivative had been isolated would seem to allow only CHM as a possibility, but the generality of this phenomenon was still open to question. Furthermore, Kekulé really wanted to establish not just CHM, but CHT, for which he offered virtually no support beyond the elegance of the formulation.

Despite the somewhat equivocal character of the evidence, the theory proved immensely appealing to Kekulé's fellow structuralists and satisfied something of a theoretical void in the emerging field of aromatic chemistry. Many of the best young organic chemists of the 1860s flocked to the new field and journals began to fill with amazing numbers of aromatic investigations, most of them guided by CHM or CHT theories.

Two of the most successful such chemists were Albert Ladenburg and Wilhelm Koerner, both former students of Kekulé. By 1869 each had independently developed *positive* arguments from cleverly designed experiments for the chemical equivalence of all six hydrogen atoms of benzene.[10] Ladenburg went on in the next seven years to devise a proof that there exist in every mono-substituted benzene two pairs of chemically equivalent hydrogen atoms (i.e., two ortho and two meta hydrogens), an argument which would seem to exclude all but CHM benzene.[11] In the meantime Koerner deduced from CHT theory a set of predictions regarding isomer numbers of tri-substituted benzenes which he then succeeded in verifying, moreover in such a way as to provide a method for determining in absolute fashion the detailed structures of the three isomers of each *di*-derivative.[12]

Koerner's absolute method of determining positional isomers, still cited as a classic experimental investigation in organic chemistry texts, was supplemented by some more circumstantial but nonetheless convincing arguments relating to positional (ortho-, meta-, and para-) isomers. Within a very few years there was no longer any real doubt that every di-derivative had three isomers. The fact that three moles of acetone can condense to form one of mesitylene (trimethyl benzene) suggested that the methyl groups are symmetrically situated, a hypothesis that was supported by the fact that only one isomer was produced when the mesitylene nucleus was further substituted. The decarboxylated oxidation product of mesitylene, isophthalic acid, must then be a 1,3 derivative. The ease of anhydride formation of phthalic acid, and the fact that it can be derived from oxidized naphthalene, suggested that the carboxyl groups are placed 1,2; the third dicarboxylic acid, terephthalic acid, must therefore be the 1,4 isomer. All of these arguments, and many more of a similar character, were developed on the basis of new experiments performed and novel aromatic isomers prepared during the first decade after Kekulé's summary article of 1866.[13] Even Kekulé's arch-enemy and principal opponent of both the structure and benzene theory, Hermann Kolbe, conceded as early as 1874 that the benzene theory (by which he meant CHT) was "accepted by the great majority of chemists".[14]

This judgement, which accords with nearly universal testimony, should not be taken to imply that the theoretical dialectic had ceased by the mid-1870s: lively debates over the precise structure of aromatic substances continued for decades. In 1869 Kolbe himself proposed an SCN benzene theory that denied hydrogen-equivalence; but Kolbe's theory was not even translatable into unambiguous structural terms, and he was unable to win any significant converts, even among his own students. An example of Kolbe's evidentiary problems was that in order to explain the non-existence of more than one benzoic acid (for instance), he was forced to assert that the missing isomers predicted by his theory were in fact formed, but were too labile to be isolated and underwent intramolecular rearrangements to the observed form. Although this hypothesis was logically impeccable, it was understandably regarded by the peer community as unacceptably ad hoc, since there was no independent evidence for such rearrangements.[15]

The episode with Kolbe's theory underlined the unprofitability of non-CHM theories generally, and after 1866 virtually all of the benzene debates centered on a half-dozen CHM variants. One CHM structure, the so-called 'Dewar benzene', was quickly ruled out due to its inability to account for observed isomer numbers of both mono- and di-derivatives. Another CHM proposed by Adolf Claus, known as the diagonal formula, made a few converts, but most chemists thought that Claus' formula implied the chemical equivalence of the ortho and para positions in mono-derivatives, resulting in the false retrodiction of only two isomers of each di-derivative. An extremely subtle and powerful but quite non-classical CHM structure was the hexagonal 'centric' formula, proposed in three different variants by Lothar Meyer, Henry Armstrong and Adolf Baeyer in the 1870s and 1880s. In 1872 Kekulé himself proposed a very non-classical CHT variant that was operationally equivalent to centric benzene, namely the symmetrized CHT theory.[16]

The only classical CHM that proved to be a real rival to Kekulé's CHT was Ladenburg's 'prism' formula, unveiled in 1869. The real power of this formula consisted in the circumstance that its isomer-number predictions in the absence of carbon-carbon bond breaking matched those of CHT precisely, which in fact was CHT's most celebrated success. Indeed, prism benzene suggested the observed *full* chemical equivalence of the ortho hydrogen pair, which Ladenburg (convincingly though not conclusively) argued was not provided by CHT. Some of the other evidence supporting CHT, such as neatly accounting for the products of oxidation of naphthalene or condensation of acetone, was more difficult to reconcile with prism benzene, but by no means impossible.[17]

As a consequence, for many years the principal arguments used in favor of the prism on the one side or CHT on the other rested on predictions of the magnitude of chemical effects of relatively subtle structural distinctions, or on the plausibility of largely hypothetical reaction mechanisms, over which reasonable scientists could differ. It was not until nearly the turn of the century that the prism formula was effectively and ultimately decisively countered by Baeyer's studies on the hydrogenation of aromatic compounds. It was even more difficult to find experimental purchase on the negligible differences between the centric formula and CHT, especially if one used Kekulé's symmetrized variant of the latter. Because of this relatively close approach to observational indistinguishability between the CHT, prism, and centric formulas during the period with which we are concerned, the discussion that follows will usually proceed as if all three CHM rivals are minor variants of a single theory (indeed, the overview just provided suggests that this is an intrinsically plausible view of the subject). Kekulé's CHT formula will be used as the most convenient representation of this single theory, not because it is the formula still preferred by many chemists to represent benzene, but because, even when it was under attack by a CHM rival, it was still by far the most popular among organic chemists during the last third of the 19th century.

*   *   *

THESIS (T2.4): *The appraisal of a theory is based entirely on those phenomena gathered for the express purpose of testing the theory and which would be unrecognized but for that theory.* Kekulé's benzene theory is in fact one of the neatest examples of what Lakatos refers to, in this context, as the 'excess corroboration' of a theory. At the time CHT was (most likely) conceived there was little existing data to check its validity, and at the time it was published the situation was only slightly improved. The number of post-1865 confirmed predictions based on CHT is legion: the non-existence of lower homologues of benzene or benzoic acid; the singularity of benzene and of mono- and penta-derivatives; the tri-isomerisms of di-, tri-, and tetraderivatives; hydrogen-equivalence in benzene; two pairs of equivalent hydrogens in mono-derivatives; Koerner's prediction of isomer numbers on converting di-to tri-derivatives; structural relations of oxidized and condensed aromatic hydrocarbons; and innumerable more of a similar character.

By the same token, other theories were relatively quickly excluded by failed predictions; examples include the Dewar and Kolbe formulas, along with all other non-CHM theories. Every time a result was published that was inconsistent with CHM, it was pounced on by one or more structuralists and was soon shown to be based on flawed experimental work.[18] Conversely, when predictions of CHM variants such as the CHT, prism, or centric formulas coincided, or at least failed to lead to *unambiguously* distinct experimental outcomes, the peer community remained indecisive and refused to commit unanimously to a single variant. In sum, one could not ask for a more classic case study of the predictive success of a theory (CHM and CHT) and, more generally, the application of predictive *criteria* for both confirming and falsifying instances, using hypothetico-deductive method. The predictive criterion and hypothetico-deductive nature of the theory was also clearly recognized at the time, as can be seen from Kekulé's 1865 letter to Baeyer, and Ladenburg's summary monograph of 1876.[19]

So it can be said that the thesis is well supported here. And yet an important qualification must be added: it is difficult in this case to make a general distinction between *predictions* (i.e., of outcomes of experiments not yet performed) and *retrodictions* (i.e., comparing data from experiments predating the formulation of the theory, or from experiments that were performed for reasons other than testing the theory). Every use of the word 'prediction' in the last two paragraphs must for historical accuracy be interpreted to include the idea of retrodiction as well. Even Kekulé's initial formulation of the theory, as hypothetical as it was in the still poorly-developed aromatic field, had some retrodictive as well as predictive power; and the later exclusion of non-CHM theories exhibited at least as much use of retrodiction as of prediction. Indeed, it is difficult to imagine how any instance of hypothetico-deduction could rely on *pure* prediction, for a hypothesis would not attract the attention of a scientist unless at least some sort of empirical justification (in the sense of retrodiction) could be adduced. Lakatos himself suggests as much in his historical exemplification of 'excess corroboration', where he implicitly includes retrodictive as well as predictive success as important for theory appraisal.[20]

THESIS (T2.6): *The appraisal of a theory is usually based on only a very few experiments, even when those experiments become the grounds for abandoning the theory.* Here it appears that philosophers of science may be overgeneralizing from the history of physics. Our examination of the appraisal of benzene theory, exhibiting what I have called the 'empirical richness' of late-19th-century organic chemistry, leads to a very different conclusion: the number of experiments that were performed to test the theory numbered in the thousands, if not tens of thousands. To give one revealing example, Koerner's investigation of the isomeric relations between di- and tri-derivatives, which resulted in a dramatic verification of predictions from CHT as well as an absolute method of determination of positional isomerism, required the preparation of no fewer than 126 new aromatic substances.[21] In a sense, the discovery of *each* of the tens of thousands of new aromatic compounds in the first generation after 1865 constituted a potential (weakly) verifying or falsifying instance. Moreover, it is unclear what constitutes a single experiment in this context, since the preparation, isolation, and purification of each new substance required a complicated sequence of unit operations (for more on this definitional problem see the discussion of (T2.9) and (T2.10) below).

The necessity for a very large number of experiments here is understandable. An important part of the evidence for CHT and other CHMs was negative, such as the non-existence of certain isomers, which could by its very character only be inductively confirmed through extensive cumulative experience. Some of the positive evidence, such as the predicted existence of specific numbers of possible isomers, was similarly slow to appear, since some of the predicted isomers proved difficult – or even impossible – to prepare. For instance, it was not until 1878 that Beilstein succeeded in accomplishing what might appear to be a relatively simple and straightforward task, namely the preparation and CHT-structure assignments of all twelve predicted chlorobenzenes.[22]

THESIS (T2.7): *The appraisal of a theory is sometimes favorable even when scientists do not fully believe the theory, specifically when the theory shows a high rate of solving problems.* This thesis is supported by the present episode. The discussion above documents the impressive rate of problem-solving which characterized CHT and other CHM variants. Kekulé's theory made immediate sense of a large amount of information regarding aromatic substances and was successfully used heuristically by nearly the entire community of organic chemists of the day; as Edvard Hjelt accurately commented regarding the theory, "in dem Vorhandenen schuf sie Ordnung, in das Zukünftige warf sie Licht".[23] In short, the appraisal of the theory in the peer community of the day was on the whole extraordinarily favorable, and this seems to have been related to its proven ability to solve chemical problems.

This was true even though no one, not even Kekulé or propounders of rival CHM variants, was fully satisfied by the verisimilitude of his brainchild. Kekulé's cautious comments in 1866 have been noted. Although he never backed away from CHT, for the rest of his life he was curiously shy about using *any* resolved structural formula for aromatics, nearly always simply writing C_6H_5 for phenyl, and so on. This cautious attitude was no doubt partly due to a prevalent air of positivism, at least in rhetorical convention, regarding structural formulas in general; but it was also due to certain difficulties which Kekulé and other CHT and· CHM advocates were never able fully to circumvent. For instance, Ladenburg's point about the non-equivalence of the ortho hydrogens of CHT was telling, and Kekulé's symmetrized CHT variant, designed to solve this problem, required an interpretation of valence that few found acceptable – or even comprehensible. Furthermore, the presumed double bonds of CHT had little in common with the chemical behavior of olefins, quite contrary to reasonable expectations. Prism benzene had different implausibilities, especially in the interpretation of the structure and oxidation reactions of condensed rings. And advocates of centric benzene had to argue for a disposition of the fourth valence bond of each ring carbon that had no analogy to any aliphatic compound.

The remarkable thing is that this contrast between the striking success and popularity of the theory on the one hand, and the seemingly insoluble problems of detail on the other, seemed not at all to disconcert the leading theoreticians. Kekulé remained cautious, but retained confidence in the scientific worth of his idea. Ladenburg also remained faithful to his prism structure, though twice in the 1870s he remarked that CHT was "at least as appropriate, if not more so, than the prism".[24] In the first major review of aromatic theory, published in 1876, Ladenburg revealed a strongly relativistic approach to scientific theory in general. He wrote:

I am in fact of the opinion that in a science logical deductions are valuable even if they are derived from weak hypotheses, if, as is the case here, they can be verified directly by the facts, and that they are retained even when their [theoretical] foundation is damaged or even destroyed.[25]

Ladenburg proved his point by citing examples from the history of physics. In his theoretical discussions he used prism and CHT formulas

side by side. A similar combination of heuristic confidence and ontological skepticism remained characteristic of summary discussions of aromatic chemistry into the present century.[26] The aromatic theory was
trusted and popular, and justly so – but not fully believed.

THESIS (T2.8): *The appraisal of a theory is relative to prevailing
doctrines of theory assessment and to rival theories in the field.* In one
sense this thesis is unexceptionable; indeed it approaches the status of a
truism, in that scientists, as human beings living within a cultural
context, cannot help but be guided by the context in general, and by the
prevailing set of scientific doctrines in particular. This thesis can be said
to be confirmed here in three ways (although we will conclude the
discussion with a crucial reservation). First, it has been noted that
explicit advocacy of the method of hypothesis, familiar to physicists for
a generation, began to replace inductivist rhetoric in chemical circles
just about the time that Kekulé was first formulating his benzene
theory;[27] Kekulé's publication of the theory was unapologetically
hypothetico-deductive in character. It is not surprising, therefore, that
the benzene theory was appraised in hypothetico-deductive terms, as is
underlined by the discussion of (T2.4) above.

Second, the benzene theory was a specific application of a more
general chemical doctrine, namely the theory of structure, and it was
necessarily appraised in that context. This theory was still quite new –
Kekulé's and Couper's papers, generally regarded as marking the birth
of the theory, appeared in 1858 – but it had already celebrated a number
of successes and by 1865 was rapidly gaining in popularity. Most of the
prominent young organic chemists of the early to mid-1860s were
structuralists: examples include Erlenmeyer, Butlerov, Crum Brown,
Baeyer, Beilstein, Fittig, Graebe, Dewar, Frankland, Ladenburg, Na-
quet, Koerner, and Heintz. Older chemists, such as Liebig, Wöhler,
Dumas, Wurtz, Williamson, and Hofmann, did not advocate the theory,
but did not oppose it, either. The number of outright opponents – Kolbe
of course comes to mind here – remained very small. Kekulé's benzene
theory, built on a structuralist foundation, handsomely repaid the debt;
it was widely regarded as the most successful early application of the
theory, and proved to be a factor in its widespread acceptance.

Finally, the debate between CHT and rival CHMs, and the exclusion
of non-CHM theories, underscores the circumstance that these were
regarded as alternatives that must compete head-to-head *vis à vis* the

empirical data. In making choices between theoretical alternatives, a scientist is of course restricted to the alternatives that exist at the time, including any new modifications he may be able to devise.

However, as much as the thesis may seem to be confirmed by the present case, the self-evident character of the last point suggests a circularity in the thesis that may be vicious; in other words, in the sense we have been discussing it the thesis may be true, but in an either trivial or tautological way. In particular, the thesis seems true only if one considers the *pragmatic* necessity for choices between existing theoretical options. The case under study suggests that there is also apparently at work an additional criterion in the minds of scientists, namely judgements as to absolute truth-value. Our discussion of the last thesis shows that the benzene theory was *both* extremely popular and successful, *and* disbelieved in absolute terms. The same could be said for the atomic theory during much of the 19th century. The present case seems to argue that scientists may embrace a theory with enthusiasm – as preferable to rivals – without at the same time being blinded by its deficiencies in absolute terms.

THESIS (T2.9) *and* (T2.10): *The appraisal of a theory occurs in circumstances in which scientists can usually give reasons for identifying certain problems as crucial for testing a theory; and it depends on certain tests regarded as 'crucial' because their outcome permits a clear choice between contending theories.* The discussion of (T2.6) above would seem to weaken the possibility of a role for 'crucial experiments' in appraising the benzene theory. Such a conclusion, however, depends *inter alia* upon one's definitions of the two words in the phrase. If 'experiment' is interpreted to apply not just to unit operations, but also to extended investigations designed to examine a predicted outcome, and if 'crucial' designates not so much the possibility of rigorous proof, as a perception in the peer community that a relatively clear choice between well defined theoretical alternatives can be made upon the results of that investigation, then the case under study provides a number of such confirming instances. To be sure, many predictions based upon CHT and other CHMs were of the 'negative' sort that could only be tested inductively, by extensive empirical evidence – as was argued in the discussion of (T2.6). Others, however, were of a more positive character that were amenable to straightforward examination.

For instance, the demonstrations by Ladenburg and others of the chemical equivalence of the hydrogen atoms in benzene and of the existence of two pairs of equivalent hydrogens in mono-derivatives rested only upon the assumptions of structure theory; these 'proofs', whether or not rigorous in an absolute sense, commanded the assent of the overwhelming majority of structuralists and excluded the possibility of non-CHM theories. Similarly, any single confirmed example of three isomers of a di-derivative, combined with the previous demonstration, excluded all but a small handful of CHMs. Koerner's absolute method of determination of positional isomerism might also be mentioned as having had a similar impact.

But perhaps the best example in this case study is an instance of a crucial experiment performed by an *opponent* of benzene theory. In 1860 Kolbe claimed to have discovered an isomer of benzoic acid, which he named 'salylic acid'; four years later Beilstein showed that this material was simply impure benzoic acid, a conclusion which Kekulé appropriated the following year for his fledgling benzene theory. Beilstein's demonstration was straightforward: when 'salylic acid' was distilled, pure benzoic acid was found in the receiver and a small residue of impurities remained in the distillation flask. Kolbe understandably did not appreciate this interpretation, as it simultaneously impugned his experimental skill, falsified his own benzene theory, and was used to support Kekulé's structuralist CHT theory – a theoretical school which he heartily loathed. He suggested that Beilstein's distillation had induced an intramolecular rearrangement from 'salylic' to benzoic acid.

In 1875 he designed an experiment that would permit a clear choice between the two alternatives: 'pure salylic acid' or 'impure benzoic acid'. His test was ingeniously simple: perform Beilstein's distillation, but then recombine the residue and the distillate. If the substance had rearranged during distillation, the recombined material should retain the properties of benzoic acid; if Beilstein were right, the properties of 'salylic acid' should reappear. The experiment proved Beilstein, and by implication, Kekulé's benzene theory, correct.[28]

To Kolbe's credit – and to the credit of the concept of crucial experiments – he published these results and admitted his error. But this instance underscores the *weakness* of crucial experiments as well, for Kolbe found no reasons to alter his *general* opposition to CHT or his advocacy of his own non-structuralist benzene theory. He had lost a battle, but not the war: just because the isomers which his own theory predicted (such as salylic acid) had not yet been found, was no reason to conclude that they would always remain hidden from the scientist.[29]

* * *

The case study just examined, the appraisal of the benzene theory during the last third of the nineteenth century, provides support for several of the theses under examination here: (T2.4, T2.7, T2.8, T2.9, and T2.10). Thesis (T2.6) is the only one of those examined here that appears to be weakened. On the other hand, it is remarkable that in nearly every instance of a supported thesis it was found necessary to add qualifications, reservations, or interpretations in order properly to test the thesis against the historical case. Do successful predictions appear to be the sole criterion for theory appraisal here? Yes, but only if retrodictions are considered to be a subcategory of predictions. Was the appraisal of the benzene theory relative to existing criteria of theory assessment and to specific rivals? Yes, but there was a sense of absolute standards as well. Can 'crucial experiments' be identified? Yes, but this identification depends on a particular definition of the term, and moreover such instances of crucial experiments, even when the result and interpretation of the experiment was not contested, did not seem to compel the conviction of determined opponents.

If the appraisal of scientific theories cannot be reduced to a kind of straightforward application of rules of theory assessment on whose results virtually all reasonable scientists can agree, neither can models of theory appraisal ever be tested against history in a fully rigorous fashion. But by the same token, the very circumstance that successful scientific theories do normally command general allegiance even in the absence of such agreed rules suggests that it is not unrealistic to hope for the sort of result sought by this paper and by others in the project: a consensus as to the greater or lesser empirical success of individual models of theory appraisal.

NOTES

[1]Francis Japp, 'Kekulé Memorial Lecture', *J. Chemical Society* **73** (1898), 97–138 (135–136). Kekulé's three initial papers are most accessible in R. Anschütz (ed.), *August Kekulé* (Berlin, 1929), 2 vols., **2**, 371–83, 388–96, and 401–53.

[2]Anschütz, *ibid.*, pp. 649–685.

[3]A. J. Rocke, 'Hypothesis and Experiment in the Early Development of Kekulé's Benzene Theory', *Annals of Science* **42** (1985), 355–381, which provides further details and full literature citations for the following discussion.

[4]The discovery of isomeric benzenes or aromatics with less than six carbons would mean that aromatic compounds do *not* all have a common (i.e., single and unique) six-carbon nucleus. The existence of isomers of mono-substituted benzene (the discovery of a second benzoic acid, or a second chlorobenzene, for example) would require that an atom or functional group could attach to *two* chemically distinct parts of the CHM ring. In CHT, all carbons are chemically equivalent, and the only CHM structure where two chemically different carbons are found is 'Dewar' benzene.

[5]For instance, dichlorobenzene can have the two chlorine atoms attached to: (a) adjacent (1,2) carbon atoms – the 'ortho' isomer; (b) carbons once removed from each other (1,3) – the 'meta' isomer; or (c) opposing (1,4) carbon atoms – the 'para' isomer. This situation would be expected not only for Kekulé's CHT structure, but also for prism and centric benzene, and arguably for the diagonal structure as well. 'Dewar' benzene would require several (probably six) isomers of dichlorobenzene. The half-dozen other possible non-CHM SCN benzene theories also fail this criterion of three di-substituted products. Beilstein's work therefore effectively limited the field to the four or five CHMs listed.

[6]Kekulé to Baeyer, 10 April 1865, August-Kekulé-Sammlung, Institut für Organische Chemie, Technische Hochschule, 61 Darmstadt, West Germany. I thank Professor K. Hafner for permission to cite this letter.

[7]Anschütz, *ibid.*, pp. 401–403, 438.

[8]A. S. Couper (in 1858) and Joseph Wilbrand (in 1861) had each suggested a diallene benzene structure ($H_2C = C = CH - HC = C = CH_2$), and Wilbrand also speculated on a highly condensed-ring structure. The best discussion is still Anschütz's: *ibid.*, **1**, 177–179 and 296–305.

[9]Namely, $C_3(CH_2)_3$ or $C_4(CH_3)_2$.

[10]Albert Ladenburg, 'Bemerkungen zur aromatischen Theorie', *Berichte der Deutschen Chemischen Gesellschaft* **2** (1869), 140–142; *idem*, 'Ueber Benzolformeln', *ibid.*, pp. 272–274; Wilhelm Koerner, 'Fatti per servire alla determinazione del luogo chimico nelle sostanze aromatiche', *Giornale di scienze naturali ed economiche* **5** (1869), 208–256. Koerner's work remained unknown in Germany until 1875.

[11]Ladenburg's accomplishment, and his degree of indebtedness to others, is well summarized in his *Theorie der aromatischen Verbindungen* (Brunswick, 1876), pp. 10–21. The experiments and theoretical argument amount to a fairly rigorous demonstration of the generality of Beilstein's empirical finding of tri-isomerism of *all* di-derivatives.

[12]Koerner, 'Studj sull'isomeria della cosi dette sostanze aromatiche a sei atomi di carbonio', *Gazzetta chimica italiana* **4** (1874), 305–446.

[13]For descriptions of these developments see Richard Meyer, *Einleitung in das Studium*

der aromatischen Verbindungen (Leipzig and Heidelberg, 1882), pp. 29–102, and Ladenburg, *Theorie*. It will be difficult for readers unacquainted with organic chemistry to follow the reasoning in the several experimental tests mentioned in this paragraph and a detailed explanation would take me far beyond the necessary space limitations that the editors have imposed. Suffice it to say there are two kinds of tests here. One is the classic isomer-counting exercise: the theory allows for *n* isomers exhibiting such-and-such a formula; how many are actually observed? A second kind of test examines structural details or possibilities by means of synthetic or analytic reaction routes.

[14]Hermann Kolbe, 'Ueber eine neue Darstellungsmethode und einige bemerkenswerthe Eigenschaften der Salicylsäure', *J. für praktische Chemie* **118** (1874), 89–112 (90). Ladenburg commented two years later that "there is almost no significant scientist in this field who does not at least to a certain degree adhere to [this theory]" (*Theorie*, p. 1).

[15]Kolbe, *Ueber die chemische Constitution der organischen Kohlenwasserstoffe* (Brunswick, 1869), pp. 9–22; discussed by Meyer, *Einleitung*, pp. 87–89.

[16]The benzene-structure controversies are reviewed, e.g., by Colin Russell, *The History of Valency* (Leicester, 1971), pp. 242–257, and by J. R. Partington, *A History of Chemistry*, **4** (London, 1964), pp. 543–565 and 802–805.

[17]V. M. Schelar, 'Alternatives to the Kekulé Formula for Benzene: The Ladenburg Formula', in O. T. Benfey (ed.), *Kekulé Centennial* (Washington, 1966), pp. 163–193.

[18]Ladenburg enumerated several such instances and cited the relevant refutations in the literature: *Theorie*, pp. 21–22.

[19]See, e.g., quotation attached to note 25 below.

[20]For instance, Lakatos cites successful retrodictions of general relativity theory as a factor in its favorable appraisal: *The Methodology of Scientific Research Programmes* (Cambridge, 1978), p. 39.

[21]H. W. Schütt, 'Guglielmo Körner (1839–1925) und sein Beitrag zur Chemie isomerer Benzolderivate', *Physis* **17** (1975), 113–125 (122, 124). Schütt argues that Koerner's work exemplifies verification rather than falsification, which he suggests supports Kuhnian normal science rather than the Popperian model.

[22]F. Beilstein and A. Kurbatov, 'Ueber die Chlorderivate des Benzols', *Annalen der Chemie und Pharmacie* **192** (1878), 228–240.

[23]E. Hjelt, *Geschichte der organischen Chemie* (Brunswick, 1916), p. 300.

[24]Ladenburg, 'Ueber das Mesitylen', *Berichte der deutschen chemischen Gesellschaft*, **7** (1874), 1133–1137; *idem, Theorie*, pp. 1 and 28.

[25]Ladenburg, *Theorie*, p. 4.

[26]e.g., Meyer, *Einleitung*, p. 102; W. Marckwald, 'Die Benzoltheorie', *Sammlung chemischer und chemisch-technischer Vorträge* **2** (1898), 1–34; V. Meyer and P. Jacobson, *Lehrbuch der organischen Chemie* 2:1 (Leipzig, 1902), pp. 46–63.

[27]Laudan, *Science and Hypothesis* (Boston, 1981), pp. 9–15, 72–127; A. J. Rocke, 'Methodology and Its Rhetoric in Nineteenth-Century Chemistry: Induction versus Hypothesis', in E. Garber (ed.), *Robert Schofield Festschrift*, unpublished manuscript.

[28]Kolbe, 'Ueber die chemische Natur der Salicylsäure', *J. für praktische Chemie* **120** (1875), 151–157.

[29]Kolbe, 'Chemische Constitution des Benzols und einiger Derivate derselben', *J. für praktische Chemie* **122** (1876), 347–355.

WILLIAM BECHTEL

FERMENTATION THEORY: EMPIRICAL DIFFICULTIES AND GUIDING ASSUMPTIONS*

1. INTRODUCTION

How do scientists react when their guiding assumptions run into empirical difficulties? This project leads us to consider six answers: (GA2.1–GA2.6). To evaluate the adequacy of these answers to the question of how scientists deal with empirical difficulties that confront their guiding assumptions, I will examine two empirical difficulties that confronted the guiding assumptions of early 20th-century biochemists working on fermentation and discuss the way the biochemical community responded to these difficulties.

Examining two cases will obviously not suffice for a definitive test of any claim about how scientists deal with empirical difficulties. We are confronted not only by the problem that single tests, confirmatory or disconfirmatory, cannot establish the correctness or incorrectness of a universal claim, but also by the problem that for these purposes scientists may not form a natural kind. Different scientists may react differently in the face of empirical difficulties confronting their guiding assumptions. Nonetheless, the cases I shall describe offer greater support for the first three theses than the latter three. I shall argue, however, that these two cases reveal a deeper problem which points to the inapplicability of all six theses. The theses are structured on the assumption that scientists are aware when their guiding assumptions confront empirical difficulties. In these two cases, however, the researchers did not seem to possess such awareness. While in retrospect it is clear that the two problems represented empirical difficulties for the guiding assumptions of early biochemistry, these difficulties were recognized as difficulties for the guiding assumptions only after the problems had been solved by new theories that departed from one or more of the guiding assumptions which had led to the original problem. Thus, although we can make the cases fit three of the theses identified above, I shall argue that this way of defining the issues misrepresents what was happening in these two cases, and perhaps others.

163

2. GUIDING ASSUMPTIONS OF EARLY 20TH CENTURY BIOCHEMISTRY

Before presenting the two cases and the scientific response to them, it is necessary to identify the guiding assumptions present in early 20th-century biochemistry. For purposes of this project, the term 'guiding assumption' refers to the "large-scale, relatively long-lived conceptual structures which different modelers refer to as 'paradigms' [Kuhn], 'global theories' [Feyerabend], 'research programmes' [Lakatos], or research traditions [Laudan]". Although guiding assumptions are sometimes referred to as theories, they are at a higher level than typical specific theories and are construed as providing the framework for developing these more specific theories. They perform this task by offering a general conception of nature in the domain in question or by offering a set of heuristics for developing specific theories.

Identifying the guiding assumptions operative in any discipline at a given time is difficult. Sometimes practitioners of a discipline may publish a "manifesto" setting out what they take to be the way the discipline ought to proceed, but such a document at best reflects the views of some practitioners. Further, it may not reveal the actual assumptions that guide their research. As biochemistry emerged in the early decades of this century, though, there was no such manifesto. The guiding assumptions of the field, however, can be at least partially elicited from the way researchers construed their research tasks and the manner in which they went about attempting to develop their explanations.

It is controversial whether one should treat biochemical investigations in the early years of this century as continuous with the physiological and organic chemistries of the 19th century (see Teich, 1965) or whether it should be construed as a new discipline (see Kohler, 1975). What is clear, though, is that the research program of biochemistry took on a distinctive character in the years after 1897. Prior to that time, a number of researchers had found chemical substances which would facilitate some of the same chemical reactions as those performed by living cells. Berzelius had coined the name 'catalysts' for these substances and Hoppe Seyler introduced the term 'enzyme' for catalysts that carried out their functions within living cells. While some researchers held out the hope that enzymes could be found to account for all the reactions associated with living organisms, in the case of some prominent reactions, including fermentation, that did not seem possible. Over a fifty-year period a number of researchers had tried, without success, to establish that fermentation could be accomplished in press juice or cell extract in which one no longer had a whole living cell. Their failure seemed to support the view, promoted by Schwann and Pasteur, that fermentation was a process that required whole living cells and could not be explained in purely chemical terms. In 1897, however, Eduard Buchner succeeded where others had failed when he demonstrated that fermentation could be accomplished in yeast press juice and did not require the presence of whole living cells. After that date there was increasingly broad agreement within the biochemical community that biological systems are chemical machines that extract their energy from their foodstuffs and use this energy to carry out their work. Hopkins (1913), for example, in making his case against protoplasm theories, contended that in studying the reactions in living organisms researchers were "confronted not with complex substances which elude ordinary chemical methods, but with simple substances undergoing comprehensible reactions". Once this assumption was made, the challenge was to figure out what were the working components of the biological machine and how they accomplished such activities as fermentation.

From this point, the research endeavor followed a pattern that is quite common in mechanistic research programs in the life sciences.[1] A common first move is to try to explain the ability of a system to perform a function by identifying a component in the system that itself performs that function. Buchner initially pursued the attempt to explain fermentation in this manner, attributing it to a single enzyme, which he call 'zymase'. He and other researchers, however, recognized quickly that fermentation was a multi-step process. At this point the attempt to develop a mechanical explanation of fermentation began in earnest as researchers tried to identify the various steps in the fermentation process and the enzyme responsible for catalyzing each reaction. Since enzymes themselves could not be identified in terms of their chemical structure but only in terms of their functional capacities, the key to developing the explanation was the attempt to identify the intermediate products formed by the various reactions leading from sugar to ethanol. The simplest assumption was that the reactions formed a linear chain leading from sugar through a series of intermediate substances, each produced by a basic chemical reaction (e.g., dehydration, decarboxylation, oxidation, etc.), to the final ethanol. The theoretical task confronting biochemists was to propose pathways that made chemical sense in that they employed generally recognized chemical reactions, while the empirical task was to provide evidence that the proposed intermediate substances were actually produced in the course of fermentation. This, however, was not an easy task, since intermediate substances formed in the course of fermentation would proceed on through the reaction cycle as rapidly as formed. To test whether these substances were intermediaries in ordinary fermentation, researchers had to rely on indirect evidence such as interrupting the reaction pathway to determine what substances would then accumulate or supplying potential intermediates to the pathway to see if they would proceed through the pathway. In using the last technique a critical assumption was often made, namely that nothing could be an intermediary unless it could be metabolized as rapidly as the original metabolite. This assumption is really a consequence of the assumption that the pathways are linear. If they are linear, then nothing that happens earlier in the pathway should affect later reactions and those should proceed independently, but if the pathway were not linear, then later reactions may depend on earlier ones and so supplying more of the intermediary may not actually lead to a normal reaction. Since researchers granted this assumption, this technique became a principal test for determining whether something was an intermediary.

From this description of the research program of early-20th-century biochemistry, we can elicit the following set of guiding assumptions:

(1) Reactions in living systems are ordinary chemical reactions,
(2) Such reactions as occur in living systems but which do not occur naturally in approximately the same physical conditions (e.g., temperature, pH) outside the cell are catalyzed by enzymes,
(3) These reactions are generally linked in linear pathways in which the product of one reactions will provide the input to the next reaction,
(4) One can identify the steps in these linear pathways by a two-pronged investigation of (a) interrupting or stimulating these processes as they occur either *in vivo* or *in vitro* so as to identify intermediary products, and (b) building models of possible reaction pathways that would adhere to the known principles of chemistry,

(5) If a potential intermediary is added to a fermenting situation and it is not fermented as rapidly as the sugar itself, then it cannot be an intermediary.

In offering these as guiding assumptions of early biochemical research, I am not arguing that individual researchers ever argued for them explicitly or even would have acknowledged them if asked. Rather, I am offering them as part of an historical reconstruction that is intended to make sense out of the research endeavor that was pursued. Moreover, they were not universally held by all researchers (e.g., some researchers continued to reject the enzyme assumption for several decades) and could be violated on occasion. Thus, they were guiding heuristics, not mandatory principles. This is particularly true of principle (3). Together with principle (5), which is really a corollary to it, principle (3) was the source of the empirical difficulties to be discussed below. As I shall show, though, while researchers held to this principle in the face of the empirical difficulties I will be discussing, they were willing to violate them in other cases.

3. EMPIRICAL DIFFICULTIES ENCOUNTERED IN EARLY-20TH-CENTURY BIOCHEMISTRY

Having identified some of the guiding principles that confronted early-20th-century biochemistry, I will now discuss two sets of empirical difficulties which researchers confronted and which eventually led them to modify some of their guiding assumptions. In each case I will consider how well (GA2.1–GA2.6) characterize the researchers' responses to these difficulties. As I have already indicated in the introduction, though, I will also go on to suggest that none of these theses fully captures what was occurring in these cases since researchers were generally not fully aware that the empirical problems they were confronting were challenges to their guiding assumptions. I will return to explain why this was the case after I have examined the difficulties themselves and researchers' responses to them.

3.1. *Empirical Difficulties Presented by Methylglyoxal*

As I have already indicated, once researchers discovered that fermentation could be performed without living cells and concluded that it was a purely chemical process, they set out to try to identify potential intermediaries in the pathway. Investigators took a clue from the results of attempts by organic chemists in the 19th century to degrade sugars with alkalis. Duclaux (1886) had shown that when strong alkalis were used, alcohol could be produced, while weaker alkalis yielded lactic acid. Other researchers had shown that various other three-carbon compounds including glyceraldehyde and methylglyoxal could also be produced. Taking this as a guide, researchers assumed that fermentation began with the scissioning of the sugar molecule into such three-carbon sugars and ended with the decarboxylation of a three-carbon compound to form alcohol. A further critical clue derived from Neubauer's work on amino acid metabolism, in which he discovered that pyruvic acid formed in that process was subsequently fermented to yield alcohol. He suggested that pyruvic acid might also be an intermediary in alcoholic fermentation (Neubauer and Fromherz, 1911). Neuberg followed up on Neubauer's proposal and provided evidence for an enzyme he labeled 'carboxylase' which could carry out the decarboxylation of pyruvic acid to form acetaldehyde, which could then form alcohol by reduction.

From these pieces of information and knowledge of what reactions were chemically possible, Neuberg put together a comprehensive scheme for alcoholic fermentation. In it he proposed a scissioning of the glucose molecule into two molecules of methylglyoxal (step (1) – see Figure 1). In the first stage of the reaction he proposed that these two undergo a Cannizzaro reaction, oxidizing one molecule to pyruvic acid and reducing the other to glycerol (step (2a)). The pyruvic acid is then

(1) $\quad C_6H_{12}O_6 \xrightarrow{\ 2H_2O\ } 2\,C_3H_4O_2$

   hexose   methylglyoxal

(2a) $2\,C_3H_4O_2 + 2H_2O \longrightarrow C_3H_8O_3 + C_3H_4O_3$

   methylglyoxal   glycerol   pyruvic acid

(2b) $C_3H_4O_2 + C_2H_4O + H_2O \longrightarrow C_3H_4O_3 + C_2H_5OH$

   methylglyoxal aldehyde   pyruvic acid alcohol

(3) $\quad C_3H_4O_3 \longrightarrow C_2H_4O + CO_2$

   pyruvic acid   aldehyde

Fig. 1. Glycolytic pathway proposed by Neuberg and Kerb (1913).

decarboxylated, yielding acetaldehyde (Step (3)). After enough acetaldehyde is produced, he proposed that it was the acetaldehyde that participates in the Cannizzaro reaction, being reduced to alcohol while the methylglyoxal is oxidized to pyruvic acid (step (2b)).

Neuberg's scheme was not the only one advanced during this period. Lebedev, Kostychev, and others proposed a variety of schemes using various three-carbon compounds as intermediaries (see Florkin, 1975). But Neuberg's found greatest favor and became the generally accepted model of fermentation for approximately thirty years. Its main virtues included the fact that besides providing a sequence of reactions that produced the output from the input through simple reactions of known types, this model could account for the methyl that had been found amongst the products of fermentation. Neuberg engaged in a number of empirical investigations to try to show that his postulated intermediaries are actually produced in biological fermentations by fixing and removing them from living cells. For example, he employed sodium sulfate to prevent the reduction of acetylaldehyde and show its occurrence in fermenting material (Neuberg and Reinfurth, 1920).

The model, however, faced a serious empirical difficulty. Methylglyoxal did not satisfy the requirement that an intermediary in a sequence of reactions should undergo the reaction at least as rapidly as the initial substance. Methylglyoxal could not be fermented in living cells. During the twenty years that the Neuberg model was widely accepted, this was a point of intense investigation, with some investigators reporting success in fermenting methylglyoxal, but most reporting failure. Neuberg himself excused the failure as an artifact of trying to add a substance to the cell experimentally rather than having it produced in the cell. He argued that the fate of naturally produced substances differed from those added experimentally to the cell, especially if the natural substance had a slightly different constitution than that used in experimental contexts. The natural methylglyoxal might therefore be fermented while the experimental methylglyoxal could not.

The failure of methylglyoxal to ferment constituted an empirical difficulty to the guiding assumptions of early biochemistry because, of all the theories advanced in response to these assumptions, only Neuberg's seemed compelling, and yet the failure of methylglyoxal to ferment violated the last of these assumptions. Let us consider how well (GA2.1 –GA2.6) account for the response of the biochemical community to this empirical difficulty. It might appear that the case supports (GA2.1) according to which scientists blame their own experimental skill rather than their guiding assumptions. The fact that experimentalists continually sought to establish that methylglyoxal could ferment or was found in fermenting cells suggests that they thought that with additional work they could resolve the difficulty. But there is no evidence that the researchers blamed their experimental work. When researchers reported negative findings, they reported them as that, not as their own shortcomings. They kept returning to the problem because it constituted a major difficulty and they continued to hope that they might find some way in which cells did manage to ferment methylglyoxal or evidence that would suggest an alternative model that avoided the problem. In accord with (GA2.2) there is a sense in which researchers were prepared to leave the difficulty unresolved for years. But this should not be viewed as a conscious choice. Without a solution, they had no alternative but to leave the problem unsolved. They did not, however, ignore the problem but continued to confront it. Again, in accord with (GA2.3) researchers did not abandon or change their guiding assumptions in response to the problem, but this too should not be viewed as a conscious and principled refusal. In fact, Neuberg's theory itself departed from the linearity assumption in a critical respect – step (2b) makes use of a product formed in step (3), thus introducing a non-linearity. The motivation for introducing this cycle was purely chemical – oxidation of methylglyoxal had to be balanced by a reduction, and one already required a reduction of pyruvic acid to lactic acid. Thus, assumption (3) was not a rigid rule, but could be bent if a non-linear model suggested itself. The problem was that no such model that solved the methygloxal problem was suggested until the 1930s.

This case offers even less support for the remaining three theses. Thesis (GA2.4) holds that researchers will stick with guiding assumptions as long as they anticipate novel phenomena. It is not clear that major success was being achieved during this period. In research on fermentation and other metabolic processes, researchers continued to amass data about possible intermediaries, and a variety of schemes were advanced to handle these phenomena, but there were few major advances. With reference to (GA2.5), it is not clear that temporal persistence of the problem was a major consideration in researchers' decisions as to whether they should maintain their guiding assumptions or reject them. Rather, the lack of any viable alternative scheme seemed to be the critical factor. Finally, in accord with (GA2.6), it might seem as if

Neuberg at least introduced an untestable hypothesis to protect himself against the result that methylglyoxal could not be fermented. There is one sense in which it is untestable – when any known form of methylglyoxal fails to ferment, he can allege that the true intermediary is a still unknown form. And yet the claim is one that could be investigated by looking for this alternative form or by devising ways of trapping it during the course of the reaction.

While the first three theses seem more applicable to this case than the others, none of them seems to characterize clearly what was happening. Part of the problem is that the scientists, as far as one can tell from their publications, were not consciously aware that their guiding assumptions confronted empirical difficulties. In part, this is a result of the fact that one of the critical guiding assumption – the linearity assumption – was not explicit but tacit in the thinking of most investigators. There was explicit commitment to its corollary, thesis 5, which maintained that nothing could be an intermediary that was not itself fermentable, but this seemed a reasonable assumption in its own right and there was no clear guidance on how to proceed if this was given up. In addition, researchers were not convinced that the empirical difficulty would be be resolved either along the lines Neuberg proposed, or by devising a somewhat different theory than Neuberg's that would not encounter the problem of methylglyoxal. Researchers continually came back to the problem, viewing it as a puzzle that had not yet yielded, but might. They could do this because of the tacit character of the guiding assumption and the distance between guiding assumptions and actual empirical tests. Empirical evidence, even if accepted as inconsistent with a particular theory, was not thereby rendered inconsistent with the guiding assumptions.

3.2. Empirical Difficulties Posed by Hexosediphosphate

The second empirical difficulty confronting the guiding assumptions of biochemical research on fermentation arose even before Neuberg developed his general scheme of fermentation. Even though Buchner had shown that fermentation could occur in a cell-free extract, such fermentation was always slow and ceased after a short period. Harden investigated this phenomenon and discovered that adding blood serum to the extract increased the rate of fermentation. Together with Young, he traced this effect to the fact that the blood serum supplied phosphates and a heat-stable, dialyzable substance which facilitated the fermentation (Harden and Young, 1906). The latter substance Harden and Young designated as a 'co-ferment'; its actual function remained a mystery, despite numerous investigations, until a much more complex account of fermentation was developed in the 1930s (see Bechtel, 1984). In their continuing work, Harden and Young focused on the role of phosphates. They showed that adding inorganic phosphate to a fermenting solution resulted in a temporary increase in this rate followed by a return to the lower level. The additional carbon dioxide produced in the rapid phase was shown to be proportional to the increased phosphate.

Harden and Young also found that the added phosphate disappeared rapidly into a form that was no longer precipitable. Young (1909) isolated the product of the phosphate reaction as an ester, hexose-diphosphate. This substance was found to ferment very slowly. Applying the maxim that nothing could be an intermediary in the fermentation pathway that was not fermentable as rapidly as sugar itself, he concluded that it was a by-product of the fermentation reaction and not an intermediary on the actual pathway of fermentation. Since the absence of phosphate stopped the reaction, though, Harden and Young considered the formation of this by-product to be essential to the fermentation reaction itself. The discovery of the need for phosphates in the fermentation process but the determination that the compounds formed from them could not figure as intermediaries in the fermentation pathway constituted the second empirical difficulty for the theory. Given the guiding assumptions used by early biochemists, no role could be identified for the phosphorylated compounds. Thus, Neuberg did not give them any role in his scheme of fermentation.[2] Yet empirical investigations showed that the formation of these phosphorylated compounds was necessary for the fermentation process.

That a difficulty existed here is shown by the continued attention given to this problem over the following three decades. Harden and Young were particularly concerned to make sense of their discovery and in 1908 they advanced a theory to explain it. They proposed (see Figure 2) that the combination of a hexose molecule and two phosphates made another hexose molecule more labile, causing it to break down into two triose molecules and then continue along the pathway. Since it appeared that the ester itself could not be fermented, however, they proposed that it had to be hydrolyzed and release its phosphates (a slow process) before the sequence could repeat itself. The rapid fermentation when

Glycolytic reaction:

$$2\ C_6H_{12}O_6 + 2\ R_2HPO_4 \longrightarrow 1\ C_6H_{10}O_4(PO_4R_2)_2$$

hexose   phosphate   hexosediphosphate

$$+ 2\ C_2H_5OH + 2CO_2 + 2H_2O$$

alcohol

Hydrolysis Reaction:

$$1\ C_6H_{10}O_4(PO_4R_2)_2 + 1H_2O \longrightarrow 2\ C_6H_{12}O_6 + 2\ R_2HPO_4$$

hexosediphosphate   hexose   phosphate

Fig. 2.   Harden and Young's conception of the role of phosphates in glycolysis.

phosphates were added to the solution was due to there being an adequate supply of inorganic phosphate for the first of these reactions to proceed. The reaction slowed down, however, when this phosphate was used up so that it could only proceed as rapidly as the second reaction released free phosphate so that it could continue. (For further discussion of this research, see Harden's (1932) recount of it and Kohler, 1974.)

Harden and Young's theory was not the only one advanced to handle the formation of hexosediphosphate. Lebedev (1912) developed an alternative account of the role of phosphates according to which it was only after the hexose molecule split into two three-carbon compounds that phosphate entered the cycle. The two triosephosphates would then reunite to form hexosediphosphate, which would in turn yield inorganic phosphate, alcohol and carbon dioxide:

hexose → 2 triose → 2 triosephosphates → hexosediphosphate → alcohol + CO_2 +inorganic phosphate

The motivation for this rather esoteric scheme was evidence that when dihydroxycetone (a triose sugar) was fermented, hexosediphosphate appeared. Lebedev proposed that dihydroxyacetone was one of the trioses formed on splitting hexose (the other being glyceraldehyde, which he thought could be decarboxylated into alcohol without phosphorylation) and that if it were not removed through formation of the hexosediphosphate, the reaction could not proceed. This explanation was disputed by Harden and Young (1912), who showed that high concentrations of dihydroxyacetone would not inhibit fermentation of glucose. Lebedev's proposal, though, was just one of several that were advanced during this period to account for the role of phosphates in alcoholic fermentation. None of them seemed to offer any advantages over Harden and Young's proposal, which remained the most widely accepted explanation until the 1930s.

Despite its general acceptance, however, Harden and Young's theory remained unsatisfying to many researchers, who continued to focus on the role of phosphates in fermentation. The evidence about hexosediphosphate became all the more confusing in the wake of research on muscle glycolysis. This reaction increasingly appeared to be a very similar reaction to alcoholic fermentation, differing primarily in having lactic acid rather than alcohol as its final product. Embden and his colleagues tried to simulate Buchner's work with alcoholic fermentation by carrying out muscle glycolysis with muscle press juice. While some lactic acid was formed by this means, Embden noted that adding additional glucose or glycogen failed to increase the reaction. Embden interpreted these results as pointing to an unknown predecessor of lactic acid in muscles, which he labeled 'lactacidogen' (Embden, Kalberlah and Engel, 1912). Beyond this, Embden not only confirmed earlier reports that phosphoric acid appeared in the course of muscle action but showed that in his experimental arrangement, phosphoric acid appeared in equimolar proportion with lactic acid. He concluded that lactacidogen was a phosphorylated substance. Knowing of Harden and Young's results with alcoholic fermentation, Embden tested hexosediphosphate and showed that it produced an increase in formation of lactic acid and phosphoric acid in his muscle press juice, suggesting that it might be lactacidogen.[3]

Embden's results were disturbing. Evidence continued to amass that alcoholic fermentation and muscle glycolysis were analogous processes, but in one hexosediphosphate seemed to play an intermediary function while in the other it could only be a by-product of the main pathway. Meyerhof tried to resolve these difficulties by suggesting that all sugar goes through a phosphorylated stage in the process of fermentation or glycolysis, forming a hexosemonophosphate. He proposed that this monophosphate was an active form, some of which then proceeded to transfer its phosphate group to another molecule of monophosphate and then scission into trioses. This reconciled the conflicting data from alcoholic fermentation and muscle glycolysis, but it seemed a rather baroque and *ad hoc* way to account for the role of phosphates. Moreover, it failed to account for how normal cells were able to maintain their fermentation rates without excessive build-ups of hexosediphosphate.

Considering again the focal theses for this paper, this case can also be interpreted in accord with (GA2.1–GA2.3), but the fit is not perfect. The fact that a variety of investigators continued to try to resolve the problems created by hexosediphosphate within the framework of their basic guiding assumptions suggests that they attributed the problem to their own skill rather than inadequacies of their guiding assumptions (GA2.1). But, once again, the deficiency in skill is not in experimental skill but in developing an adequate theoretical model that conforms to the guiding assumptions. In conformity with (GA2.2), investigators left the problem of hexosediphosphate unresolved for years, but once again the reason was that no alternative solution was available, not that they were not concerned about the problem. In accord with (GA2.3), they did not change their guiding assumptions, but that was not a principled decision. In one respect, all the proposals to deal with hexosediphosphate introduced a non-linearity into the process since, with Harden and Young, they posited a side-reaction needed to activate the main reaction, or with Meyerhof, proposed a loop at the beginning of the pathway.

The remaining theses again seem unhelpful in attempting to understand this case. Thesis (GA2.4) seems only marginally applicable since there were no major novel phenomena being anticipated by the acceptance of the guiding assumptions, only accumulation of further data about reactions that can or do occur in living tissue. With respect to (GA2.5), the temporal duration of the problem does not seem to have been a major factor in how serious the problem was regarded. Finally, while proposals like Meyerhof's have an *ad hoc* character, they do not seem to be untestable hypotheses designed simply to save the guiding assumptions, as (GA2.6) holds. Rather, they are empirical proposals designed to accommodate rather confusing data.

As with the previous case, the reason why none of the theses seems particularly applicable is that the difficulties raised by hexosediphosphate were not consciously recognized as empirical difficulties confronting the guiding assumptions of early biochemistry. The formation of hexosediphosphate was a puzzling phenomena insofar as it did not satisfy the maxim that any true intermediary should be fermented as rapidly as sugar itself. Thus, biochemists were left to try to figure out what role hexosediphosphate could play, a challenge they could not meet as long as they accepted the linearity assumption. But they did not view the problem posed by hexosediphosphate as raising a difficulty either for assumption (3) or (5). Once again, the distance between guiding assumptions and empirical data provided a fair amount of maneuvering space for developing alternative empirical theories, so that the difficulties for the guiding assumptions were not readily apparent.

4. CONCLUSION

I have examined the response of early biochemists to two empirical difficulties that confronted the guiding assumptions of early-20th-century biochemistry. There is an interesting contrast between these two problems. In both cases the potential intermediary, methylglyoxal or hexosediphosphate, failed to meet the test of being fermentable as rapidly as sugar itself. In the case of methylglyoxal, biochemists nonetheless continued to include it in the main fermentation pathway anyway, since the simplest scheme of fermentation that met their other assumptions, that proposed by Neuberg, required it. They then tried to overcome the difficulty that this violated assumptions by trying to find evidence that it was fermentable, but perhaps in a different form. In the case of hexosephosphate, it was empirical evidence that showed that phosphate was required for fermentation. The ester formed from it, hexosediphosphate, however, also failed to meet the maxim. In this case biochemists adhered to the maxim and excluded hexosediphosphate from the main fermentation pathway. The problem then became that of explaining what function phosphates and the formation of hexosediphosphate played in the fermentation process.

Both of these empirical difficulties resulted from an oversimplified view of fermentation which stems from the linearity assumption (assumption (3)). The linearity assumption seems entirely reasonable when one looks at fermentation simply as a chemical process producing alcohol from sugar. But in adopting such a view one entirely overlooks the biological function of fermentation, which is to liberate the energy from the sugar and make it available to the organism. The method by which nature accomplishes this function, however, introduces a number of non-linearities into the pathway. As investigators discovered during the 1930s, it was through the phosphate bonds that the energy released from fermentation was stored and transported to other locations where it could be liberated to perform cellular work. The chief vehicle for storing and transporting this energy was adenosine triphosphate (ATP). As was recognized by Meyerhof shortly after ATP was discovered in the late 1920s, the third phosphate bond (as well as to second) released a great amount of energy in hydrolysis. In 1940 Lipmann labelled these bonds "high energy phosphate bonds" and they became generally recognized as vehicles of energy transport.

During the same decade, moreover, researchers came to recognize that the intermediaries in the fermentation pathway were themselves phosphorylated and that it was in the phosphate bonds of these intermediaries that the energy released in the oxidation reaction was stored until it was transferred to ATP. The first breakthrough to this discovery occurred in 1933 when Embden offered evidence that the triose compounds resulting from scissioning of the sugar were themselves phosphorylated and proposed a scheme wherein all the intermediaries in fermentation were phosphorylated substances. Subsequent research established that one of the phosphate groups that figured in the formation of hexosediphosphate also came from ATP and provided the energy for the scissioning of the sugar molecule as well as other steps in which phosphates were added into or removed from the fermentation pathway. By the end of the decade it was clear that a phosphate cycle figured centrally in the process of fermentation and that the transfer of phosphates between different intermediaries and ATP served to integrate reactions in the fermentation process. (Another coenzyme, NAD, was also discovered which linked the oxidation event with the later reduction of pyruvic acid.)[4]

These discoveries undercut the linearity assumption and helped to demonstrate that various reactions in the fermentation pathway are linked to others earlier or later in the pathway so that one is dealing with a highly integrated system, not one that can be decomposed into a simple chain of events. Once these mechanisms of integration were recognized, both the empirical difficulties to which I have attended were readily explained. Methylglyoxal, it turned out, was not itself an intermediary, and was removed from the pathway in favor of the phosphorylated three carbon sugars. These phosphorylated substances did ferment, thus not violating the critical maxim. Hexosediphosphate, on the other hand, is an intermediary in the pathway. The reason it did not ferment in the test solutions was that production of these solutions destroyed the ATPase that was necessary to break down the ATP formed during the reaction, leaving only a slow breakdown of ATP to ADP (adenosine diphosphate) and inorganic phosphate. As a result, all ADP available in the solution was transferred into ATP in the rapid phase of the Harden and Young reaction, leaving none available to take up the phosphate bonds at latter steps in the fermentation pathway and no available phosphate to create more of the hexosediphosphate ester. This caused the reaction to slow down. Once it was recognized that in fact what one was dealing with in a normal cell was such an integrated system, it no longer seemed surprising that in cell extracts where part of this system was removed that true intermediaries would not ferment.

With regard to the theses indicated at the outset, we have seen that the first three do seem partially applicable to these two cases, but the later three do not. Further, I have argued that even the first three do not really get at what was going on. Rather, I have contended that while the problems surrounding methylglyoxal and hexosediphosphate can be seen in retrospect to constitute empirical difficulties for the guiding assumptions of early biochemistry, it was not clear at the time that they did so. They were difficulties for specific theories constructed under the guiding assumptions, and as such they provoked intensive inquiry amongst biochemists. But at the time it was not clear that one could not develop alternative theories that could overcome the difficulties without modifying the guiding assumptions.

The failure of researchers to recognize that these difficulties were in fact due to the guiding assumptions with which researchers were working is in part a result of the fact that guiding assumptions were not explicitly stated, but were tacitly assumed. But, perhaps more importantly, it was due to the fact that there is a significant distance between guiding assumptions and actually testable empirical hypotheses. Guiding assumptions are general principles that do not themselves give rise to empirical predictions that can be used to confirm or refute them. Rather, they provide a framework in which to develop more specific theories which can be put to such empirical tests. Such tests only count for or against these more specific theories. The relationship between the guiding assumptions and the more specific theories is not a deductive one, so one cannot reason from the fact that an empirical test of a more specific theory resulted in disconfirmatory data to the claim that the guiding assumptions are wrong. It therefore remains possible that an alternative specific theory developed under the same guiding assumptions might be empirically adequate. The only way empirical data could itself undermine guiding assumptions would be if it was clear that no theory could be constructed under the guiding assumptions which would fit the data. This is generally nearly impossible to prove.

Instead, what happened in this case, and may be true of other cases as well, is that other factors led researchers to consider theories that do not conform to their guiding assumptions, and in such an indirect manner researchers are led to abandon their guiding assumptions. Once they were abandoned, researchers were then in a position to recognize that some of the empirical difficulties that were confronted by earlier theories were due not just to the theories, but to the guiding assumptions which directed their construction. Prior to that point, there was no recognition that the guiding assumptions were in trouble, but only that one had empirical difficulties that needed to be solved in some way or other. Thus, only after the linearity assumption was rejected and the problems surrounding methylglyoxal and hexosediphosphate resolved, was it clear that these problems resulted from the guiding assumptions. The difficulty with all six theses that have been offered by theorists of science is that they assume that researchers possess a greater degree of awareness as they are carrying out their research that guiding assumptions are facing empirical difficulties than may be the case.

NOTES

*Research on this paper has been supported by a grant from the National Endowment for the Humanities, which I hold jointly with Robert C. Richardson.

[1]For further discussion of this pattern and an account of how it figures in the research program of numerous disciplines in the life sciences, see Bechtel and Richardson, forthcoming.

[2]Neuberg dismissed the evidence about the need for phosphates and the production and accumulation of hexosediphosphate in yeast juice preparations as an artifact of the experimental circumstances and contended that phosphates did not figure in fermentation in living cells. He argued that they could not figure since hexosediphosphate is not fermented in living cells (Neuberg and Kobel, 1925). This argument is particularly interesting since Neuberg was quite willing to include methylglyoxal in the pathway even though it too could not be fermented by living cells.

[3]In actuality, the process is more complex than Embden could anticipate. The phosphoric acid actually results not directly from hexosediphosphate, but from the breakdown of ATP.

[4]I discuss the discovery of the phosphate cycle in greater detail in Bechtel, 1986a and 1986b, and Bechtel and Richardson, forthcoming.

REFERENCES

Bechtel, W. (1984), 'Reconceptualization and Interfield Connections: The Discovery of the Link Between Vitamins and Coenzymes', *Philosophy of Science* **51**, 265–292.

Bechtel, W. (1986a), 'Developing Chemical Explanations of Physiological Phenomena: Research on Intermediary Metabolism', in W. Bechtel (ed.), *Integrating Scientific Disciplines*, Dordrecht: Martinus Nijhoff.

Bechtel, W. (1986b), 'Building Interlevel Theories: The Discovery of the Embden–Meyerhof Pathway and the Phosphate Cycle', in P. Weingartner and G. Dorn (eds.), *Foundations of Biology*, Vienna, Austria: Holder–Pichler–Tempsky, pp. 65–97.

Bechtel, W. and Richardson, R. C. (forthcoming), *A Model of Theory Development: Localization as a Strategy in Scientific Research*.

Buchner, E. (1897), 'Alkoholische Garung ohne Hefezellen (Vorlaufige Mittheilung)', *Berichte der deutschen chemischen Gesellschaft* **30**, 117–124.

Duclaux E. (1896), 'Études sur l'action solaire', *Annales de l'Institut Pasteur* **10**, 129–168.

Embden, G., Kalberlah, and Engel (1912), 'Über Milchsaurebildung im Muskelpressaft', *Biochemische Zeitschrift* **45**, 45–62.

Florkin, M. (1975), *A History of Biochemistry, Part III*, volume 31 of Florkin, M. and Stotz, E. H., *Comprehensive Biochemistry*, Amsterdam: Elsevier.

Harden, A. (1932), *Alcoholic Fermentation*, 4th edn., London: Longmans, Green.

Harden A. and Young, W. J. (1906), 'The Alcoholic Ferment of Yeast-Juice', *Proceedings of the Royal Society (London), Series B* **77**, 405–520.

Harden, A. and Young, W. J. (1908), 'The Alcoholic Ferment of Yeast-Juice. Part III – The Function of Phosphates in the Fermentation of Glucose by Yeast-Juice', *Proceedings of the Royal Society (London), Series B* **80**, 299–311.

Harden, A. and Young, W. J. (1912), 'Der Mechanismus der alkoholischen Garung', *Biochemische Zeitschrift* **40**, 458–478.

Hopkins, F. G. (1913), 'The Dynamic Side of Biochemistry', *Nature* **92**, 213–223.

Kohler, Robert (1973) 'The Enzyme Theory and the Origin of Biochemistry', *Isis* **64**, 181–196.

Kohler, Robert (1974), 'The Background to Arthur Harden's Discovery of Cozymase', *Bulletin of the History of Medicine* **48**, 22–40.

Lebedev, A. (1912), 'Über den Mechanismus der alkoholischen Garung', *Biochemische Zeitschrift* **46**, 483–489.

Neubauer, O. and Fromherz, K. (1911), 'Über den Abbau der Aminosauren bei der Hefegarung', *Zeitschrift für physiologische Chemie* **70**, 326–350.

Neuberg, C. and Kerb, J. (1913), 'Über zückerfreie Hefegarungen. XII Über die Vorgänge bei Hefegarung', *Biochemische Zeitschrift* **53**, 406–419.

Neuberg, and Reinfurth, (1920), 'Ein neues Abfangverhahren und seine Anwendung auf die alkoholische Garung', *Biochemische Zeitschrift* **106**, 281–291.

Teich, M. (1965), 'On the Historical Foundations of Modern Biochemistry', *Clio Medica* **1**, 41–57.

Young, W. J. (1909), 'The Hexosephosphate Formed by Yeast-Juice from Hexose and Phosphate', *Proceedings of the Royal Society* **81B**, 528–545.

ARTHUR M. DIAMOND, JR.

THE POLYWATER EPISODE AND THE APPRAISAL OF THEORIES*

1. BACKGROUND AND DATA

Instead of only examining how often a thesis about scientific change is confirmed in cases of scientific advance, it would be better to examine both how often the thesis is true in cases of scientific advances *and* how often it is true in cases of scientific mistakes. (For the purpose of this paper, a 'mistake' is defined as a scientific claim that is initially accepted by a sizable proportion of specialists, but is shown to be false before being integrated into the prevailing interpretation of the research area in question.) One could then conclude that the higher the success to failure ratio (out of the total number of successful and mistaken episodes in which the thesis was true), the greater the warrant for affirming the thesis as a normative standard for how good science should be done. The main criterion for choosing an episode for this study was that the episode be an exemplary mistake. For the purposes of this study, the polywater episode has several advantages over other mistakes that could be studied: (a) the episode occurred recently enough for biographical and citation data to be available, and (b) the number of identifiable scientists involved was larger than for most other mistakes that could be studied.

Some current commentators are ready to label the polywater episode as mistaken *ex ante* (see Eisenberg, 1981; Franks, 1981, pp. 180–182). Zuckerman, for instance, discusses the polywater episode as a 'classic' case of a 'disreputable error' because polywater's proponents failed "to live up to the cognitive norms prescribing technical procedures designed to rule out even the most favored hypotheses if they are in fact unsound" (1977, pp. 111, 112). But other scientists (e.g., Deryagin, 1983; Gould, 1981, p. 15; and Pethica, 1982) seem to view polywater more generously as one of those "mistakes of good men earnestly seeking the truth" (Metzger, 1970, p. 9).

Franks' *Polywater* (1981) is by far the most extensive account of the episode, but other useful accounts are also available (Everett, Haynes and McElroy, 1971; Hasted, 1971; Howell, 1971; Gingold, 1973; Gingold,

181

1974; Klotz, 1986, pp. 67–95). Anomalous water was first discovered in 1961 by a Russian scientist named Fedyakin who found that in very small capillaries he could condense a form of water that has a lower freezing temperature and a higher boiling temperature than ordinary water. The discovery was taken up by the prestigious Russian scientist Deryagin[1] who brought his research to the attention of English-speaking scientists in a series of lectures from 1966 to 1968. In the West the surge of articles on anomalous water began in 1969. The term 'polywater' was introduced by Lippincott and his co-authors to refer to a new polymeric form of water that they claimed exhibited the anomalous phenomena. In what follows we sometimes adopt the convention of subsequent writers by using the word 'polywater' more broadly to refer to anomalous water regardless of whether it is due to impurities in ordinary water or to a new polymeric structure. By 1973, when Deryagin admitted that the polywater phenomena were probably due to impurities in ordinary water, over 400 publications had appeared that referred to the subject (see Bennion and Neuton, 1976).

I began my study of the polywater episode by collecting biographical and career information for a sample of scientists from the West who wrote on polywater and a comparable sample of scientists from the West who did not write on polywater (the control group). Gingold's (1973) nearly exhaustive bibliography on polywater was the basic source for the sample of scientists who wrote on polywater. For more details on the sample selection criteria and on the sources of biographical and career information, consult Diamond, (1987a, b). After the elimination of all excluded entries, the final sample of research publications on polywater consisted of 112 entries.

These remaining entries were examined in order to judge whether their author's attitude toward polywater was pro, con or neutral. An article was judged to be pro-polywater if on balance it seemed to support the position that the anomalous phenomena were due to some different structure of water, while an article was judged to be con-polywater if on balance it seemed to support the position that the anomalous phenomena were due to impurities in the water. Articles that were so heavily qualified that they did not seem to lean toward either position were judged to be neutral. For the polywater case, I evaluated 66 entries as pro, 4 entries as neutral and 42 entries as con.

A scientist was judged as having been pro-polywater if he had ever authored or co-authored a research article that was judged to be pro-polywater. Thus Allen and Kollman, who wrote an important pro article in 1970 and an important con article in 1971, were classified as 'pro'. A scientist was judged as having been con-polywater if he had never written a pro research article but had authored a con research article. A scientist was judged to have been neutral on polywater if he had never authored either a pro or con research article, but had authored an article that was judged to have been neutral.

2. RELEVANCE OF POLYWATER TO THE THESIS CLUSTERS

Which of the canonical set of theses is the polywater episode most relevant to? The episode has some of the sociological characteristics that have been associated with revolutions, so one might be tempted to focus on the theses that refer to revolutions. One revolution-like feature of the episode, for instance, was the enormous popular controversy that was generated.

For the purposes of this volume, a 'revolution' has been defined as "the replacement, abrupt or gradual, of one set of guiding assumptions by another". In the polywater episode most of the participating scientists accepted one or the other of two mutually inconsistent guiding assumptions. The first was that the polywater phenomena were due to some new structure of water, while the second was that the polywater phenomena were due to some impurity in ordinary water. Some of those who accepted the first guiding assumption attempted to develop theories of a new structure of water that would explain the properties of polywater in a manner consistent with the laws of physics and chemistry. Some also attempted experimentally to produce polywater under conditions in which every precaution against contamination had been taken.

Some of those who accepted the guiding assumption that the polywater phenomena were due to impurities in ordinary water developed theories concerning which impurities were producing the phenomena. Some of them also attempted to show how mixtures of the suggested impurities with water would produce the phenomena. Finally, some of them attempted to demonstrate that such impurities would in fact be produced by following the experimental procedures of those who accepted the competing guiding assumption.

Although some aspects of the polywater episode can be interpreted in terms of divergent guiding assumptions, the scientists on both sides of the controversy shared, from beginning to end, far more assumptions than did scientists in many of the examples of revolutions that are traditionally discussed. We will, therefore, consider first, and focus most of our attention upon, the theses on the appraisal of theories. In conclusion we will consider the relevance of the polywater episode to one thesis from the set of theses on scientific revolutions, viz., the thesis on age and the acceptance of theories.

Since most of our attention will be focused on theses concerning theory appraisal, we should note that we reject a common view that theory appraisal constitutes a phase of the scientific decision-making process distinct from problem choice. Ample evidence exists that appraisal is an on-going process, intimately involved in a scientist's choice of problems for research.

3. THESIS (T2.8)

Thesis (T2.8) states that "the appraisal of a theory is relative to prevailing doctrines of theory assessment and to rival theories in the field". One question of interpretation that must be answered before the thesis can be evaluated is whether the thesis applies only to the ultimate appraisal or also to the on-going appraisal that occurs while the issue is still in doubt. Here we focus mainly on the ultimate appraisal. Ultimately the theories that proposed new structures for polywater were rejected in favor of the theories that proposed that the phenomena were due to impurities.

In what follows we also give disproportionate attention to the appraisals made by the 'elite' where that term is operationally defined as those scientists who either play a substantial role in the accounts of the episode given by authorities such as Franks and Klotz or else who were important by some independent, objective measure such as by the Institute for Scientific Information's co-citation clusters. Much work, largely in the sociology of science, could be cited to justify the decision to focus on the elite (cf. Cole and Cole (1973), *passim*; and Mulkay (1980), pp. 28–42).

Using citations made during 1970 and 1971, the cluster method identifies 11 distinct articles as having been important.[2] Of the 11 (all published in either 1969 or 1970) 8 are pro polywater and 3 are con. Of the 8 pro articles, 4 are focused on reporting experimental results, 2 are focused on presenting new theoretical structures of water and 2 incorporate both experimental results and a significant discussion of new theoretical structures. Of the 3 con articles, all are focused on reporting experimental results.

The theories that proposed new structures of water in order to explain the polywater phenomena can be divided into two broad categories. One group consisted of structures derived largely in analogy with other known chemical structures. Allen and Kollman characterize this first group of theories by saying that the theories employ "intuitive physical arguments" (1971a, p. 461). The second, and much smaller, group consisted of structures derived largely from computer models based on quantum mechanics. The former group of theories can be called more 'casual' than the latter both in the sense that less time and effort went into the construction of such theories, and also in the sense that such theories tended to be given less probative weight. Referring to this group of theories, Pimental and McClellan claim that "'back-of-the-envelope' proposals for novel H-bonded structures to account for the properties became a popular luncheon sport" (1971, p. 367). First-hand testimony to the casualness of these theories is provided in the admission of Jerry Donohue (not to be confused with the F. J. Donahoe who proposed in *Nature* that polywater might end life on earth) that his own theory was "concocted one evening while watching TV, between commercials, with tongue in cheek".[3]

For the most part, those scientists appraising the theories of a new structure of water did not make much effort to distinguish the various theoretical accounts from one another. Some rough evidence of this is that only one purely theoretical paper, that of Allen and Kollman, was heavily enough cited to qualify for the 10-article polywater citation cluster for 1971. One reason for this might have been that the constraints imposed by the known regularities of chemistry (such as Gibb's law) combined with the constraints imposed by the then-current observations may have been thought to underdetermine theory choice in the straightforward sense that more than one theory might be consistent with the corpus of chemical knowledge and the specific evidence on the polywater phenomena. Another possible reason might have been that although relative assessments would have been possible, the greater part of the interested scientific community did not consider such assessments worth the time and effort required until the broader issue of the presence of impurities was resolved.

Allen and Kollman, at the June 1970 Lehigh symposium, held near the end of the polywater episode, made a systematic attempt to assess the relative merits of several of the most prominent pro theories on the basis of quantum mechanical calculations of their stabilities (1971a). The authors found that their own earlier structure is clearly superior to the others. *Then*, in a companion article in the same issue of the same journal (1971b), Allen and Kollman report that on the basis of new calculations, again making use of quantum mechanics, they no longer believe that their own earlier model is stable. They finally conclude that, since their theory is the best one available and since their theory is unstable, polywater does not exist. Near the end of this article, the authors express the fear that their complete reversal on the stability issue "will cause many chemists to laugh and some to discount the efficacy of theoretical chemistry" (1971b, p. 480).

Another noteworthy reversal by a pro theorist in the small group using calculations from quantum mechanics was that of O'Konski. O'Konski had claimed that his theory was superior to various particular alternative theories in that it accounted for (1) how polywater could be in equilibrium with ordinary water and (2) how polywater could be stable. (O'Konski, 1970, pp. 1089–1090) He argued that polymeric theories are superior to allotropic theories because they account for (1), while his particular polymeric theory is superior to other polymeric theories because it accounts for (2).

The following year, using sophisticated (and at the time, computationally expensive) techniques, O'Konski concluded that his own theory cannot account for (2). On the basis of this result "together with", reports of numerous impurities found in samples of anomalous water O'Konski reached a negative appraisal of his own theory and expressed doubts that any polymer theory would turn out to explain the phenomena (O'Konski, 1971, p. 551).

It is not clear whether the theoreticians changed their stand because they did their calculations more carefully, as is implied by some accounts, or rather, because they saw how the experimental work was going and changed the interpretation of their results so as to be in line with the expected scientific outcome. In their thorough, 30-page, 1971 survey article on polywater, Everett *et al.* (1970) devote only slightly more than a page to the theories of a new structure for water. They conclude that:

It is now clear that present theoretical techniques are quite inadequate to assess reliably the stability, relative to monomeric water, of postulated polymer species. Whether polywater is eventually proved to exist or not there is no doubt that by an appropriate selection of models and parameters, a theoretical basis could be found for whichever turns out to be the experimentally established situation: theoretical calculations cannot have any reliable predictive value in the present case (p. 301).

Lest these strong claims be discounted because their source was a group of con scientists, it should be noted that one of the co-authors of one of the main pro theoretical articles has retrospectively remarked of the evidence from the theoretical calculation that "one could weight the evidence to either support or not support the existence of polywater". He goes on to say: "Early on we chose to weight the positive evidence more, since one could construct such a seemingly self-consistent picture of the phenomena" (Kollman as quoted by Franks (1981), pp. 94–95).

Apparently the plethora of theories concerning which impurities caused polywater did not prevent the formation of a consensus that the phenomena were due to impurities of some sort. A turning point in the development of a consensus (see Franks (1981), pp. 103–104) was the paper by Lippincott presented at the Lehigh symposium on polywater in June 1970, in which he reported that he had been able to reproduce the polywater spectra from ordinary water contaminated with impurities, but was unable to reproduce it when the impurities were rigorously excluded. Lippincott's reversal carried more weight than papers by other researchers because he had the reputation of being a careful experimentalist (see Franks (1981), p. 68) and because his early research in favor of the existence of a new structure of water had been a key in the growth of polywater research in the West (see Klotz (1986), p. 85).

Although the Lippincott reversal may have been a turning point, the accumulation of additional evidence of impurities was necessary before the consensus was complete. In order to understand why the theories that proposed a new structure ultimately received a negative appraisal, it may be worthwhile to focus on the three con articles identified as most influential by the Institute for Scientific Information's co-citation cluster method. The three con articles identified by the cluster method as important all appeared in 1970 and were authored by Everett, Haynes and McElroy; Rabideau and Florin; and Rousseau and Porto.

One of the striking features of the three key con articles is that they propose what appear to be mutually inconsistent impurity theories of the polywater phenomenon. Rousseau and Porto conclude that "in this complex mixture of impurities only trace amounts of silicon have been found, ruling out the speculation that the anomalous properties result from a silica gel" (p. 1718). Everett *et al.* conclude to the contrary that "as the properties outlined above appear to be consistent with those of a silicic acid sol, it is not necessary on the basis of the above evidence to seek an explanation in terms of 'polywater'" (p. 1035).

Everett *et al.* claim to have shown that spectroscopic and other evidence indicates that the polywater phenomena are what one would expect from a mixture of ordinary water and silica. They do not, however, present the details of the precautions they took to avoid contamination by other impurities (such as those reported by Rousseau and Porto) nor do they report spectroscopic or other tests indicating that such impurities were absent.

At one point in their paper (p. 1036) they refer explicitly (and positively) to Occam's razor, a reference that is instructive about the structure of their argument. They do not claim to have shown that other explanations (such as the polymer theories) are inconsistent with the phenomena. Rather they claim that we should not assert a new structure unless the phenomena cannot be explained by mixtures of already known chemical structures. Further, they appear to be totally unconcerned with appraising their silica impurity theory relative to other impurity theories. In fact, near the conclusion of the paper they offer conciliatory remarks to those accepting other impurity theories (p. 1037).

Consider finally the paper by Rabideau and Florin in which they find that the main impurities are sodium and boron. Everett *et al.* had also found significant quantities of sodium, but only a trace of boron – a trace that only appeared using one of the four analytical techniques reported. Rabideau and Florin mention the Everett and Porto study (which had appeared a few months earlier) only as a source of a potential mechanism for the contamination, but they do not speculate on why the earlier study had not turned up significant quantities of boron. Perhaps even more surprising is the absence of any mention, let alone appraisal, of the rival impurity theory that emphasized silicon as the key impurity. Instead, the authors only mention in passing that "chlorine and silicon were found, but these were small weight fractions of the residue" (p. 50).

Some of the pro-polywater scientists argued that the diversity of impurities theories undermined the credibility of the guiding assumption that the phenomena were caused by impurities. But this would be so only to the extent that the characteristics of the substance obtained in different samples by the pro-polywater experimentalists had been consistent. Although there was considerable consistency in some broad qualitative characteristics, such as high viscosity, other characteristics varied considerably (see Howell (1971). p. 666; Hasted (1971) p. 136). The pro-polywater experimentalists explained the diversity of reported characteristics in two ways: (1) some experimentalists may have insufficiently guarded against impurities, and (2) all samples contain mixtures of polywater and ordinary water in differing relative proportions. On the other hand, the diversity of characteristics reported may explain why the con experimentalists frequently did not appraise their theories of what impurity was important relative to other theories of what impurity was important. The inconsistency in the characteristics reported from sample to sample was interpreted by many as indicating that different impurities were present in different samples. Apparently those characteristics that were consistently reported from sample to sample were not sufficient to distinguish between several different impurity explanations. For example, Everett *et al.* (in a useful summary article published in the year following the article we have just discussed) summarize the physical properties of anomalous water (1971, pp. 295–297) and suggest that:

Many of the properties of 'anomalous water' have now been shown not to be specific but common to dilute hydrosols of a variety of materials: it seems highly probable that the many 'anomalous waters' described in the literature represent but a few of the possible forms of this phenomenon, and deliberate attempts to prepare anomalous water in contact with other solids have in some cases been successful (pp. 298–299).

The initial disagreement among scientists about whether to give a positive or a negative appraisal to the theories of a new structure for water occurred largely because standard precautions against contamination turned out not to be sufficient in the polywater case and because the quantities of the anomalous water produced were so small that the usual techniques for detecting impurities were taxed to the limits of their sensitivity. Many, though perhaps not all, of the pro-polywater experimentalists had taken considerable precautions to avoid contamination of their water with impurities. Very early on, for instance, Deryagin substituted quartz capillaries for Fedyakin's glass capillaries because quartz was known to be less susceptible than glass to surface contamination. In addition, rinsing the capillaries with distilled water was routine. Other precautions against contamination varied from study to study but were often quite elaborate. Finally, however, the cleaning techniques that were previously considered sufficient were judged *not* to be sufficient in the polywater case. Rousseau notes that "new cleaning techniques must be devised to eliminate all surface contaminants" (p. 1718).

In addition to the unusual difficulty of preventing contamination, a second reason for the initial disagreement over how to appraise the polywater theories was the difficulty in overcoming the limitations of standard analytic techniques. On this issue Franks (1981) notes that:

With never more than few micrograms available, ingenious analytical methods had to be devised if definitive and quantitative results were to be obtained that would themselves stand up to scrutiny. The instrumentation required for much of this work was of the most advanced kind, only available at a few centers in the United States. In the end the analysts carried the day, because they showed to their own and most other scientists' satisfaction, that polywater really contained little water . . . (p. 87).

If we are to assign a date to the time by which the scientific community had reached its ultimate negative appraisal of the theories of a new structure of water, no better candidate can be found than 5 March 1971 when "Polywater Drains Away" appeared in the unsigned "News and Views" column of *Nature*. Howell (p. 666) calls this editorial an 'obituary' of polywater, although in tone it reads more like a self-satisfied homily. Deryagin, however, was not quite ready to concede his theory's demise. Five months later, in the same journal, in a note co-authored with Churaev, Deryagin attempted to respond to the claim that the phenomena were due to impurities. If due to impurities, he argued, then liquid water introduced into the capillaries should absorb the same impurities as the vapor condensate associated with the polywater phenomena. He found that it did not, from which he concluded that the impurities theories were "at odds with the facts" (1971, p. 131).

After the passage of two more years, however, Deryagin conceded (again in a note in *Nature* co-authored by Churaev) that the polywater phenomena were due to impurities in ordinary water. In the end Deryagin concluded that something in the process of vapor condensation increased the likelihood of dissolving impurities from the surface of the capillaries. He found that in samples demonstrably free of impurities, the polywater phenomena did not occur and in samples where the polywater phenomena did occur, impurities could always be detected. In justification of his earlier scepticism toward the impurities theories, he again mentioned the lack of agreement about what the impurities were and what the mechanism was for producing the contamination. In his final sentence he suggested that the mechanism is a subject requiring further research, a 'requirement' that apparently has never been filled (1973, pp. 430–431).

The temporary disagreement over the appraisal of the polywater theories consisted, not so much in disagreement over the broadly defined doctrines of theory assessment, but rather in matters of judgement over two more concrete issues. The first was the extent to which the various theories had satisfied the prevailing standards, while the second was the extent to which the standards had to be met before the theories should be appraised positively. Based on the experiments that had been performed, the pro-polywater scientists (for a time) attached a higher probability to the thesis that the polywater phenomena could be produced without impurities. Also, the pro-polywater scientists seemed to believe that the level of probability required to justify a positive appraisal was lower than the level that the con scientists believed was required. Eventually, with the accumulation of sufficient sophisticated evidence of impurities, the appraisals of the scientific community converged on the conclusion that the polywater phenomena were due to impurities rather than to a new structure of water.

All of the scientists on both sides of the polywater issue accepted, in broad terms, the prevailing standards for theory appraisal. In particular they agreed that the theories should be consistent with the laws of physics and the current knowledge of molecular chemistry. They agreed that empirical support for the theories would require that the polywater be produced under conditions meticulously excluding impurities and that application of the techniques of analytic chemistry should show no impurities sufficient to explain the phenomena. I conclude that, in broad terms, (T2.8) is consistent with what occurred in the polywater episode.

4. THESES (T2.5) AND (T2.6)

Now that we have a general account of theory appraisal in the polywater episode before us, we are able to succinctly evaluate (T2.5) and (T2.6), which treat the role of empirical evidence in theory appraisal. Thesis (T2.5) states that "the appraisal of a theory is based on phenomena which can be detected and measured without using assumptions drawn from the theory under evaluation". In the case of polywater this thesis is clearly true. The phenomena were observed first, the theories came later. The original discoverer of the phenomena was Fedyakin, an experimentalist who was interested in the behavior of water condensed in small capillaries. Unfortunately, as Klotz notes (p. 79), Fedyakin is an elusive figure. But from Deryagin's account we can infer that Fedyakin did not have a new structure of water in mind when he performed his initial experiment. Deryagin reports (1969, p. 64) that: "Fedyakin had set himself the task of studying how the properties of liquids change when they are closed in extremely narrow glass capillaries".

The basic phenomena on which the theory was based consisted of the creation under certain experimental conditions of water that had certain properties. The basic apparatus had not been designed with any theory of polywater in mind (although it was refined during the course of the episode so as to reduce the likelihood of contamination, increase the quantity produced and decrease the time required to produce it). In addition, the equipment used to determine the physical and chemical properties of the substance (most notably spectroscopy) was not designed specifically for use in the polywater episode and was used by those on both sides of the fundamental issue of whether the polywater phenomena were due to a new structure of water.

Thesis (T2.6) states that "the appraisal of a theory is usually based on only a very few experiments, even when those experiments become the grounds for abandoning the theory". Two related, yet distinguishable types of experiments were performed during the polywater episode. One was designed to see whether polywater could be created using techniques that scrupulously avoided contamination by other substances. The second was designed to see if spectra and physical properties corresponding to those found in the first kind of experiment could be replicated using mixtures of known substances with ordinary water. Neither experiment could be interpreted as a crucial test because the results of both (no matter how they turned out) could be interpreted in ways consistent with either a pro or con position on polywater. I conclude that (T2.6) is not consistent with what occurred in the polywater episode.

5. THESIS (GA4.5)

Thesis (GA4.5) states that "during a change in guiding assumptions, younger scientists are the first to shift and then conversion proceeds rapidly until only a few elderly holdouts remain". The view that age matters in acceptance of new theories is not unusual. Perhaps the most extreme and oft-quoted articulation of that view was given by Max Planck: "a new scientific truth does not triumph by convincing its opponents and making them see the light, but rather because its opponents eventually die, and a new generation grows up that is familiar with it"[4]. Kuhn (1962, p. 151) explains Planck's principle by noting the costs of intellectual retooling when a new theory is adopted. Those who are older will, *ceteris paribus*, have more human capital invested in the old theories and, hence, will have more to lose by the adoption of a new theory.

We are interested here in learning the effect of age on the acceptance of polywater. We have data on the year of receipt of the PhD for more of the scientists involved than we have data on year of birth. So in addition to looking at the effect of biological age on the acceptance of polywater, we also look at the effect of professional age (which is equivalent to the economist's operational definition of "experience"). Even apart from data considerations, professional age may be the more appropriate concept if we accept the 'costs of retooling' explanation for (GA4.5).

Since the dependent variable is equal to one when the scientist accepted polywater and equal to zero when he did not, an ordinary least squares regression would have several undesirable features, inefficiency most notable among them (see Judge *et al.* (1980), pp. 586–587). So instead of ordinary least squares a binary logit is estimated.[5]

Polywater scientists differ in the extent to which their papers were mistaken. According to Franks (1981, p. 72), Lippincott's *Science* paper "is one of the rare examples in the chronicle of polywater where a group of authors took a definite stand based on their reading of the experimental evidence". Those who added more qualifications, it can be argued, turned out to have been less mistaken when polywater was shown to be due to impurities. Stigler (1982, p. 234) persuasively argues, however, that:

Unless an author explicitly sets out to refute a theory, one should characterize his attitude toward that theory as favorable, or at worst neutral, if he actually refers to the theory. For he is reviving its currency and advertising its existence.

Since the effects of professional age on the acceptance of polywater might differ depending on the form of 'acceptance', we estimate separate logit regressions using three different definitions of what it means to have accepted polywater.

The relationship between a scientist's age and his position on polywa-

TABLE I.

Logit estimates of the effect of biological and professional age on the probability that a scientist was 'pro' on polywater

Definition of 'pro-polywater' used	Regression number					
	(1)	(2)	(3)	(4)	(5)	(6)
	1=highly visible pro; 0=control group or wrote con		1=wrote on polywater; 0=control group		1=wrote pro; 0=control group or wrote con or wrote neutral	
Biological age in 1968	−0.036 (0.87)	—	−0.005 (0.04)	—	−0.008 (0.14)	—
Professional age in 1968 (i.e., 1968 minus year of Ph.D.)	—	0.011 (0.11)	—	0.003 (0.02)	—	0.019 (0.79)
Constant	−0.083 (0.00)	−1.674 (16.85)	1.403 (1.94)	1.111 (14.61)	0.198 (0.05)	−0.428 (2.79)
Sample size	55	71	87	107	87	107
No. of obs. where pro poly. = 1	11	12	67	81	41	46
No. of obs. where pro poly. = 0	44	59	20	26	46	61
− 2 log likelihood	55.04	44.50	93.81	118.66	120.18	146.22

The absolute value of the asymptotic t-statistic is reported in parenthesis. A value of 1.96 or more would indicate statistical significance at the usually used 0.05 level. As a measure of goodness-of-fit for the logit regression as a whole, we follow the convention of reporting minus two times the log likelihood. The statistic cannot easily be given a straightforward interpretation for the uninitiated, but we should at least note that, *ceteris paribus*, the higher the statistic's value, the better the fit.

ter is explored in the logit regressions presented in Table I. The dependent variable (i.e., the variable to be explained) has a value of 1 if the scientist is 'pro' polywater and a value of 0 if the scientist is not 'pro'. We will define the various senses of 'pro' shortly. Following the convention in the statistics literature, a variable that has only two possible values is called a 'dummy' variable.

In the sentences that follow 'favorable' will mean believing that the polywater phenomena are due to a new structure of water, while 'unfavorable' will mean believing that the polywater phenomena are due to impurities in ordinary water. In logit regressions 1 and 2 the pro-polywater dummy variable was equal to 1 if the scientist had written a highly visible favorable article on polywater and was equal to zero if the scientist either had not written on polywater or else had written an unfavorable article. Scientists who had written neutral or less visible favorable articles were excluded. In logit regressions 3 and 4 the pro-polywater dummy variable was equal to 1 if the scientist had written any research article on polywater, whether favorable, unfavorable, or neutral. The dummy was equal to 0 if the scientist had not written on polywater. In logit regressions 5 and 6 the pro-polywater dummy variable was equal to 1 if the scientist had written a favorable article (whether highly visible or not) and was equal to 0 if the scientist had either not written on polywater or had written a neutral or unfavorable article.

Regressions 1, 3, and 5 use biological age as the measure of age, while regressions 2, 4, and 6 use professional age. A negative sign on the coefficients on the age variables would indicate that older scientists are less likely to be 'pro' polywater while a positive sign would indicate that older scientists are more likely to be 'pro'. The signs of the biological age coefficients in regressions 1, 3 and 5 are all negative, while the signs of the professional age coefficients are all positive. However the asymptotic t-statistics (given in parentheses underneath each coefficient) are never even close to being statistically significant for any of the age variables in any of the regressions. The appropriate conclusion is that neither biological age nor professional age seems to have mattered in predicting whether a scientist would be 'pro' polywater in any of the three senses of 'pro'. In other regressions estimated, but not reported in the table, this conclusion is further confirmed when scientists' pre-polywater quality is controlled for by including as a variable in the regression the number of citations that a scientist received in 1968. I conclude that (GA4.5) is not consistent with what occurred in the polywater episode.

NOTES

*The research reported in this paper was partially supported by a grant from the National Science Foundation and a grant-in-aid from the College of Social and Behavioral Sciences of the Ohio State University. I am grateful to Gregory Armotrading, Mark Chapinski, Jack Julian, Lisa Knazek, Bret Mizer, Steven Oetken, Maureen Ogle, Christopher Smith, James Thomas, William Tisch, Ann Wertz and Kathryn Williams for able research assistance. Henry Small of the Institute for Scientific Information graciously provided me with 1970 and 1971 co-citation cluster data. Irving Klotz suggested to me the appropriateness of the polywater episode as an example of a mistake. I have had useful conversations on the subject of this paper with my colleague Gonzalo Munevar. I also appreciate comments on an earlier version by the participants of the VPI conference and by an anonymous referee. Finally, I am grateful to many of the participants in the polywater episode for sharing with me their accounts of what happened.

[1]We have adopted a uniform transliteration for Deryagin's name. The most common alternative in the literature on polywater is 'Derjaguin'. When we quote authors who use an alternative transliteration, we change the spelling to 'Deryagin' without adding 'sic' or other scholarly distractions.

[2]For more details on the co-citation cluster method and illustrations of its use, see, e.g., Garfield (1983) and Small and Griffith (1974). For more details on the particular articles in the polywater clusters, including some biographical and career data on the authors, see Diamond, (1987a).

[3]As quoted in Franks, p. 95. Of course, a cynic might note that after the ultimate negative appraisal of polywater, those who had written theories of a new structure might have an interest in downplaying the resources they had invested in the mistaken research.

[4]Planck (1949), pp. 33–34. Nearly all those who quote Planck's principle accept it as true. For extensive bibliographies listing sources that have referred to Planck's principle, see footnotes 1 and 3 in Diamond (1980) and, for more recent sources, Diamond (1987b). For systematic tests of the importance of a scientist's age as a determinant of theory acceptance see: Hull *et al* (1978), Diamond (1980) and Gieryn and Hirsh (1983). All three studies found a statistically significant effect of age in the direction predicted by Planck, although the first two found that the magnitude of the effect was small.

[5]A justification for using the logit cumulative distribution function can be found in: Judge *et al.* (1980) p. 591. The estimates of the constants may be subject to choice-based sample bias as discussed by Manski and Lerman. But since the values of the constants are irrelevant to the test of (GA4.5), the potential problem need not concern us.

REFERENCES

Allen, Leland C. and Kollman, Peter A. (1970), 'A Theory of Anomalous Water', *Science* **167**, 1443–1454.

Allen, Leland C. and Kollman, Peter A. (1971a), 'Comparison of Theoretical Models for Anomalous Water', *J. Colloid and Interface Science* **36**, 461–468.

Allen, Leland C. and Kollman, Peter A. (1971b), 'What Can Theory Say about the

Existence and Properties of Anomalous Water?', *J. Colloid and Interface Science* **36**, 469–482.

Bennion, Bruce C. and Neuton, Laurence A. (1976), 'The Epidemology of Research on "Anomalous Water"'. *J. Amer. Society for Information Science* **27**, 53–56.

Cole, Jonathan R. and Cole, Stephen (1973), *Social Stratification in Science*, Chicago: University of Chicago Press.

Deryagin, Boris (1969), 'The Equivocal Standard', *Saturday Review* (Sept. 6, 1969), pp. 54–55.

Deryagin, Boris (1970), 'Superdense Water', *Scientific American* **223**, 52–71.

Deryagin, Boris (1983), 'Polywater Reviewed', *Nature* **301**, 9–10.

Deryagin, Boris and Churaev, N. (1971), 'Anomalous Water', *Nature* **232**, 131.

Deryagin, Boris and Churaev, N. (1973), 'Nature of "Anomalous Water"', *Nature* **244**, 430–431.

Diamond, Arthur M., Jr. (1980), 'Age and the Acceptance of Cliometrics', *J. Economic History* **40**, 838–841.

Diamond, Arthur M., Jr. (1987a), 'The Career Consequences for a Scientist of a Mistaken Research Project', (preliminary draft).

Diamond, Arthur M., Jr. (1987b), 'The Determinants of a Scientist's Choice of Research Projects', (preliminary draft).

Eisenberg, David (1981), 'A Scientific Gold Rush', *Science* **213**, 1104–1105.

Everett, D. H., Haynes, J. M. and McElroy, P. J., (1970), 'Colligative Properties of Anomalous Water', *Nature* **226**, 1033–1037.

Everett, D. H., Haynes, J. M. and McElroy, P. J. (1971), 'The Story of Anomalous Water', *Science Progress, (Oxford)* **59**, 279–308.

Franks, Felix (1981), *Polywater*, Cambridge, Mass.: MIT Press.

Garfield, Eugene (1983), 'Computer-Aided Historiography – How ISI Uses Cluster Tracking to Monitor the "Vital Signs" of Science', in *Essays of an Information Scientist*, vol. 5, Philadelphia: ISI Press. pp. 473–483.

Gieryn, Thomas F. and Hirsh, Richard F. (1983), 'Marginality and Innovation in Science', *Social Studies of Science* **13**, 87–106.

Gingold, Marcel Pierre (1973), 'Anomalous Water: General Review', *Société Chimique de France, Bulletin*, Pt. 1, 1629–1644.

Gingold, Marcel Pierre (1974), 'L'eau Anomale: Histoire d'un Artefact', *La Recherche* **5**, 390–393.

Gould, Stephen Jay (1981), 'Ice Nine, Russian-Style', *The New York Times Book Review* (30 August 1981), pp. 7 & 15.

Hasted, J.B. (1971), 'Water and "Polywater"', *Contemporary Physics* **12**, 133–152.

Howell, Barbara F. (1971), 'Anomalous Water: Fact or Figment', *J. Chemical Education* **48**, 663–667.

Hull, David L., Tessner, Peter D., and Diamond, Arthur M. (1978), 'Planck's Principle', *Science* **202**, 717–723.

Judge, George G., Griffiths, William E., Carter Hill, R., and Lee, Tsoung-Chao (1980), *The Theory and Practice of Econometrics*, New York: John Wiley.

Klotz, Irving M. (1986), *Diamond Dealers and Feather Merchants*, Boston: Birkhauser.

Kollman, Peter A. and Allen, Leland C. (1972), 'The Theory of the Hydrogen Bond', *Chemical Reviews* **72**, 283–303.

Kuhn, Thomas S. (1970), *The Structure of Scientific Revolutions* (2nd edn., enlarged), Chicago: University of Chicago Press, (1st edn. published 1962).

Manski, Charles F. and Lerman, Steven R. (1977), 'The Estimation of Choice Probabilities from Choice Based Samples', *Econometrica* **45**, 1977–1988.

Metzger, Norman (1970), 'Polywater Boils', *Chemical and Engineering News* (9 Nov. 1970), p. 9.

Mulkay, Michael (1980), 'Sociology of Science in the West', *Current Sociology* **28**, 1–116, 133–184.

O'Konski, Chester T. (1970), 'Covalent Polymers of Water', *Science* **168**, 1089–1091.

O'Konski, Chester T. and Levine, S. (1971), 'Instability of Water Polymers', *J. Colloid and Interface Science* **36**, 547–551.

Pethica, Brian A. (1982), 'Review of *Polywater*', *J. Colloid and Interface Science* **88**, 607.

Pimentel, George C. and McClellan, A. L. (1971), 'Hydrogen Bonding', *Ann. Rev. Physical Chemistry* **22**, 347–385.

Planck, Max (1949), *Scientific Autobiography and Other Papers*, New York, pp. 33–34.

Rabideau, S. W. and Florin, A. E. (1970), 'Anomalous Water: Characterization by Physical Methods', *Science* **169**, 48–52.

Rousseau, D. L. and Porto, S. P. S. (1970), 'Polywater: Polywater or Artifact?' *Science* **167**, 1715–1719.

Small, H. G. and Griffith, B. C. (1974), 'The Structure of Scientific Literatures, I: Identifying and Graphing Specialties', *Science Studies* **4**, 17–40.

Stigler, George J. (1982), 'The Literature of Economics: The Case of the Kinked Oligopoly Demand Curve', in *The Economist as Preacher and Other Essays*, Chicago: University of Chicago Press, pp. 223–243.

Zuckerman, Harriet (1977), 'Deviant Behavior and Social Control in Science', in Edward Sagarin (ed.), *Deviance and Social Change*, London: Sage Publications, pp. 87–138.

3. NINETEENTH-CENTURY PHYSICS

JAMES R. HOFMANN

AMPÈRE'S ELECTRODYNAMICS AND THE
ACCEPTABILITY OF GUIDING ASSUMPTIONS

1. INTRODUCTION

The primary target of the following essay is an important historical thesis (GA 1.3) promoted in various forms by Imre Lakatos[1] and, subsequently, by Elie Zahar and John Worrall.[2] A closely related claim also to be scrutinized, is (GA1.4).

In order to test (GA1.3) in the case of Ampère's electrodynamics, I begin by stipulating a meaning for the term 'novel prediction'. This concept has generated a considerable variety of interpretations due to conflicting opinions about the most relevant way to apply it in the analysis of actual scientific debate. Ampère's attempts to promote acceptance of his program in electrodynamics provide a clear test for the historical significance of the type of novelty Michael Gardner has labelled 'use novelty'.[3] My conclusions thus bear most directly upon John Worrall's revisions of Lakatos's ideas, in that Worrall incorporated 'use novelty' into his concept of 'genuine empirical support'. My assessment of (GA1.3) thus addresses Worrall's rendition of this thesis in terms of 'use novelty' more directly than any of Lakatos's numerous earlier formulations.

With this understanding of 'novel prediction' at stake, I present evidence that Ampère did call attention to the importance of novel predictions prior to the beginning of his research in electrodynamics in 1820. After a summary of the guiding assumptions of Ampère's electrodynamics, I test (GA1.3) by an analysis of the arguments Ampère offered for the acceptability of his program. To avoid a potentially misleading reliance upon a single episode, I assess Ampère's prolonged apologetics during the formative and highly competitive period between 1820 and 1823.

My general conclusion is that, when 'novel prediction' is understood in terms of 'use novelty', Ampère's argumentation does not rely upon a significantly clear-cut stipulation of this concept. His language is sometimes ambiguous or general enough to include the criterion cited in (GA1.3), but we certainly cannot conclude that he wanted his program

201

to be judged largely according to this criterion. It is clear that Ampère relied heavily upon other types of evidence, one of which is cited in (GA1.4). In this respect (GA1.3) is not confirmed by Ampère's argumentation.

The present essay is almost entirely limited to the criteria operative in Ampère's own attempts to promote acceptance of his program. A more complete analysis would have to investigate a wide variety of scientists to determine what factors were responsible for the stance they took with respect to Ampère's program. In my concluding remarks I present some evidence that (GA1.3) is not substantiated by the criteria employed by Ampère's colleagues.

2. NOVEL PREDICTIONS

The starting point for Imre Lakatos's analysis of scientific change is the recognition that sequences of closely related scientific theories evolve through an application of the much more general sets of guiding assumptions he called research programs. According to Lakatos, these assumptions include both a "hard-core" of primarily ontological or metaphysical commitments and a 'positive heuristic' intended to guide the creative process of theory revision within a program. A scientific revolution involves a widespread shift in allegiance from one research program to another; the assessment of the guiding assumptions promoted by rival research programs thus becomes a central issue in Lakatos's analysis of scientific debate.

In particular, Lakatos insisted that scientists expect a well-founded research program to generate theories that are not simply created by ad hoc adjustments of earlier theories due to confrontation with anomalous phenomena. That is, scientists are claimed to prefer 'progressive' research programs in which theories should not only account for the anomalies that plagued their predecessors, but should also successfully 'predict novel facts'. Lakatos originally formulated this criterion in terms of strictly temporal predictions of hitherto unobserved phenomena.[4] Realizing that this requirement was too strict to reflect accurately most actual scientific practice, Lakatos repeatedly revised his concept of 'novel fact' before finally agreeing to a definition proposed by Elie Zahar.[5] According to Zahar[6]

A fact will be considered novel with respect to a given hypothesis if it did not belong to the problem-situation which governed the construction of the hypothesis.

Unfortunately, as Michael Gardner has argued,[7] it is difficult to see how this rendering of novel fact could function very readily in scientific debate. The potential domain of applicability originally envisioned for a theory may perhaps be kept private by its originator and, thus, not be available to other scientists attempting to judge the merits of the research program concerned.

John Worrall's response to this problem was to reconceive novelty in the sense subsequently labelled 'use novelty' by Gardner.[8] Following Lakatos's terminology, Worrall still holds that one criterion for the comparison of rival research programs involves determining which program has been most "progressive". But for Worrall a program is relatively progressive insofar as its theories progress beyond their predecessors by being 'genuinely supported' by more empirical facts. That is, a theory and its research program are only genuinely supported by an empirical fact when the theory entails a description of the fact, and when that fact did not contribute to the production of the theory through a modification of its predecessor within the program; instead, the theory must be generated from its predecessor by an application of the positive heuristic of the program without relying upon the information provided by the fact in question. On the other hand, the mere explanation of a fact through a *post hoc* modification of a prior theory in direct response to the fact to be explained does not provide genuine support for the theory so generated; the research program concerned is stigmatized as 'degenerate' if this tactic has become its sole means of increasing its empirical scope.

The importance of the guiding assumptions making up a positive heuristic leads Worrall to cite heuristic strength as a second criterion for the appraisal of research programs. The basic touchstone for heuristic strength is the degree to which the assumptions of a positive heuristic provide a clear-cut strategy for the creation of a progressive sequence of theory revisions, that is, a strategy that specifies techniques for theory revisions that result in an increase in genuinely supportive empirical scope. In physics, for example, a program is being guided by a strong positive heuristic when the 'hard-core' conceptual, dynamical and onto-logical assumptions of the program are expressed in a mathematical structure appropriately complex as to suggest specific tactics for progressive theory revision in anticipation of the inevitable discovery of anomalous new experimental evidence.

The task of the present essay thus is complicated by the fact that, in contrast to the formulation used in theses such as (GA1.3) and (GA1.4), Worrall's version of the methodology of scientific research programs does not claim that any one of the criteria cited in these theses is to be singled out as the one that is 'largely' appealed to in scientific debate. While I shall argue that debates in early-19th-century electrodynamics did not rely predominantly upon the progressive criterion cited in (GA1.3), this does not necessarily impugn Worrall's less exclusive attribution of importance to both progressiveness and heuristic strength. But this argument requires a brief preliminary summary of the guiding assumptions of Ampère's program.

3. THE GUIDING ASSUMPTIONS OF AMPÈRE'S ELECTRODYNAMICS

Oersted's 1820 discovery that the orientation of a suspended bar magnet could be altered by an electric current initiated an intensely competitive debate over the new domain Ampère soon labelled 'electrodynamics'. By November 1820, Ampère had made the experimental discoveries and conceptual interpretations that inspired his adoption of the guiding assumptions for a new research program.[9] These assumptions can be briefly summarized as follows.

Based upon Oersted's discovery and his own discovery of attractions and repulsions between current-bearing conductors, Ampère held that all electrodynamic phenomena, including all purely magnetic effects, are produced by forces acting between particles of two electric fluids moving within electric currents. Although the exact trajectories of these motions could not be specified, Ampère asserted that the most fundamental electrodynamic forces were central forces acting along the line joining pairs of electric fluid particles. He also believed that electrodynamic effects were propagated due to the transmission of a state of polarization within the electric fluids constituting an all-pervasive ether.

Secondly, Ampère claimed that all magnetic phenomena are produced by electrodynamic forces arising from electric currents confined to circuits within the neighborhoods of individual molecules of magnets.[10]

Although Ampère's most fundamental ontological assumptions about electric fluids were not directly testable, these claims were supplemented by additional heuristic assumptions about how the observable effects of electric fluid particles were to be shown to be consequences of empirically testable laws pertaining to electric currents. These laws were to be formulated without reference to the electric fluid micro-structure of electric currents. That is, Ampère assumed that all electrodynamic phenomena could be given a preliminary explanation through the application of a law stating the force between electric circuit segments small enough to be considered 'infinitesimal', but only in comparison to measurable magnitudes. If this law could be determined, it was assumed that mathematical integrations of it would be applicable both to directly observable circuits and to the molecular circuitries of magnets.

The sequence of theory revisions generated by Ampère's program thus was expected to provide increasingly accurate magnetic circuitries and more detailed expressions for the electrodynamic force between circuit elements. In order for his program to be judged 'progressive' in Worrall's sense, these theories would have to achieve genuine empirical support from phenomena not used to invent them. Thus, when (GA1.3) is interpreted in terms of 'use novelty', the thesis implies that the acceptability of Ampère's guiding assumptions was assessed largely on the basis of the genuine empirical support garnered by the sequence of theory revisions produced by his program.

4. AMPÈRE AND NOVEL PREDICTIONS PRIOR TO 1820

Ampère's study of electrodynamics came at a relatively late stage in his unconventional career. By 1820 he had reached the age of forty-five without making any systematic contributions to physics. The majority of his professional activities were performed as a mathematician and were supplemented by erstwhile excursions into chemistry;[11] his primary interests were metaphysics, psychology and epistemology. Although Ampère never brought his voluminous notes on these subjects into a publishable condition, they nevertheless provide valuable insights into his methodological views prior to 1820.

In particular, it is clear that Ampère originally held the successful prediction of new phenomena to be the most important type of confirmatory evidence for a theory. For example, in some notes pertaining to his earliest speculations about electricity in 1801, Ampère claimed that in proposing a *système physique* as a causal foundation for a given domain,[12]

. . . one has all the more means of verifying it as one has used fewer in discovering it. One has only to imagine some new experiments and predict their outcome; it then suffices to try them in order to see if the conjectures one has made are confirmed or not.

In this early passage, Ampère clearly has in mind a strictly temporal prediction pertaining only to phenomena hitherto unknown. Furthermore, he seems to restrict confirmatory evidence to these successful predictions just as was required by Lakatos's original definition of 'novel facts'.

By 1814, however, Ampère had liberalized his attitude toward confirmation so as to approach Worrall's conception of genuine empirical support. In that year he commented as follows concerning his hypothesis that all gases at a given temperature and pressure have the same inter-particle separation.[13]

Whatever may be the theoretical reasons which seem to me to support it, one can only consider it as a hypothesis. However, when comparing the consequences, which are the necessary implications of it, with the phenomena or the properties that we observe, if it is in agreement with all the known results of experiment, and if one deduces consequences from it that are found to be confirmed by later experiments, it will acquire a degree of probability which will approach what, in physics, is called *certitude*.

Although in this passage Ampère does not attempt to specify the relative weighting of the two types of confirmatory evidence he cites, he clearly continued to give predictions of new phenomena an important status.

In 1816 he made some additional methodological remarks in terminology strikingly comparable to Worrall's. According to Ampère, when a scientist is confronted by a choice between alternative theoretical explanations, he should[14]

. . . definitely adopt the one explication that seems the most probable only after having verified it by drawing some consequences from it which are a necessary implication of it and which are confirmed by some facts that one did not have recourse to during the research which led to it and which these consequences themselves indicate to the observer in the case where he would not yet have noticed them.

Although the qualification appended in the last phrase of this passage is not of optimum clarity, Ampère may very well have been trying to express a conception of genuine empirical support. He seems to be putting his emphasis upon the fact that decisive evidence is provided by all entailed facts that were not used to create the theory in question; temporal prediction of new phenomena does not appear to be essential.

Prior to 1820, Ampère's methodological remarks thus provide some preliminary evidence that he was predisposed to judge the guiding assumptions of a research program according to how successfully its theories make novel predictions. But be this as it may, in order to address (GA1.3), we must investigate the extent to which this predilection was put into practice in actual debate in electrodynamics.

5. AMPÈRE'S ASSESSMENT OF CONFIRMATORY EVIDENCE DURING 1820 AND 1821

When (GA1.3) is interpreted in terms of 'use novelty', we should expect that, when Ampère began to argue for the acceptability of his guiding assumptions late in 1820, he would rely 'largely' upon the degree to which his theories achieved genuine empirical support from phenomena not used to construct them. In fact, however, we find that he also relied upon other factors and did not restrict himself to genuine empirical support as his primary mode of appraisal. Thesis (GA1.3) thus is not confirmed by the period prior to 1822.

First of all, in some of his earliest comments on the superiority of his program in contrast to Biot's rival magnetic fluid program, Ampère conceded that many phenomena could be explained in a non-quantitative manner by both programs; he made no effort to specify which of these phenomena had been used to invent the explanatory theories in question. Instead, he stressed that his own general assumptions offered the twofold advantage of ontological simplicity and conformity to the longstanding French penchant for central forces.[15] Ampère thus was quite willing to promote his program on the basis of standards quite independent of empirically successful prediction.

A second reason for denying the truth of (GA1.3) during 1820 results from a scrutiny of how Ampère publicized the extensive set of new electrodynamic effects explained by his program during November and December. During this period Ampère did indeed successfully confirm some important predictions drawn from his assumption that magnetic effects are due to internal electric circuits. Most importantly, he duplicated many of the effects of bar magnets and magnetized needles using electric circuits in the form of large suspended loops, tightly-wound spirals, and helices of small cross section.[16] But many of these explanations were only made possible by an important new theoretical assertion Ampère proposed in order to explain what had originally been a seriously anomalous phenomenon.[17] Nevertheless, whenever Ampère had occasion to tabulate favorable evidence, he made no effort to exclude this particular experimental fact just because he had used it to make an important theory revision. Ampère thus did not employ the 'use novelty' distinction in the manner (GA1.3) implies that he should have.

During the subsequent year of 1821, Ampère's argumentative strategy was influenced by two constraints. First, he still could not claim superiority over Biot's program either on the basis of genuine empirical support (assuming that he would have wanted to) or on the basis of a more inclusive conception of empirical confirmation. Although he repeatedly attempted to perform conclusive 'crucial experiments' to decide between his program and Biot's, most of the resulting observations turned out to be equally well explained by both programs. In fact, in January 1821 the one experiment that did not fall into this category turned out in favor of Biot's program and was tactfully ignored by Ampère for over two years until an important theory revision allowed him to explain it.[18]

Secondly, Ampère encountered serious calculational difficulties when he attempted to apply his force law to any circuits more complicated than two parallel wires.[19] This meant that he could not upstage Biot by providing quantitative explanations for the phenomena they both had explained less rigorously.

Ampère summed up this situation in an expository memoir written in March 1821. After summarizing the known facts to be explained, Ampère pointed out how he had been able to derive an expression for his electrodynamic force law solely from a study of phenomena pertaining to experimentally-manipulated electric currents; he had not needed to rely upon phenomena produced by magnets and their alleged internal currents. Ampère's important concluding comment is worth quoting in full.[20]

These results, based on necessary deductions from the data of experience, cannot be placed in doubt as long as they concern only the mutual action of conducting wires discovered by M. Ampère.; he extends them not only to what M. Oersted has made known to us concerning a wire conductor and a magnet but to those actions of two magnets on each other in conformity, as we have said above, to the manner in which he conceives that electricity arranged in magnets produces all the magnetic phenomena. This extension given to the mathematical laws of the mutual action of voltaic conductors will decide the question of the identity of electricity and magnetism; this question is reduced by M. Ampère's work to a question which is solely within the province of mathematical analysis, since it is only a matter of calculating, according to the formulas which he has given, all the circumstances of the mutual action of two magnets which physicists have observed or measured up to the present and of seeing if the results of these calculations constantly accord with the data of experiment.

The first point to be noted in this passage is that Ampère does not make any explicit claim that magnetic effects and phenomena produced by interactions between magnets and electric currents are domains which provide particularly important evidence for the accuracy of the mathematical expression he had determined for his electrodynamic force law between current elements. He does not attempt to foster acceptance of his guiding assumptions about this force by appealing to the criteria cited in (GA1.3, GA1.4) and (T2.2).

Secondly, in his remarks about the quantitative application of his force law to the two domains where magnets are concerned, Ampère foresees a decisive confirmation of his assumption about the electric basis for magnetism. Nonetheless, contrary to (GA1.3), he does not foresee any need to isolate those facts used to stipulate magnetic circuitries from the more general set of all facts to be explained by a subsequent application of the force law to these circuitries. Ampère's comments are conspicuously void of any reference to the "use novelty" distinction.

During the remainder of 1821, Ampère's rhetoric shifted slightly in response to further unexpected developments. Inquisitive scientists throughout Europe continued to discover new electrodynamic phenomena which they sometimes cited as anomalous counterexamples to Ampère's program. Ampère typically responded to such discoveries and allegations by pointing out that, although he could not yet calculate the magnitude of the new effects, a non-quantitative study of the implications of his force law could at least account for their existence. Furthermore, these *post hoc* explanations were important confirmatory evidence because his latest theory "could have been used to predict them in advance".[21] But although Ampère's remarks during this period do indicate that he did not restrict his concept of relevant evidence to successful predictions of new phenomena prior to their experimental confirmation, he gave no indication that his program should be judged 'largely' upon the basis of 'use novelty'.

6. AMPÈRE AND NOVEL PREDICTION DURING 1822

Early in April 1822, Ampère concluded a discourse at the annual public session of the Académie with the following contrast between his program and Biot's:[22]

It is in this way that, of two hypotheses serving to explain a certain number of phenomena, the one which can account for them only by striving to agree with them is ordinarily refuted by other phenomena for which time eventually brings the discovery; on the other hand, the one which is simply, so to speak, the expression of the true relations of the facts which it explains, is found confirmed each time experiment makes new ones known to us.

Ampère thus implied that Biot's program had degenerated because it had relied upon *ad hoc* tailoring of theory in response to anomalous phenomena. However, Ampère made no attempt to explain how, prior to the rotary effect discoveries, his own program could have been distinguished as "the expression of true relations between the facts it explains". Indeed, in an earlier remark he had admitted that during 1820 and 1821 he had rejected Biot's program "more due to the general order of the facts than by basing myself upon direct proofs".[23] Ampère did not rely upon novel predictions or 'use novelty' as criteria that could have supported his program prior to Faraday's discoveries.

Ampère's silence on this issue is not too surprising when we consider that he himself was making major theory revisions in response to new experimental phenomena. He responded to De La Rive's and Faraday's experimentation by revising his latest theory of magnetic circuitry. More fundamentally, during March 1822, Ampère produced a new variation of Faraday's rotary effects and discovered that this implied that one of the parameters in his force law should have the value $-\frac{1}{2}$ rather than the negligibly small value he had previously assigned it.[24] This was a major theory revision for Ampère, and it was partially responsible for an important set of experiments he performed in Geneva in September 1822. Although one of these experiments involved the detection of an effect brought about by electromagnetic induction, Ampère did not carry out the additional research required to achieve an accurate understanding of induction; Faraday only accomplished this nine years later. As I have explained in detail elsewhere,[25] one reason for this neglected opportunity was the far greater importance Ampère assigned to another experiment performed on the same day. This experiment confirmed a prediction Ampère had drawn from his revised force law. He had publically announced this prediction to the Académie on 24 June 1822, and he naturally provided some enthusiastic publicity when experiment confirmed his prediction. For example, the relevant passage from the text for Ampère's verbal report to the Académie concluded with the words: "Voilà donc un fait tout à fait inattendu, conclu d'avance de mes formules et vérifié ensuite par l'observation et de la manière la plus frappante."[26]

As had been the case throughout his career, Ampère obviously still held the successful prediction of a new phenomenon to be an impressive piece of evidence. Nevertheless, although Ampère promptly placed his experimental discovery in a prominent position in an axiomatically-structured publication of 1822, he did not emphasize the fact that he had predicted it. Thus, as had been the case in 1821, during 1822 Ampère did not make any methodological proclamations that can be interpreted as evidence in favor of (GA1.3). Rather than highlighting novel phenomena as the decisive source of evidence, Ampère devoted extensive attention to a more general critique of Biot's static magnetic fluid program due to the problem posed by the rotary effects. The distinguishing factor Ampère cited most often was the fact that he could explain these effects on the basis of *moving* electric fluids. In conformity to (T1.1) and (T2.2), Ampère stressed his successful solution of problems that had stymied Biot; but there is no indication that Ampère considered "use novelty" to be the deciding issue.

7. 1823: SAVARY'S APPLICATION OF AMPÈRE'S PROGRAM TO MAGNETISM

Early in 1823 Félix Savary at last carried out the application of Ampère's force law to magnetic circuitries. Ampère had stressed the importance of these calculations almost two years earlier, and he immediately proclaimed Savary's accomplishment to mark "une sorte d'epoque dans l'histoire de l'électricité dynamique".[27] Savary described his primary concern to be an application of Ampère's force formula so as to study "l'analogie qui existe entre les aimans et des assemblages de courans électriques".[28] More precisely, in a subsequent abstract of his memoir, Savary summarized his derivation of Ampère's force law from a set of non-magnetic phenomena and then added that:[29]

... M. Ampère's opinion on the constitution of magnets can acquire the type of proof demanded by the present state of physics only when, starting from this formula and by applying it to the circular currents he imagines within magnets, one will find by calculation the same results as those given by experiment, both with respect to the mutual action of two magnets and with respect to that between a magnet and a voltaic conductor.

After considerable labor, Savary was indeed able to derive, at least approximately, the following impressive set of previously established empirical laws: the Biot-Savart law for the effect of a current on a suspended magnet; Coulomb's law for the forces between magnetic poles; a hitherto unexplained effect produced by Ampère and Depretz in January 1821; laws pertaining to rotary effects established by Claude Pouillet; and a law of terrestrial magnetism studied by Biot and Bowditch. Savary's concluding remarks about all this were rather low–keyed.[30]

The conclusion from these various results is that electrodynamic cylinders act, at least at distances not greatly exceeding their diameters, as do magnets for which the poles would be situated at the very extremities of these cylinders.

Savary made no attempt to draw a distinction between the effects that Ampère had relied upon to create his theory of magnetism and other effects not employed in this way. His discussion of his results shows no sign of any concern about the issue of 'use novelty'. On the other hand, Savary was more sensitive to the criteria cited in (GA1.4) and (T1.1); in particular, he referred to the rotary effects and pointed out that only Ampère's program could explain why some of these phenomena could not be produced using magnets.[31]

Similarly, in his own comments on Savary's memoir, Ampère was more concerned about the superior explanatory scope of his program in comparison to Biot's than he was in drawing fine distinctions based upon 'use novelty'.[32] It is true that in Ampère's private correspondence he took considerable pride in the fact that at a very early stage in his research he had foreseen the eventual quantitative confirmation of his magnetic theory. In 1820 he had published a set of 'conclusions' stipulating the basic assumptions of his reduction of magnetism to the action of electric currents. Writing to De La Rive in 1823, Ampère looked back on this summary as "a kind of prevision of all that has developed since". Furthermore,[33]

It is above all in comparing these conclusions with the results of all the experiments performed and all the calculations performed since that I congratulate myself in having written it; this is because I have always thought that there is all the more merit in having been the first to conceive truths which subsequently discovered facts confirm insofar as one had, at the time of this type of divination, as M. Biot calls it, less data with which to achieve this.

Ampère's pride is evident, but whatever he may have meant by "merit" in this passage, he did not make the clarifications that would have been required if he had wanted to alert De La Rive to the importance of the 'use novelty' criterion.

Similarly, when confronted by resistance to his program by Faraday, Ampère simply returned to the rhetoric he had used in 1821. For example, in a published version of an 18 April 1823 letter to Faraday, Ampère made the following claim about developments subsequent to the 1820 'Conclusions' memoir he had cited in his letter to De La Rive the previous month:[34]

Subsequently, new phenomena, which I was not able to predict, have been discovered by various physicists; far from being found in opposition to my theory, they have all provided new proofs of it, or rather necessary consequences which it would have been able to predict in advance.

In spite of Ampère's claim to the contrary, in 1820 he obviously could not have predicted *all* subsequent discoveries. Some of these discoveries originally had been seriously anomalous and had required important revisions in both his force law and his magnetic theory. Secondly, Ampère made no attempt to stipulate which facts he had used to revise his theories; the criterion of 'use novelty' was no more operative during 1823 than it had been during the preceding period.

8. CONCLUSION

My conclusions can be summarized under four points. First, although in arguing for the acceptability of the guiding assumptions of his electrodynamics Ampère did maintain a robust respect for the successful prediction of new phenomena, he never pointed out in any clear way that such predictions should be recognized as special cases of 'use novelty'. He surely would have done so if he had wanted 'use novelty' to be the primary mode of appraisal for his program.

Secondly, Ampère also relied upon other modes of appraisal such as ontological simplicity, restriction to central forces and overall empirical scope. He did not want his guiding assumptions to be judged 'largely' on the basis of 'use novelty'. Ampère's own promotion of his program thus presents evidence against (GA1.3).

Thirdly, however, John Worrall's version of the methodology of scientific research programs is not necessarily impugned by the forego-ing conclusion. In addition to progressiveness with respect to the genu-ine empirical support provided by 'use novelty', Worrall also allows judgement of the 'heuristic strength' of a program to play an important role: in the present case it remains to be seen whether this notion can be rendered sufficiently precise to be tested.

Finally, the present essay obviously is not an exhaustive study of the modes of appraisal operative in early-19th-century electrodynamics. I have shown that Ampère did not argue in the way (GA1.3) would lead us to expect. This leaves open the possibility that other scientists, or even the majority of them, did assess rival programs largely according to the criterion in question. I shall close with some evidence that 'use novelty' was not the primary mode of appraisal among Ampère's colleagues; criticism was aimed primarily at Ampère's mathematical manipulations of current elements irrespective of his acknowledged explanatory success on the empirical level. For example, Ampère's close friend Maurice relayed an extensive critique from a correspondent who wished to remain anonymous. The objections included[35]

> . . . the large number of entirely gratuitous hypotheses, the abuse of the use of infinitely small magnitudes of which one can say whatever one wishes, and the mixing of various dynamical ideas for which the introduction is not sufficiently motivated and the influence is not precisely characterized.

These criticisms are motivated by a wide range of worries; no one criterion is cited as primary. I suspect that further research will confirm that this attitude was common among Ampère's critics and that the distinctions required to employ 'use novelty' as a decisive standard were simply too delicate to play a major role in the decisions of Ampère's ·contemporaries. Furthermore, it is clear that (GA4.5) is a far more accurate rendition of this period in the history of French electrodynam-ics than (GA4.4). Although most French physicists gradually conceded the superiority of Ampère's program over Biot's, 'elderly holdouts' such as Poisson and Biot himself never abandoned research within the magnetic fluid program.

NOTES

[1]Lakatos [24], [25], and [26].
[2]Zahar [33] and Worrall [31] and [32].
[3]Gardner [19].
[4]Lakatos [24].
[5]Zahar [33] and Lakatos [26], p. 185.
[6]Zahar [33].
[7]Gardner [19].
[8]Worrall [31], p. 50.
[9]Blondel [17], Caneva [18], Hofmann [21] and [23] and Williams [29].
[10]Ampère tentatively adopted the molecular aspect of this assumption in January 1821 and it became an increasingly central assumption during the following few months; see Hofmann [22] and Williams [30].
[11]Caneva [18].
[12]Ampère [1]. This and all other translations are my own.
[13]Ampère [2], p. 47.
[14]Ampère [3]. Ampère's remark conforms closely to the wording of (T2.4).
[15]Ampère [4], p. 549.
[16]Williams [29] and Hofmann [23].
[17]Hofmann [23]. In conformity to (T2.1), Ampère clearly was greatly encouraged about the status of his theory when he resolved this serious anomaly.
[18]Ampère [17] and Hofmann [21]. Ampère's strategy here offers a striking confirmation of (GA2.2) and (GA2.3); when his assumptions were confronted with empirical difficulties, he not only refused to change those assumptions but was quite willing "to leave the difficulties unresolved for years".
[19]See Ampère [7]. Also, throughout this period Ampère was using an incorrect value for one of the parameters in his force law. He did not correct this until April 1822.
[20]Ampère [6], p. 319.
[21]Ampère [8], p. 128. Similar remarks appear in Ampère [9], pp. 306 and 308.
[22]Ampère [11], p. 66.
[23]Ampère [11], p. 64.
[24]Hofmann, [21].
[25]Hofmann, [22].
[26]Ampère, [12], p. 331.
[27]Ampère, [16], vol. 2, p. 624.
[28]Savary [27], p. 1.

[29]Savary [28], pp. 94–95.
[30]Savary [28], p. 97.
[31]Savary [28], pp. 91–92.
[32]Ampère [14], pp. 354–364.
[33]Ampère [16], vol. 2, p. 627.
[34]Ampère [15], p. 398.
[35]Ampère [16],vol. 3, p. 924.

REFERENCES

[1] Ampère, A. M. (1801), 'Recherches sur les causes de l'attraction et de la répulsion électrique', manuscript collection: Archives de l'Académie des Sciences, carton 10, chemise 205 bis.

[2] Ampère, A. M. (1814), 'Lettre de M. Ampère à M. le comte Berthollet sur la détermination des proportions dans lesquelles les corps se combinent d'après le nombre et la disposition respective des molécules sont composées', *Annales de chimie* **90**, 43–86.

[3] Ampère, A. M. (1816), 'Traité de psycologie', manuscript collection: Archives de l'Académie des Sciences, carton 17, chemise 269.

[4] Ampère, A. M. (1820), 'Mémoire sur l'expression analytique des attractions et répulsions des courans électriques. Lu le 4 décembre 1820', *Annales des Mines* **5**, 546–553.

[5] Ampère, A. M. (1820), 'Conclusions d'un Mémoire sur l'action mutuelle de deux courans électriques, sur celle qui existe entre un courant électrique et un aimant, et celle de deux aimans l'un sur l'autre; lu á l'Académie royale des Sciences, le 25 septembre 1820', *Journal de physique, de chimie, d'histoire naturelle et des arts* **91**, 76–78.

[6] Ampère, A. M. (1821), 'Exposé sommaire des divers Mémoires lus par Mr. Ampère à l'Académie des Sciences de Paris, sur l'action mutuelle de deux courans électriques, et sur celle qui existe entre un courant électrique et le globe terrestre ou un aimant', *Bibliothèque universell des sciences, belles-lettres, et arts* **16**, 309–319.

[7] Ampère, A. M. (1821), 'Calcul de l'action qu'exerce sur un aimant cylindrique . . . un fil conducteur . . .', manuscript collection: Archives de l'Académie des Sciences, carton 8, chemise 166.

[8] Ampère, A. M. (1821), 'Lettre de M. Ampère à M. Arago', *Annales de chimie et de physique* **16**, 119–129.

[9] Ampère, A. M. (1821), 'Lettre de M. Ampère à M. Erman', *Journal de physique, de chimie, d'histoire naturelle et des arts* **92**, 304–309.

[10] Ampère, A. M. (1821), 'Notes relatives au mémoire de M. Faraday', *Annales de chimie et de physique* **18**, 370–379.

[11] Ampère, A. M. (1822), 'Exposé sommaire des nouvelles expériences électromagnetiques faites par differens physiciens depuis le mois de mars 1821, lu dans la séance publique de l'Académie royale des Sciences le 8 avril, 1822', *Journal de physique, de chimie, d'histoire naturelle et des arts* **94**, 61–66.

[12] Ampère, A. M. (1822), 'Notice sur quelques expériences nouvelles . . .' *Collection de Mémoires relatifs à la physique*', vol. 2, *Mémoires sur l'électrodynamique*, J. Joubert (ed.), Paris: Société de Physique 1885–1887, pp. 329–337.

[13] Ampère, A. M. (1823), *Exposé des phénomènes électro-dynamiques et des lois de ces phénomènes*, Paris: Bachelier.

[14] Ampère, A. M. (1822), *Recueil d'observations électrodynamiques* . . . 2nd edn., Paris.

[15] Ampère, A. M. (1823), 'Extrait d'une Lettre de M. Ampère à M. Faraday', *Annales de chimie et de physique* **22**, 389–400.

[16] Ampère, A. M. (1936–1943), *Correspondance du Grand Ampère*, L. de Launay (ed.), 3 vols, Paris: Gauthier Villars.

[17] Blondel, C. (1982), *A.-M. Ampère et la création de l'électrodynamique* (1820–1827), Paris: Bibliothèque nationale.

[18] Caneva, K. (1980), 'Ampère, the Etherians and the Oersted Connection', *British J. History of Science* **13**, 121–138.

[19] Gardner, M. R. (1982), 'Predicting Novel Facts', *British J. Philosophy of Science* **33**, 1–15.

[20] Faraday, M. (1821), 'On some new Electro-Magnetical Motions, and on the Theory of Magnetism', *Quarterly J. Science* **12**, 74–96.

[21] Hofmann, J. (1982), 'The Great Turning Point in André-Marie Ampère's Research in Electrodynamics: A Truly "Crucial" Experiment', PhD dissertation, University of Pittsburgh.

[22] Hofmann, J. (1987), 'Ampère, Electrodynamics, and Experimental Evidence', *OSIRIS* **3**, 45–76.

[23] Hofmann, J. (1987), 'Ampère's Invention of Equilibrium Apparatus: A Response to Experimental Anomaly', *British J. History of Science* **20**, 309–341.

[24] Lakatos, I. (1968), 'Criticism and the Methodology of Scientific Research Programmes', *Proc. Aristotelian Society*, **69**, 149–186.

[25] Lakatos, I. (1970), 'Falsification and the Methodology of Scientific Research Programmes', in I. Lakatos and A. Musgrave (eds.) *Criticism and the Growth of Knowledge*, Cambridge: Cambridge University Press.

[26] Lakatos, I. (1978), *The Methodology of Scientific Research Programmes*. J. Worrall and G. Currie (eds.), Cambridge: Cambridge University Press.

[27] Savary, F. (1823), 'Mémoire sur l'application du Calcul aux Phénomènes électro-dynamiques', *Journal de physique, de chimie et d'histoire naturelle* **96**, 1–25.

[28] Savary, F. (1823), 'Extait d'un Mémoire lu à l'Académie des Sciences, le 3 février 1823', *Annales de chimie et de physique*, **22**, 91–100.

[29] Williams, L. P. (1983), 'What were Ampère's earliest discoveries in electrodynamics?', *ISIS* **74**. 492–508.

[30] Williams, L. P. (1986), 'Why Ampère did not discover electromagnetic induction', *Amer. J. Physics* **54**, 306–311.

[31] Worrall, J. (1978), 'The ways in which the methodology of scientific research programmes improves on Popper's methodology', in G. Radnitzky and G. Andersson (eds.), *The Structure and Development of Science*, Dordrecht: D. Reidel, pp. 45–70.

[32] Worrall, J. (1978) 'Research programmes, empirical support, and the Duhem problem', in G. Radnitsky and G. Andersson (eds.), *The Structure and Development of Science*, Dordrecht: D. Reidel, pp. 321–338.

[33] Zahar, E. (1973), 'Why did Einstein's Research Programme Supercede Lorentz's?', *British J. Philosophy of Science* **24**, 95–123, 223–262.

DEBORAH G. MAYO

BROWNIAN MOTION AND THE APPRAISAL OF THEORIES

1. INTRODUCTION

Brownian movement (discovered by the botanist Brown in 1827) refers to the irregular motion of small suspended particles in various fluids which keeps them from steadily sinking due to gravitation. Attempts to explain this phenomenon are linked with the atomic debates from the late 19th century to the early 20th. In particular, the testing of the Einstein–Smoluchowski theory of Brownian motion by Perrin[1] is considered to have provided the long sought-after evidence in favor of the molecular-kinetic theory of gases against classical thermodynamics. While this episode has often served as a case study for philosophers,[2] such case studies, as remarked generally by Laudan et al. (1986, p. 159) "are not 'tests' of the theory in question at all, but applications of the theory to a particular case". The aim of my discussion will be to employ this episode to carry out genuine tests, focusing on the following set of interrelated theses on theory appraisal: (T2.3, T2.4, T2.5, T2.6, T2.8 and T2.10). As many of these pertain to evaluating guiding assumptions, I shall suggest briefly how my results bear on a thesis on the appraisal of guiding assumptions (GA1.1), and a thesis on scientific revolutions (GA4.1). The atomic debate as a whole is too broad to be taken up here. I restrict my focus to theories of Brownian motion, those put forward both prior and subsequent to those explicitly linked with the kinetic theory in the 1870s.

But what justifies taking my analysis as a genuine test of the theses, in contrast to the mere 'applications' mentioned earlier? Does not our lack of a theory of science open any alleged test of rival theories of science to the charge of circularity? The answer, I claim, is no. So long as the theses aim to accord with actual scientific inquiry – as the theorists under consideration all do – non-circular criteria for testing them exist. It needs only to be assumed that the criticism raised against 'mere applications' is indeed a criticism. The complaint, as put by Laudan et al. (1986, p. 159) is that "such applications, by treating the model in question as unproblematic, fail to be probative . . .". If the model is assumed,

219

clearly the study has very little, if any, power to reveal its inadequacies. To deem this a criticism is tantamount to requiring that tests of theses have sufficient power to reveal correctly where the theses fail to describe science, as well as where they succeed. Otherwise the theses will fail to be constrained by how science actually functions.

This plausible demand alone provides the basis for two (meta-level) criteria for testing these theses. I shall frame them in the (attempted neutral) language of my test results, that is, in terms of assertions about whether a test case *indicates* (or is a good or poor indication of) a thesis H.[3] They may be stated as:

Two (Meta-level) Criteria: A good test of a thesis H must have little chance of incorrectly indicating thesis H, as well as a good chance of correctly indicating H.[4]

(A quantitative construal of this occurs in Note 4.) My intention in explicitly setting out the criteria I am using is not only to facilitate carrying out my tests, but to facilitate their critical scrutiny by others.

A shared thrust of many of these theses is a downplaying of the importance of empirical data methods in appraising theories and guiding assumptions. In addition to testing these theses, my study serves to test, or at least uncover, the presuppositions underlying this post-positivist tendency. In the final wrap-up I shall remark on the bias that these presuppositions introduce into the testing program and suggest how it might be avoided. I begin by testing (T2.3):

The appraisal of a theory is based on the success of the guiding assumptions with which the theory is associated.

Sets of guiding assumptions, which I shall abbreviate throughout as GAs, are supposed to include not only specific theories, but ontological claims about what exists and methodological claims about how they should be studied. The debates I am considering as to the cause of Brownian motion correlate with a number of opposing views which may be seen as sets of GAs (e.g., molecular *vs.* energeticist ontologies, mechanical *vs.* phenomenological explanation, atomic *vs.* continuous metaphysics, statistical *vs.* non-statistical models, the method of hypotheses *vs.* positivistic methodologies, realism *vs.* instrumentalism, and others). Thus, if there are genuine causal connections between appraising theories and their associated guiding assumptions, as (T2.3) alleges, there is a good chance they would be operating in this case. But that just means it is easy for an analysis of this episode to have a high chance of indicating such interconnections, as a number of discussions of this episode show. Such an analysis satisfies only one of the two criteria I am assuming are required for a good meta-test. A second requirement is that a meta-test does not construe the episode as indicating (T2.3) too readily. That is, the test must have a reasonably good chance of *counter*-indicating (T2.3) if the theory appraisal actually was *not* based on the success of GAs. This requires getting clear on what might be intended in denying (T2.3).

Thesis (T2.3) appeals to many contemporary theorists of science because its denial is typically assumed to be an endorsement of positivist views of theory appraisal, in which theories were thought to be evaluated by some sort of logical comparison with the evidence, in isolation. Granted, this episode counts against that sort of positivistic theory appraisal. However, my test of this episode shows the supposition that *either* a theory is tested in isolation (along positivistic lines), or else (T2.3) is correct (and its appraisal is based on the success of its associated set of GAs), is a false dilemma. For, neither need be the case.[5] As in the rest of my study, I shall focus on the early theories positing causes of Brownian motion (around 1827–1877) as well as the evaluation of theories explicitly based on the kinetic theory of gases (e.g., the theory of Einstein and Smoluchowski).

Thesis (T2.3) is counter-indicated by the early theories about the cause of Brownian motion, because they were vigorously debated and appraised from its initial discovery by Brown in 1827 – even before it was known whose problem it was. (It could have been chemistry, biology, physics or something else.) Brown tested and rejected his own initial view that the particles were self-animated (by finding that the motion occurred in inorganic as well as organic substances). Brown's experiments were recognized by Faraday (in an 1829 lecture) and others to have ruled out all of the causes of the motion suggested up until that time (e.g., unequal temperatures in the water, evaporation, air currents, heat flow, capillarity, motions of the observer's hands, and others). (See, for example, Jones (1970, p. 403).) Thus, theory appraisal took place, but without knowing to which set of GAs (or even which science) it belonged.

It might be countered that (T2.3) refers only to appraising more

full-blown theories, such as those explicitly based on the kinetic theory
of gases, e.g., by Delsaulx in 1877. Even with this restriction, (T2.3), as
I understand it, misdescribes the episode. For starters, it is not clear that
there is a set of 'GAs with which the theory is associated'. Some critics
of the atomic-kinetic accounts (Ostwald, Duhem,) endorsed an energeticist ontology; others, like Mach and Planck, did not. A number
of scientists (e.g., Clausius, Kelvin, and Born) worked within the school
of the phenomenological ('pure') theory of thermodynamics, whose
validity did not depend on any theory about the nature of matter, while
they also contributed to atomic theory in different publications. Secondly, conflicting appraisals of the kinetic account were urged by
scientists who held the *same* assessment of the success of associated
GAs. This was so in what was considered at the time to be the most
important challenge to the kinetic explanation, provided by Nageli, a
German cytologist. Nageli shared with Delsaulx the belief in the success
of the atomic-kinetic viewpoint, admitting that "the idea that the
molecules of a gas travel past each other with large velocities has
entered physics and, because of its irrefutable proof, has found general
agreement . . .".[6] Ironically, from the point of view of (T2.3) Nageli's
(and his followers') denial of Delsaulx's theory (attributing Brownian
motion to molecular collisions) was based on *accepting* the molecular-
kinetic model of gases. For they use the molecular magnitudes given by
the kinetic theory (e.g., Avogadro's number N) to argue against
molecular collisions as their cause. (Nageli calculated that a million
water molecules would have to hit Brownian-particles at the same
moment from the same direction to account for a single jerk of the
particle. But since the collisions would come from all different directions, he argued, the particle should not appear to jerk at all. He was in
error.) The main outlines of the statistical argument by which Smoluchowski answers Nageli's objection will arise later.

     Thesis (T2.3) comes closest to describing portions of the dispute in
which certain anti-atomists (e.g., Ostwald, Mach (at times) and the
mathematician Zermelo), insisting on the absoluteness of the Second
Law of Thermodynamics, base at least part of their opposition to the
molecular-kinetic theory on the fact that it would allow exceptions to
this principle. But even here there is evidence that the disputants on
both sides did not regard such criticisms as an adequate basis for
appraising the kinetic theory. (That this is so for anti-atomists like

Ostwald will be indicated in testing, and affirming, (T2.10) on crucial experiments.) Perrin makes it clear that without experimental arguments, the disputants were not yet at the level of scientific theory testing. While "the beautiful discoveries that we owe to molecular hypotheses" make Perrin hesitant "to support this radical opinion" of the anti-atomists, Perrin continues:

But since one has no direct proof of the existence of molecules, he can only go by *esthetic reasons* if he has not succeeded in proving by *experimental arguments* that the second law does not have the character of absolute rigor, in the name of which one would sacrifice the molecular theories (Perrin, 1950, p. 57 emphasis added).

It is not that our episode indicates that theory appraisal takes place without *reference to facts* from associated sets of GAs. It indicates only that theory-testing need not appeal to the overall success of the GAs. And where there is such an appeal to GAs it does not seem to count as the basis for an adequate theory appraisal. Theory appraisal, in this episode, seems to be based neither on appraising GAs nor on positivist-style algorithms (relating statements of evidence and theories). Instead, it is based on statistical and experimental reasoning. Examining this reasoning shows this episode is seriously at odds with thesis (T2.6):

The appraisal of a theory is usually based on only a very few experiments, even when those experiments become the grounds for abandoning the theory.

Lakatos' endorsement of (T2.6) is expressed as a contrast between his methodological falsificationism and claims espoused by Popper. Lakatos (1978, p. 35) claims his own view "denies that 'in the case of a scientific theory, our decision depends upon the results of experiments' . . . It denies that 'what ultimately decides the fate of a theory is the result of a test, i.e., an agreement about basic statements'." It is true that actual scientific practice shows the inadequacy of this version of a Popperian test, as of other positivist models of testing. But in so doing, these episodes show that theory appraisal requires many *more*, not less, experiments than positivist models suppose. For, the fact that theory appraisal depends on intermediate theories of data and experiment, and that data is theory-laden, inexact, and 'noisy', only underscores the necessity for numerous experiments, shrewdly interconnected, if one is to learn from data. From the initial discovery of Brownian motion by Brown in 1827, each inquiry into its cause is a story of hundreds of experiments. They may be grouped into two main classes: (1) experiments to (arrive at and) test theories attributing Brownian motion either to the nature of the particle studied, or to factors external to the liquid medium in which the Brownian particles were suspended (e.g., temperature differences in the liquid observed, vibrations of the object glass); (2) experiments to test the quantitative theory of Brownian motion put forward by Einstein and (independently) by Smoluchowski. I shall abbreviate the Einstein–Smoluchowski Theory of Brownian motion as the *ES Theory*.

Each molecular-kinetic explanation of Brownian motion (first qualitatively proposed by Wiener in 1863) spurred a flurry of experiments by biologists and physicists aimed at refuting it. Each non-atomic theory tried, say by devising a new way to view Brownian motion as due to temperature differences, triggered a new set of experiments to refute the challenge. The enormous variety of organic and inorganic particles studied is too numerous to list in full, but it includes sulfur, cinnabar, coal, urea, india ink, and gamboge. Equally numerous were the treatments to which such particles were subjected in the hope of uncovering the cause of Brownian motion – light, dark, country, city, red and blue light, magnetism, electricity, heat and cold, even freshly drawn human milk.

Counter-indicating (T2.6), our test case shows experiments to be the main basis for checking predictions, as well as for learning how better to run experiments. The major scientists working on this problem (e.g., Brown, Wiener, Ramsay, Gouy, Perrin and Smoluchowski) would begin by carrying out experiments to exclude all exterior causes – checking and rechecking even those suspected factors that had already been fairly well ruled out. (Even after Perrin and the general acceptance of molecular theory, experiments continued, using ever-improving methods to observe hundreds of thousands of microscopic grains.[7]) By the end of the 19th century the most favored explanations attributed the cause to heat in some way (e.g., theories of Exner, Dancer, Quincke). Ironically, finding that the same Brownian particles could be used over and over again, sometimes conserved on slides for 20 years, compelled experimenters who had been searching for *non*-kinetic explanations to admit this as strong evidence *for* the kinetic explanation. It indicated the motion was eternal and spontaneous, in accordance with the kinetic account. But most required many more quantitative experiments before abandoning non-kinetic explanations.

It is only by keeping in mind that a great many causal factors had been ruled out by experiments prior to Perrin's tests (around 1910) of the ES theory of Brownian motion, that the work his experiments performed can be understood. Because his experiments are so central in testing all of the other theses I plan to consider, I shall discuss a few points in some detail. The central problem with appraising the ES theory was that for many years experimenters were simply measuring the wrong thing. What they thought had to be checked was whether the molecular effects on the *velocity* of Brownian particles accorded with that hypothesized by the kinetic theory. But this mean velocity had been ascertained by trying to follow the path of a Brownian particle, inevitably yielding a measured path much simpler and much shorter than the actual path, which changes too fast. An important advantage of the ES theory was that it provided a testable prediction that made no reference to this unmeasurable velocity (and at the same time explained why attempts to measure it had failed[8]). Instead, it was put in terms of the expected *displacement* of particles. The displacement of a Brownian particle is the total distance it travels in any direction (say along the x-axis of a graph) as it weaves its zig-zagged path, and this could be measured using the microscopes of the day. As is usual – even in testing theories not themselves statistical – the testable prediction was of a *statistical distribution law*. Here the distribution law specifies how frequently Brownian particles would be expected to be displaced along the x-axis by certain amounts over a given time t. If molecular agitation (as described by the kinetic theory of gases) causes Brownian movement, then the displacement of a Brownian particle is Gaussian distributed about its mean (which by symmetry is 0) with variance equal to $2 Dt$ (where D is the *coefficient of diffusion* and t is the time). (Simply put, this tells us that displacements near 0 are most likely to occur, while those further from 0 are increasingly less likely.) Since Avogadro's number N is a function of D, the prediction of the kinetic theory (for a given type of particle) can be stated as a predicted *standard deviation*.[9] Once D is estimated, Avogadro's number N can be calculated and compared to values hypothesized by the kinetic theory. So the crux of Perrin's appraisal of the ES theory is evaluating the *statistical hypothesis*:

H: The experimental displacement distribution is from a population distributed according to the Gaussian distribution M with parameter value a function of N^*, where N^* is the (probable) value for N hypothesized by the kinetic theory. However, the sample data can be used to this end only if it *can* be seen (i.e., modelled) as the results of observing displacements from the hypothesized Gaussian process. Being a good experimenter, Perrin realized that appraising H requires two broad steps which I shall call *Step (I)* and *Step (II)*. *Step (I)* consisted in checking whether the results of the experiment actually performed follow the given statistical distribution M, and *Step (II)* involved using estimates of D to estimate (or test) values of N. While statistical theory was not fully developed at the time, the methods (e.g., *chi-square* tests) employed by Perrin were fairly routine. Step (I) involved a (*chi-square*) test of the claim

> (j) The data approximates a (random) sample from the (hypothesized) Gaussian process M.

In other words, the *denial* of j is j^1:

> (j^1) The assumptions of the experimental model M are violated; the sample displacements are characteristic of systematic effects.

Being able to rule out j^1 indicates that violations are sufficiently negligible for the purpose of estimating the parameter D. The necessity of doing so introduces the need for a large number of experiments. The aim is to learn enough about potential violations to either generate data that avoids them regularly, or to compensate them in the analysis. Step (II) calls for numerous experiments for a different reason. Essentially it is because the statistical prediction *is* an assertion of what would be expected in a large number of experiments. As well as indicating the importance of large numbers of experiments, the statistical nature of the theory appraisal indicates well (T2.5):

The appraisal of a theory is based on phenomena which can be detected and measured without using assumptions drawn from the theory under evaluation.

This is the case because the statistical data analysis concerns only low level, independently checkable theories. The availability of the data analyses (thanks to Perrin) and the clarity of the derivation of the Gaussian model by Einstein and others[10] is such that there is a very good chance of uncovering such question-begging assumptions, were Perrin's theory appraisal to require them. Thus, this is a good test case of (T2.5).

Without actually getting down to the nitty-gritty of the data analysis in Step (I), however, Perrin's avoidance of such assumptions is likely to be overlooked, and the reason why the results of Perrin's Step (II) were so convincing overlooked as well. (Feyerabend (1978, e.g., p. 39) is a case in point.) What made Perrin's results (at Step (II)) so telling was his ability to ensure the particles observed had approximately the same size, that they could be counted, weighed, and a host of extraneous factors controlled or 'subtracted out' – even Einstein expressed surprise. The general argument in ruling out possible external factors was that were Brownian motion the effect of such a factor, then neighboring particles would be expected to move in approximately the same direction; the movement of a particle's neighbors would not be independent of its own. Thus, the object of Perrin's inquiry was to determine whether the movement of Brownian particles exemplified a type of random phenomenon, known from simple games of chance. As Perrin (p. 112) shows, Einstein's derivation of the displacement distribution depends on "making the *single* supposition that the Brownian movement is completely irregular . . . ." Checking this single supposition was a matter of testing if the Brownian movement exemplifies the type of random phenomenon known from simple games of chance. Thus, this supposition could be checked 'without using assumptions drawn from the theory under evaluation' as (T2.5) requires. These checks made use of a completely independent statistical theory of the *random walk* phenomena. The displacement of a particle may be seen as the result of k steps, where at each step the particle has an equal chance of being displaced by a given amount in either a positive or negative direction, as described in Note 9. (This is called a *simple* random walk.) Were it *in*correct to assume j, there would be an unequal chance of being displaced by an amount and these dependencies would be detectable.[11] This single supposition is precisely what is checked in Step (I), and doing so avoids question-begging assumptions.

Consider one of Perrin's checks, based on observed displacements of 500 grains of gamboge (a type of microscopic vegetable particle). The positions of each are recorded every 30 sec on paper with grids of squares, as shown in Figure 1. The values are then shifted to a common origin in order to count the number found various distances from this point. As Perrin reasons,

Fig. 1. Tracing of horizontal projections of lines joining consecutive positions of a single grain every 30 sec (Perrin, p. 115).

The extremities of the vectors obtained in this way should distribute themselves about that origin as the shots fired at a target distribute themselves about the bull's-eye. (p. 118)

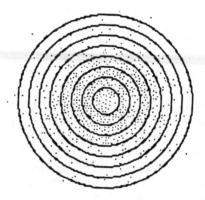

Fig. 2. Observed distribution of displacements of 500 gamboge grains.

Here again we have a quantitative check upon the theory; *the laws of chance enable us to calculate how many points should occur in each successive ring* (p. 118, emphasis added).

The number observed in each ring is close to the hypothesized number. That is, the observed displacements do not differ in a *statistically significant* way from what 'should occur' according to hypothesis *j*. (They are not improbable under *j*.) This indicates that *j* holds, only because the experiments were deliberately designed so that were the model inadequate, it would almost always yield differences that *were* statistically significantly different from what is expected according to j. On such grounds, j^1 is ruled out by Perrin and others, and j is upheld:

In short, the irregular nature of the movement is quantitatively rigorous. Incidentally *we have in this one of the most striking applications of the laws of chance* (Perrin, p. 119, emphasis added).

The multiple experiments for which Perrin stresses the need (e.g., Perrin, p. 96) are those deliberately designed so that if one misses a bias, another is likely to find it. The bias of concern here is that some regularity of molecular motion has been concealed. Such a regularity was the main alternative to Perrin's causal hypothesis H.[12] He rules it out by finding that the Brownian motion is totally irregular (as described in M). But does the theory appraisal in this episode seem correctly described as relative "to rival theories in the field", as asserted in (T2.8)?

The appraisal of a theory is relative to prevailing doctrines of theory assessment and to rival theories in the field.
Focusing on the latter part of this thesis, as I understand it, the answer seems to be no. A test claim H always has a rival but it may be a vague 'something other than H'. Since Brown's tests in the 1820s, scientists talked of testing theories as to the cause of Brownian motion (as traced in testing (T2.6)) where the only 'rivals' were assertions denying the hypothesized causal factor. As in checking whether experimental assumptions are met or not (j vs. j^1), the only rivalry is between a claim that a given model adequately captures certain aspects of the Brownian movement, and a claim that the motion is so far from what the phenomenon described in the model is likely to produce, as to indicate it is not so produced. Here the testing follows the logic of Fisherian statistical significance testing, in use at the time. Discrepancies either are or are not detected from a single hypothesis. (It is more like a Popperian refutation.[13])

But it might be countered that these low level theories do not provide a good test of what is asserted in (T2.8). After all, testing Einstein's theory was regarded as a test telling against the rival Classical Thermodynamics. My response is twofold. First, to the extent that Perrin's tests tell against Classical Thermodynamics – by showing that Brownian motion violates a non-statistical version of the Second Law – that counter-evidence is already provided by the simple statistical tests at

Step (I). (This is discussed in my test of (T2.10).) Second, even in Perrin's tests of (the parameters of) Einstein's theory of Step (II), the testing strategy is by local statistical hypotheses. Having checked that the assumptions of the experimental model have been met, by ruling out j^1 in Step (I), Perrin remarks

> To verify Einstein's diffusion equation, it only remains to see whether the number [obtained for N by substituting the estimate of D into the equation $D = RT/N6\pi a\varsigma$] is near 70×10^{22} (p. 132).[14]

The 'alternatives' at this stage of appraisal (i.e., Step (II)) are the possible values of the parameter space of N or D. This is still hardly more of a rival assessment than in testing j and j^1 in Step (I).[15] More than this is typically meant in asserting (T2.8).

How *does* the test of the predicted value for Avogadro's number (Step (II)) proceed? Nearly all the estimates of N from several experiments were (statistically) insignificantly far from that predicted by the kinetic theory, N^* (i.e., 70×10^{22}). Perrin declares:

> It cannot be supposed that, . . . values so near to the predicted number have been obtained by chance for every emulsion and under the most varied experimental conditions (p. 105).

The many instances this case affords of such (so-called) 'arguments from coincidence'[16] recommend it as a good test of theses on the role of 'surprise' and 'novelty' in science, such as (T2.4):

Theory appraisal is based entirely on those phenomena gathered for the express purpose of testing the theory and which would be unrecognized but for that theory.

This statement is interestingly ambiguous, and whether it is indicated by this test case depends on its interpretation. It is counter-indicated if understood as first intended by Lakatos (e.g., Lakatos, 1978, p. 38). On this *temporal view* of novelty, as Lakatos asserts, (p. 184) "The best novel facts were those which might never have been observed if not for the theory which anticipated it." Brownian motion, of course, had been recognized since 1827. But Zahar's (1973, p. 103) definition of novel prediction, which Lakatos subsequently endorses (e.g., Lakatos, 1978, p. 192), would not be forced to deny that Brownian motion was novel. For it presumably did not belong to the problem situation which generated the molecular-kinetic theory of gases. The gas theory, according to Einstein (1949, pp. 46–49), arose from a different aim, also of interest in this connection:

My major aim in this was to *find facts* which would guarantee as much as possible the existence of atoms of definite finite size. In the midst of this I discovered that, according to atomistic theory, there would have to be a movement of suspended microscopic particles open to observation, without knowing that observations concerning the Brownian motion were already long familiar (emphasis added).

But what about Einstein's theory of Brownian motion that Perrin appraised? In one sense it certainly seems that Brownian motion *was* part of the problem situation that yielded the specific theory that became testable. As mentioned in testing (T2.6), it was to avoid the anomalies for the kinetic theory due to attempting to measure the mean velocity of Brownian particles that Einstein formulated his theory of Brownian motion without any reference to this velocity. So even if construed using Zahar's redefinition of novel fact, it is not clear that (T2.4) would count experiments on Brownian motion as grounds for appraising the ES theory. Moreover, as Musgrave (1974) argues, Zahar's account suffers from being too 'psychologistic'. Worrall's (1985) reformulation of novelty attempts to avoid this charge by making novelty, not a matter of whether the theory was devised to explain the fact, but whether the fact was used to construct the theory. So long as evidence e was not used to construct theory H, and H entails e, e satisfies Worrall's criterion of *use novelty* for testing H. But for Worrall, this is not only a necessary but also a sufficient criterion for e to provide support for H. (See for example, Worrall, 1985, p. 318.) As a result, even if Worrall is right, and determining if this criterion is satisfied is an objective historical matter, the resulting account seems counter-indicated by the episode under consideration.

For, if it is granted that estimates of kinetic parameters, such as Avogadro's number, constituted novel facts in Worrall's sense (which it must be for Perrin's experiments to be taken to support the ES theory), successful predictions of these parameters would also seem to have warranted the kinetic theory, long before Perrin's tests. As Brush (1968, p. 380) notes:

Several independent methods of determining these parameters had been known since 1870 or before, to say nothing of the many successes of kinetic theory in predicting the properties of gases.

Worrall's account would fail to explain why scientists accorded a great deal more weight and importance to Perrin's experiments than to these earlier appraisals. Similarly, since any fact satisfying Worrall's criterion for novelty is on par – all providing 'severe tests' for H – his account would render inexplicable both the scientific criticisms of the experiments prior to Perrin's and the pains scientists took in constructing and interpreting their experiments so as to avoid such criticisms. So all of these senses of novelty appear to render (T2.4) a poor description of this episode. However, this episode itself, I think, indicates a more appropriate construal of the requirements for a good test under which (T2.4) accords well with this episode. What it indicates is that what mattered for an adequate appraisal was not an issue of when, why, or how theories were constructed, but rather, of the manner in which a theory was *tested*. The aim was to *deliberately construct* or design a test that had the ability to generate evidence e that would be highly surprising (extremely improbable) if (and only if) a given theory were an incorrect account of the origin of e. Then the occurrence of such evidence is a good indication that the theory was *not* an incorrect account of e's origin.[17] This is our (meta-level) notion of a good test now operating on the object-level. Successful fits with kinetic predictions are highly probable given the correctness of the kinetic explanations; but, contrary to the criterion of 'use novelty', this was not a sufficient warrant for their acceptance. Such successful predictions were disparaged (by pro-and anti-atomists) when such good fits (between H and e) were seen to be probable even if the kinetic theory did not give the correct causal account (of the origin of phenomena such as Brownian motion). Perrin's experiments avoided just such disparagement.

Consider again Perrin's Step (II), where the predicted value for Avogadro's number was tested against observed estimates. It was not the concordance of estimates itself that mattered; it was that experiments were *deliberately designed* (using what was learned at Step (I)) so that such concordance would be overwhelmingly *surprising* (so inexplicable) if there were a real discord between the hypothesized value, N^*, and the 'true' value for N (where the 'true' value is the mean of N in a population of experiments[18]). It was this assurance that was lacking in experiments prior to Perrin's.

It should be noted that the rationale for deeming novel predictions as better evidence than non-novel ones, in the existing senses of novelty, runs counter to the major theories of statistical testing favored by

philosophers (Bayesian and Likelihood schools).[19] For these theories
are based on the principle that given the same evidence e and hypothesis
h, the same evidential weight (that e affords h) should be assigned.[20] If
this principle is strictly adhered to, when or how the data or theory is
generated cannot matter in assessing the evidential support provided.
To the extent that the historical record indicates the importance of
novelty (in any of the senses mentioned here), it simultaneously
counter-indicates these statistical philosophies.

The same notion of novelty that renders (T2.4) a good description of
this episode underlies the importance often attributed to crucial experiments. It provides a good test case for (T2.10):

*The appraisal of a theory depends on certain tests regarded as 'crucial'
because their outcome permits a clear choice between contending theories.*

With regard to causal theories appraised before and after the appraisal of the ES theory, it has already been noted that fully-fledged
rival theories may be absent. Nevertheless, (T2.10) seems well indicated, if understood in a statistical sense suggested by the notion of
'novelty' arrived at above. A test is crucial between theories H and H^1 to
the extent that the test has a high probability of yielding one sort of
answer (i.e., one set of experimental outcomes) if H is the case and
another set if alternative H^1 is. Letting the former set of experimental
outcomes be just the ones taken to indicate H (as against H^1) gives a test
with a very low chance of indicating H erroneously. At the same time
the test will have a good chance of detecting H^1 if H is not the case.
Again, these are just the criteria for a good meta-test defined at the
outset.[21] In the extreme case, the test correctly discriminates between H
and H^1 with probability one. This corresponds to the strongest type of
crucial test: one whose result discriminates unequivocally between H
and H^1. The constant experimentation to appraise proposed causal
factors reflected the importance attached to evidence that would fairly
unambiguously discriminate between causal factors, in this sense. A test
result may not logically compel rejecting a theory (as Duhem and others
point out), but once the only available explanations that could be
appealed to are fairly well ruled out, there are excellent grounds for
doing so. Granted, Perrin's work does not provide a crucial experiment
between kinetic theory and classical thermodynamics taken as whole
sets of guiding assumptions in the sense often attributed to Popper (as is

consistent with its counter-indicating (T2.3)). But that does not mean it is not an instance of a crucial experiment in the sense meant by Einstein, Perrin and others.

Mach and Ostwald attacked the molecular-kinetic theory alleging that the theory was not needed, that a phenomenological description such as thermodynamics contains sufficient information, and avoids various problems that plagued atomic theory (e.g., the use of entities deemed hypothetical). Responding to this allegation, Einstein (1926) begins by stressing that the two theories give conflicting predictions about Brownian notion:

> If the movement discussed here can actually be observed (together with the laws relating to it that one would expect to find), then classical thermodynamics can no longer be looked upon as applicable with precision to bodies even of dimensions distinguishable in a microscope: an exact determination of actual atomic dimensions is then possible. On the other hand, had the prediction of this movement proved to be incorrect, a weighty argument would be provided against the molecular-kinetic conception of heat (Einstein, 1926, pp. 1–2).

Perrin viewed the aim of his experiments as providing such a crucial test:

> I have sought in this direction for crucial experiments that should provide a solid experimental basis from which to attack or defend the Kinetic Theory (Perrin, p. 89).

The kinetic theory, in contrast to the classical theory, views dissolved molecules as differing from suspended particles only in their size; their motion would be the same. If Brownian motion could be explained as caused by something outside the liquid medium or something within the particles themselves, then it would not be in conflict with the classical theory. If, on the other hand, it could be shown that Brownian motion was caused by a molecular motion in the liquid medium, as given in the kinetic theory, then it would be in conflict. Moreover, it would show a statistical process was responsible (and that the Second Law requires a statistical rendering). So Perrin's appraisal does test the ES theory against a non-statistical version of the Second Law of Thermodynamics, while its result is still consistent with the Second Law statistically rendered. Perrin states:

> Thus the practical importance of Carnot's principle *for magnitudes and lengths of time on our usual dimensional scale* is not affected (p. 107).[22]

That the kinetic theory indicated a non-statistical version of the Second Law was argued by Boltzmann around 1870. That Brownian motion, if indeed spontaneous, would be an exception to the (non-statistical version of the) Second Law was also recognized before the ES theory was formulated. For, if it is true that without temperature differences in the system, a Brownian particle denser than water rises spontaneously, then it constitutes a case in which part of the heat of the medium is transformed into work. This recognition is explicitly discussed by Gouy around 1890. Poincaré, persuaded by Gouy's arguments, declares:

. . . but we see under our eyes now motion transformed into heat by friction, now heat changed inversely into motion and that without loss since the movement lasts forever. This is the contrary of the principle of Carnot (Poincaré, 1904, p. 610).

In what I referred to as Step (I) (see test (T2.6)), Perrin showed that one can generate at will an observable process due to an agitation *not* attributable to the particles or to external energy sources. Thus, in carrying out Step (I), Perrin demonstrates the existence of violations to the (non-statistical version of) the Second Law. Perrin often describes his experiments as methods for generating such violations:

Briefly, we are going to show that sufficiently careful observation reveals that at every instant, in a mass of fluid, there is an irregular spontaneous agitation which cannot be reconciled with Carnot's principle *except just on the condition of admitting that his principle has the probabilistic character suggested to us by molecular hypotheses* (Perrin, 1950, p. 57, emphasis added).

Even ardent anti-atomists (probably excepting Mach) construed Perrin's experiments as providing this crucial evidence. On the basis of such experimental evidence, even Ostwald came to reverse himself on the atomic-kinetic energy theory in 1909:

I have convinced myself that we have recently come into possession of experimental proof of the discrete or grainy nature of matter . . . this evidence now justifies even the most cautious scientist in speaking of the *experimental* proof of the atomistic nature of space-filling matter.[23]

3. SUMMARY OF RESULTS AND IMPLICATIONS FOR THESES ON APPRAISING AND CHANGING GA s

My test results have indicated (T2.4, T2.5 and T2.10), and counter-indicated (T2.3, T2.6 and T2.8). These results point to the important role of experimentation (as against T2.6) in which novel facts (T2.4) may be deliberately used to construct crucial tests (T2.10) without question-begging assumptions (T2.5). The overall thrust, then, points to the importance of empirical methods. To the extent that accepting the

ES theory is linked to a change in GAs, the results of my testing the above theses on theory appraisal are relevant for testing theses on the importance of empirical methods in appraising and changing GAs. In particular, my results appear to indicate thesis (GA1.1):

The acceptability of a set of GAs is judged largely on the basis of empirical accuracy.

In asserting (GA1.1), Kuhn correctly notes that the discrepancies between theories and data and between rival theories are typically quite small. I suggest that *that* is why statistical considerations are important; it is not because scientists in this episode are interested in which theory is most probable or best supported (according to some probabilistic measure) that statistics enters. It is because in order to make what is a small difference, on our measuring scale, big enough to matter, they needed to consider what would result in a large number of experimental outcomes (contrary to (T2.6)). As with the examples Kuhn (1977) gives as illustrations, I believe examples from this episode show

how difficult it is to explain away established quantitative anomalies, and to show how much more effective these are than qualitative anomalies in establishing unevadable scientific crises (Kuhn, 1977, pp. 210–211).

Their being 'unevadable' indicates the concern scientists had with getting the cause of Brownian motion correct. There is no doubt that the dispute involved disagreement over whether mechanical or other (e.g., energeticist) explanations are preferable in science and in corresponding differences in world-views (GAs). But because the final evaluation turned on quantitative experimental results, it counter-indicates Kuhn's incommensurability claim in (GA4.1):

During a change in GAs (i.e., a scientific revolution) scientists associated with rival GAs fail to communicate.

As this case shows, through shared methods of experimental design, quantitative analysis was robust across rival GAs. Yet Kuhn rejects probabilistic testing across GAs:

If, as I have already urged, there can be no scientifically or empirically neutral system of language or concepts, then the proposed construction of alternate tests and theories must proceed from within one or another paradigm-based tradition (Kuhn, 1962, p. 146).

So Kuhn's endorsement of (GA4.1) is based on a positivist assumption that this episode shows to be in error. Members of any of the rival GAs one cares to identify can be seen to have agreed that the argument first raised by Nageli (based on the popular assumption that the molecules would have to move in a coordinated fashion to cause Brownian motion) posed a real challenge to the kinetic account. Nor did their allegiances prevent them from being convinced that the statistical argument Smoluchowski elucidated answered Nageli's cricitism.[24] Representatives of rival GAs could similarly agree that Perrin's experiments, if successful, would show Smoluchowski's statistical argument was instantiated. The arch anti-atomist Ostwald tried out kinetic ideas for his own purposes and, though he was disappointed, what matters for *our* purpose is that even he had no difficulty in using notions from incompatible GAs.

The appraisal and acceptance of the ES theory of Brownian motion corresponded to a change in beliefs about fundamental entities, about the existence of molecules and in the particulate nature of matter. It also led to a change in scientific methodology; a new limit to experimental accuracy due to Brownian fluctuations and 'noise' in measuring systems was introduced. The entry of statistical validity into physics also conflicted with the cherished philosophical conception that physics discovers wholly exact laws. This episode is a good illustration of how empirical tests constrain beliefs in each of these GA components, of how a reasoned appraisal of GAs themselves is possible. For example, in appraising the ES theory, methods of theory assessment were used intersubjectively (i.e., across sets of GAs), although these very methods led to a change in testing methods (by leading to that theory's acceptance)! Perhaps this illustrates how something like what is suggested in Laudan's (1984) *reticulated model* takes place.

4. A POINT ON SYSTEMATIC BIAS IN META-METHODOLOGY

As Laudan *et al.*, (1986, p. 155) note, one of the main points that the theorists here considered share is the rejection of positivistically inspired logics of induction and confirmation. At the same time, however, my tests consistently revealed the theses on theory testing to be rooted in positivist testing models. These models sought to provide quantitative measures of support (confirmation, truth-likeness, etc.) that evidence affords hypotheses. Any complete theory of testing must involve statistical ideas of data generation, modelling, and analysis. But there are statistical theories of testing very different from those of the positivist models that seem more adequate (e.g., those developed by Giere and other philosophers including the present author, based on current statistical methods). Nevertheless, even if I am correct to allege that the theses on theory testing are based on positivist models, *this* does not create any bias for the present meta-level testing program. At worst, the theses would misdescribe scientific episodes, and good tests should detect this, as I believe some of my tests did. What is worrisome for the program is positivistic assumptions about experimental methods in *carrying out* tests of these theses. For they are likely to create a systematic bias that will prevent such meta-level tests from being good tests.

Suppose a tester fails to find positivistic views of theory testing playing an important role in a given historical case – which seems likely, given it is generally agreed that such views inadequately capture actual science. A tester who operates with a positivistic conception of experimental testing must take this failure to indicate that what is important is something *other* than experimental testing. Then, when the historical study *does* show correlations between the holding of rival theories and various extra-scientific differences (e.g., in subjective beliefs, pragmatic interests, cultural background, etc.), these are assumed to be a major basis of theory appraisal. The danger, then, is that starting out with a positivistic view of theory testing is likely to lead to severe bias in testing theses against historical cases – one which will not be detectable afterwards. Such tests have little if any chance of *not* pointing to the importance of non-empirical or non-scientific factors. They thereby violate the meta-criteria for a good test I set out at the start.

Acknowledgment: I would like to thank Alan Musgrave for useful suggestion on an earlier draft.

NOTES

[1]All page number references to Perrin, except where noted, refer to Perrin (1923).
[2]In addition to being discussed by the theorists considered here, and in discussions concerning these models, e.g., by Clark, Gardner, and Worrall, it has served as a key illustration for views held by Cartwright, Hacking, Harman, and Salmon.
[3]My aim here is to avoid the assumptions that go along with viewing evidence or instances as providing *degrees of support* or confirmation for or against assertions.
[4]This can be expressed more formally by considering two possible results of a test of a thesis as follows:
Results:
(1) Test T indicates thesis H holds in the test case e [i.e., T indicates $H(e)$]
(2) Test T counterindicates a thesis H holds in the test case e [i.e., T counterindicates $H(e)$].

Then T is a good test of H to the extent that both
(i) Prob {Result (1) given not $H(e)$} is very low
and
(ii) Prob {Result (2) given $H(e)$} is very low.
Such probabilistic considerations enable one to talk of degrees of goodness as well.
[5]In this respect our episode accords with a main thesis of Giere's model of science.
[6]This quote is from a passage of Nägeli's cited and translated by Brush (1968), p. 356. The source is Nägeli (1879), "Über die Bewegungen kleinster Körperchen", *Sitzungsberichte der mathematisch-physikalischen Classe der K. Bayerischen Akademie der Wissenschaften zu München* **9**, pp. 389–453.
[7]An excellent sourcebook detailing these modern experiments is Wax (1954).
[8]Values obtained for the mean velocity of agitation by attempting following as nearly as possible the path of a grain, gave the grains a kinetic energy 100,000 times too small. According to Einstein's theory, mean velocity in an interval t is inversely proportional to root t; it increases without limit as the time gets smaller. Thus, the meaningless results were just what would be expected. Brush (1968, p. 369) claims "One can hardly find a better example in the history of science of the complete failure of experiment and observation, unguided (until 1905) by theory, to unearth the simple laws governing a phenomenon." While Einstein's theory did in fact serve that role, it did not require that theory's acceptance. Moreover, the statistical point that was needed here was already made in 1854 by Thomson (Lord Kelvin) with respect to the problem of laying the Atlantic Cable. Thomson defends his theoretical prediction against apparent experimental refutation, by explaining the source of the diversity of measured values of the velocity of electricity to the time required in making the measurements. Brush (1968, pp. 369–370) remarks:

> Apparently the scientists who attempted to measure the velocity of particles in Brownian movement later in the nineteenth century had not followed the dispute about Thomson's law of squares in the electric telegraph problem, and they obtained a similar collection of wildly varying results (none of them in agreement with [what would be expected according to the kinetic theory]).

[9]The standard deviation (square root of the variance) is the displacement in the direction of the x-axis which a particle experiences on average (root mean square of displacement). The importance of this statement of variance, for the experimental determination of D, is that it states that the mean square displacement of a Brownian particle is proportional to the time t. This suggests, for example, that a model for Brownian motion is provided by viewing a particle as taking a *random walk*. Since it has the same chance of being displaced a given amount in the positive and the negative direction, on average, after k steps, the displacement would be 0. Occasionally, more steps will be to one direction than to the other, yielding a non-zero total displacement. That the variance is proportional to the time (the number of steps) corresponds to the fact that the more steps taken, the larger the value that this non-zero total displacement can have.

[10]In addition to Einstein, see Parzen (1960) and Wax (1954) for such derivations.

[11]In addition to deriving the statistical distribution in the ES Theory from random walk models, it was analogously derived from the recognition that the displacement of Brownian particles is distributed like the winnings of a gambler who stands to win or lose a fixed amount x with equal probability on each trial. Clear derivations of this distribution occur in Einstein (1926, Chapter V), and Parzen (1960, pp. 374–376).

[12]Perrin had to run a number of separate experiments to justify his use of Stokes's formula, whose applicability to the fluid at hand had been open to question. See, for example, Perrin (1923), p. 129.

[13]Indeed, Popper's approach is seen by some statisticians, such as Oscar Kempthorne, as an embodiment of the Fisherian logic.

[14]ς is the viscosity of the fluid, T, its absolute temperature, R, the gas constant, a, the radius of the particles.

[15]Admittedly, because Step (II) tests a predicted value of a parameter such as N against rival values (in a parameter space), it is more like (what later came to be called) a Neyman – Pearson test than the Fisherian test at Step (I). But this still does not involve the sort of fully-fledged rival presumably meant in (T2.8).

[16]Examples of philosophers who identify it as such are Cartwright and Salmon. I discuss Cartwright's and my own treatment of Perrin's experiments as arguments from coincidence in Mayo (1986).

[17]This can often be made quantitatively rigorous: One constructs a test to generate a type of outcome e that would be highly surprising (improbable) if a theory differed from a correct account of some parameter by more than a specifiable amount ∂. Then the occurrence of e is a good indication that the theoretical value of the parameter does not differ by more than ∂ from the correct value. This was indeed the case for Perrin's test of N at Step (II). See Note 18.

[18]That is, N is itself a parameter (a mean) of a population distribution from which observed estimates of N are random samples. While actual experiments do not give the precise true value of this parameter, they do tell us if estimates are discrepant from the true value by certain amounts. They do this by ruling out discrepancies which, with overwhelming probability, could not have produced observed estimates.

[19]Popper stresses this, calling such approaches 'justificationist'. Neyman–Pearson theory contrasts with these other schools on this point, for error probabilities *are* influenced by the manner of data and hypotheses generation. Interestingly, this is one of the main reasons Neyman–Pearson statistics is criticized by opponents.

[20]This is expressed formally in the *Likelihood Principle*.

[21]This notion of a statistically crucial test seems to be congenial to Laudan's idea of solving empirical problems approximately. A statistical test might be seen as specifying a rule whereby a given test outcome either deems the theory a solution or not a solution (or a fairly good, or poor one). Then what is provided by a 'good' statistical test (i.e., one with sufficiently low error probabilities) are assurances that a theory will not often be deemed a solution to the problem when it is far from being one, and not often be deemed a poor solution when it is close to solving the problem. This also suggests that we can consider a test (statistically) crucial between any two claims for which the test is a good one, according to my statistical criteria.

[22]Perrin calculates, for example, that to have a better than even chance of seeing a brick weighing a kilogramme suspended by a rope rise to a level by virtue of its Brownian motion, one would have to wait more than 10^{10} billion years. So the statistical version involves a classical frequentist interpretation of probability.

[23]This translated quotation is given in Brush (1968), p. 381. The source is Wilhelm Ostwald (1909) *Grundriss der allgemeinen Chemie*. Leipzig: Verlag von Wilhelm Engelmann, p. 4. Quotation is from the "Vorbericht".

[24]The error in Nageli's challenge was not experimental but mathematical-statistical. Gouy was first to come close to answering this objection by citing the 'law of large numbers'. Smoluchowski gave a more rigorous explanation based on a statistical argument of which Nageli and others had been unaware. The argument essentially shows how a gambler can lose a great deal of money, even with an even chance of winning or losing a fixed amount on each trial, provided he plays long enough (See Note 11). Analogously, unlike what Nageli supposed, the jerks of Brownian particles do not need coordinated motion to explain them; with enough hits, the average displacement can be large, even when each hit has an equal chance of moving the particle to the right or left a given amount.

REFERENCES

Brush, S. (1968), 'A History of Random Processes I. Brownian Movement From Brown to Perrin', *Archive for the History of the Exact Sciences* 5, 1–36, (as Reprinted in *Studies in the History of Statistics and Probability*, Vol. II (1977), edited by Sir M. Kendall and R. L. Plackett. New York: Macmillan, pp. 347–382; page references are to this reprinting.)

Brush, S. (1968), 'Mach and Atomism', *Synthese* 18, 192–215.

Brown, R. (1828), 'A Brief Account of Microscopical Observations Made in the Months of June, July and August, 1827, on the Particles Contained in the Pollen of Plants; and on the General Existence of Active Molecules in Organic Bodies', *Philosophical Magazine* 4, 161–173.

Cartwright, N. (1983), *How the Laws of Physics Lie*, Oxford: Clarendon Press.

Chandrasekhar, S.: (1943), 'Stochastic Problems in Physics and Astronomy', *Reviews of Modern Physics* 15, 1–89, (as reprinted in Wax (1954), pp. 3–91; page references are to this reprinting.)

Clark, P. (1976), 'Atomism Versus Thermodynamics', in C. Howson (ed.), *Method* and Appraisal in the Physical Sciences, Cambridge: Cambridge University Press, pp. 41–105.

Delsaulx, J. (1877), Thermodynamic Origin of the Brownian Motion,' Monthly Microscopical J. **18**, 1–7.

Einstein, A. (1926), Investigations on the Theory of the Brownian Movement, (edited by R. Furth, translated by A. D. Cowper), London: Methuen (as reprinted in the Dover Edition (1956), New York: Dover Publications; page references are to this reprinting.)

Einstein, A. (1949), 'Autobiographical Notes', in Albert Einstein: Philosopher-Scientist (ed. P. A. Schilp). Library of Living Philosophers, Tudor Pub. Co. 1949, reprinted by Harper & Brothers, New York, 1959; the quoted passage is from pp. 46–49, (German with English translation on facing pages.)

Feyerabend, P. (1978), Against Method, London: Redwood Burn.

Gardner, M. (1979), 'Realism and Instrumentalism in 19th Century Atomism', Philosophy of Science **46**, 1–34.

Giere, R. (1984), 'Toward a Unified Theory of Science', in Science and Reality (eds. T. Cushing, C. F. Delaney, and G. Gutting), Notre Dame: University of Notre Dame Press, pp. 5–31.

Gouy, L. (1888), 'Note sur le mouvement brownien', Journal de Physique **7**: 561–564.

Hacking, I. (1983), Representing and Intervening, Cambridge: Cambridge University Press.

Harman, G. H. (1965), 'Inference to the Best Explanation', Philosophical Review **74**, 88–95.

Jones, B. (1870), The Life and Letters of Faraday, vol. 1, London: Longmans, Green.

Kuhn, T. (1962), The Structure of Scientific Revolutions, Chicago: Chicago University Press.

Kuhn, T. (1977), The Essential Tension, Chicago: University of Chicago Press.

Lakatos, I. (1978), The Methodology of Scientific Research Programmes, Cambridge: Cambridge University Press.

Laudan, L. (1977), Progress and Its Problems, Berkeley: University of California Press.

Laudan, L. (1984), Science and Values, Berkeley: University of California Press.

Laudan, L., Donovan A., Laudan, R., Barker, P., Brown, H., Leplin, J., Thagard, P., and Wykstra, S. (1986), 'Scientific Change: Philosophical Models and Historical Research', Synthese **69**, 141–223.

Laymon, R. (1978), 'Feyerabend, Brownian Motion, and the Hiddenness of Refuting Facts', Philosophy of Science **44**, 225–247.

Laymon, R. (1988), 'The Michelson-Morley Experiment and the Appraisal of Theories', this volume, pp. 245–266.

Mayo, D. (1986), 'Cartwright, Causality, and Coincidence', in A. Fine and P. Machamer (eds.), Philosophy of Science Association 1986, vol. 1, East Lansing: Philosophy of Science Association, pp. 42–58.

Musgrave, A. (1974), 'Logical versus Historical Theories of Confirmation', British J. Philosophy of Science **25**, 1–23.

Nye, M. J. (1972), Molecular Reality, London: Macdonald.

Ostwald, W. (1907), 'The Modern Theory of Energetics', Monist **17**, 481–515.

Parzen, E. (1960), Modern Probability Theory and Its Applications, New York: Wiley.

Perrin, J. (1923), Atoms, translated by D. L. Hammick, London: Constable.

Perrin, J. (1950), *Oeuvres Scientifiques de Jean Perrin*, Paris: Centre National de la Recherche Scientifique.

Poincare', J. (1905), 'The Principles of Mathematical Physics', in *Congress of Arts and Science, Universal Exposition, St. Louis, 1904*: vol. 1, Boston and New York: Houghton, Mifflin, pp. 604–622.

Popper, K. (1959), *The Logic of Scientific Discovery*, New York: Basic Books.

Ramsay, W. (1882), 'On Brownian or Pedetic Motion, *Proc. Bristol Naturalists' Society* **3**, 299–302.

Salmon, W. C. (1984), *Scientific Explanation and the Causal Structure of the World*, Princeton: Princeton University Press.

Uhlenbeck, G. E. and Ornstein, L. S. (1930), 'On the Theory of the Brownian Motion', *Physical Review* **36**, 823–841 (as reprinted in Wax (1954), pp. 93–111.)

Von Mises, R. (1957), *Probability, Statistics and Truth*, New York: Dover.

Wax, N. (ed.) (1954), *Selected Papers On Noise and Stochastic Processes*, New York: Dover.

Worrall, J. (1985), 'Scientific Discovery and Theory-Confirmation', in J. C. Pitt (ed.), *Change and Progress in Modern Science*, Dordrecht: D. Reidel, pp. 301–331.

Zahar, E. (1973), 'Why Did Einstein's Programme supersede Lorentz's? *British J. Philosophy of Science* **24**, 95–123.

RONALD LAYMON

THE MICHELSON–MORLEY EXPERIMENT AND THE
APPRAISAL OF THEORIES

1. INTRODUCTION

I shall focus on the question whether we ever need to assume the very theory to be tested or one of its rivals in order to interpret an experimental result as a test of the theory. In terms of the set of theses being tested, the thesis to be examined is that:

> The appraisal of a theory is based on phenomena which can be detected and measured without using assumptions drawn from the theory under examination (T2.5).

I shall refer to this thesis as the *Independent Evidence Thesis*. The episode to be examined will center around the Michelson–Morley experiment and some historical responses to Michelson's analysis of that experiment. I shall close with a brief consideration of an underlying assumption of the historicists and a recommendation for its replacement.

2. ON THE INDEPENDENT EVIDENCE THESIS

In order to avoid unnecessary ambiguity and proliferation of philosophical possibilities, I shall introduce some interpretive conventions. I also have as an aim the maximization of the testability of the Independent Evidence Thesis by means of historical example. Let the set of procedures and theories used in the production of the value of a measurement of some phenomenon be called *the means of measurement production*. Measurement values produced by such means will count as measurements of experimental phenomena only if these values are in fact used by scientists in their analyses of the experimental phenomena, that is, in or as tests of their scientific calculations. In other words, measurement values can occur as input conditions for calculations or as targets for

245

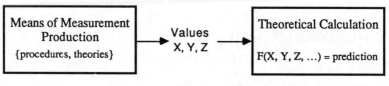

Fig. 1.

calculated outputs or predictions.[1] What I have in mind is illustrated in Figure 1.

The Independent Evidence Thesis requires then that the means of measurement not include the very theory that is to be tested, that is, of the theory used in the calculation of the prediction. If the Independent Evidence Thesis is false, then there will be cases where there exists a theory to be tested, say T_1, that is *fundamental* in the sense that the measurement procedures needed to determine initial condition values will depend on T_1, that is, on the very theory to be tested. The importance of the Independent Evidence Thesis is seen most clearly when two or more theories are being compared and appraised with respect to their ability to account for some experimental phenomenon. If the competing theories are fundamental, then it appears that comparative testing will require a comparison of the coherence of each theory's account of its own internally specified evidence.[2] Consider now the following possibility. Say the measurement of some initial condition value v was made in a way that depended on theory T_1, and that this value v was then conjoined with some competitor theory T_2, and used (perhaps with other initial condition values) to derive some prediction about the phenomenon under question. In such a case, the appraisal of theory T_2 would not strictly speaking be a counter example to the Independent Evidence Thesis since the measurement value v does not depend on T_2, but on some competitor T_1. But such a case would seem clearly to violate the spirit of the Independent Evidence Thesis since T_2 is not being tested against relatively neutral data but against data colored by the competitor T_1. Therefore I suggest modifying the Independent Evidence Thesis to read: The appraisal of a theory is based on phenomena which can be detected and measured without using assumptions drawn from the theory under examination *or from competitors to that theory*.[3]

Some readers will want to object that the proposed modification is unnecessary on the grounds that the envisioned possibility is not genuine because it would be absurd to combine competing and inconsistent theories in the way considered. Let us call the assumption that such combinations are not to be allowed the *Fair Play Principle* on the grounds that it would be unfair to test some theory T_1 using measurements obtained by assuming some competitor T_2. The Fair Play Principle can be understood in either a normative or a descriptive sense, that is, as prescribing or describing scientific practice. My claim in this paper will be that the Michelson–Morley experiment violates the Independent Evidence Thesis *and* the Fair Play Principle, where the latter is understood in both normative and descriptive senses. In fact, I shall show that the Michelson–Morley experiment illustrates an even stronger violation of fair play. Assume that theory T is to be tested by means of some calculation C of the anticipated outcome of some experiment. Assume also that the necessary input conditions (as well as the value of the experimental outcome) are supplied by some acceptable means of measurement production. By the *Extended Fair Play Principle* I shall mean the requirement (descriptive or normative) that the calculation C not utilize assumptions formally inconsistent with T. More simply put, T should not be tested against predictions derived using assumptions inconsistent with T. The Extended Fair Play Principle is a presupposition, I believe, of virtually all existing philosophical theories of confirmation and scientific acceptability, including those of the historicists. My claim then is that this fundamental philosophical principle is violated in the case of the Michelson–Morley experiment.

Before going on to the historical case, I need to introduce an additional modification of the Independent Evidence Thesis because, as formulated above, it is immune to refutation by historical example. This is due to the modal used in its formulation. The thesis does not assert that historically made comparisons are made with respect to phenomena that are *in fact* measured in ways that do not assume the very theories to be tested, but only that scientific theories 'can be tested' in this way. This means that while historical data can be supporting (since actuality implies possibility), the absence of such historical data does not show by itself that the required sorts of experimental comparisons cannot be constructed as philosophical exercises. To show such philosophical reconstruction impossible requires significantly more than historical data. To overstate somewhat, the Independent Evidence Thesis is primarily a philosophical thesis and not one *per se* about the practice of actual science. In keeping with the spirit of these proceedings, it would seem appropriate therefore to modify the Independent Evidence Thesis in a way that will allow it to be both confirmed *and* disconfirmed by historical data alone. Replacing the modal 'can be' with 'are' is sufficient.

> The appraisal of a theory is based on phenomena which are detected and measured without using assumptions drawn from the theory under examination or from competitors to that theory.

In other words, the measurement data that *are explicitly* reported and used in the appraisal *are* obtained using a means of measurement production that does not assume the truth of either the theory to be tested or any of its alternatives. Henceforth, by the Independent Evidence Thesis I shall mean this more immediately testable version.

3. THE MICHELSON–MORLEY EXPERIMENT

The arms of the interferometer of the Michelson–Morley experiment are typically described as being equal. For example, in the analysis given in their famous paper of 1887, Michelson and Morley let 'D' represent the 'distance ab or ac,' where ab and ac are the different interferometer arm lengths. Their use of 'or' here does not indicate that an option exists for the value of D. Rather, as their analysis clearly shows, the assumption is that the interferometer path lengths are equal and that D is their common length. For convenience I reproduce this analysis. The velocity of the interferometer is assumed to be parallel to the arm ac (see Figure 2).

Let V = velocity of light
       v = velocity of the earth in its orbit
       D = distance ab or ac [interferometer arm length]
       T = time light occupies to pass from a to c
       T_1 = time light occupies to return from c to a_1.

Then

$$T = \frac{D}{V - v} \qquad T_1 = \frac{D}{V + v}$$

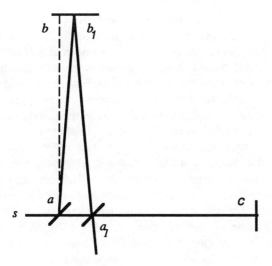

Fig. 2.

The whole time of going and coming is

$$T + T_1 = 2D \ \frac{V}{V^2 - v^2}$$

and the distance travelled in this time is

$$2D \ \frac{V}{V^2 - v^2} = 2D \left(1 + \frac{v^2}{V^2} \right)$$

neglecting terms of the fourth order. The length of the other path is evidently $2D\sqrt{1 + (v^2/V^2)}$, or to the same degree of accuracy, $2D(1 + (v^2/2V^2))$. The difference is therefore $D(v^2/V^2)$. If now the whole apparatus be turned through 90°, the difference will be in the opposite direction, hence the displacement of the interference-fringes should be $2D(v^2/V^2)$. Considering only the velocity of the earth in its orbit, this would be $2D \times 10^{-8}$. If, as was the case in the first experiment, $D=2 \times 10^6$ waves of yellow light, the displacement to be expected would be 0.04 of the distance between the interference-fringes.

We have therefore what is an apparently straightforward deduction of a prediction (the fringe shift) from various initial conditions and the Fresnel aether theory. Since Michelson's original intention was to use the interferometer as an aethereal speedometer and not as a test of the Fresnel aether theory, the prediction based on the estimated absolute velocity of the earth was meant only to demonstrate the feasibility of the interferometer conceived as a measuring instrument.[4] Of course, the experiment eventually came to be seen as a test of aether theory and not as a measuring instrument designed to determine a parameter value of aether theory. But since length determinations and the detection and measurement of interference fringes depend on fundamental optical assumptions, it seems clear that in order to use the interferometer to test a theory as fundamental as the Fresnel aether theory, it will be necessary to assume the very theory to be tested when measuring initial conditions and effect. My initial focus will be on the question of the determination of the length and equality of the interferometer arms. Since this length plays a central role in the *historically given* calculation of the anticipated effect, it seems reasonable to identify interferometer arm length as one of the components of the 'phenomenon' whose 'detection and measurement' provides the 'basis' for an experimental 'appraisal' of aether theory and its competitors. Assuming then this identification, our problem is to ascertain the measurement procedures used to determine interferometer arm lengths and the equality of those lengths.

Since the Michelson–Morley experiment was designed to measure a second-order effect (i.e., of the order v^2/c^2) one would expect there to be a careful specification of the procedures used to determine D. In fact, nothing more sophisticated than an ordinary ruler seems to have been employed, and no explicit report of measurement procedure appears in *any* of the published experimental reports.[5] If ordinary rulers were the basis of the measurement then the procedure is open to the objection that constancy of ruler length depends on assuming the non-existence of any sort of Lorentz–Fitzgerald contraction effect. But the assumption of the existence or non-existence of such contraction is a central difference among the set of theories that the experiment was historically seen as testing. To assume therefore that rulers can be used to measure the actual interferometer arm lengths is obviously to assume the very theory to be tested, namely, the Fresnel aether theory. But this means that the Independent Evidence Thesis stands refuted.

That thesis is not helped by the historical observation that the *equality* of interferometer arm lengths was determined not by ordinary rulers but by a somewhat sophisticated optical test. The interferometer mirrors were adjusted until a central dark band appeared when using a white light source. Such bands will appear only when the optical paths are "equal to the fraction of a wavelength" (Miller, 1933, p. 214). The move however from equality of optical path to equality of physical path depends essentially on the presence or absence of contraction. Therefore, while obviously more accurate than using ordinary rulers, optical tests for equality of interferometer arm lengths are dependent on the very theory to be tested in virtually the same way as is the ordinary use of laboratory rulers.

Some philosophers will be inclined to object at this stage that what the Michelson–Morley case shows is not that the Independent Evidence Thesis stands refuted, but that the experimental situation needs to be *redescribed* in terms that are neutral with respect to the competing theories. That is, we are to reject my initial identification of interferometer arm length as a component of the 'phenomena' whose measurement is to be the basis for theory appraisal. So, for example, one might try to redescribe the situation in terms of *apparently* measured interferometer arm lengths. This is, or so it is commonly claimed, the correct response to the observation that the very meaning of *length* and its operational measures are theory dependent.[6] I do not deny that such redescription may be possible as an exercise in *philosophical* reconstruction. But, as we shall see, an interpretation in terms of relatively neutral measurements was not embarked upon by the historical principals associated with the Michelson–Morley experiment. This indicates that these scientists were either philosophically naive *or* that some rationale other than appeal to relatively theory-neutral observations was at work. Going the reconstructive route therefore may cause us to overlook a rationale *already existing* in scientific practice that justifies the use of measurements that are dependent on to be tested theories. As noted above, the Independent Evidence Thesis as originally formulated is not refutable by historical data. It was in part to avoid the detour of philosophical reconstructibility that I suggested we eliminate the modal in the formulation of the thesis. What justifies this elimination is the fact that in the Michelson–Morley case historical practice offers insight into the nature of science that would be lost if there were an immediate move to philosophical reconstruction in terms of neutral-observation procedures. Therefore, we ought not to try to save the thesis by restoring the modal. But this is merely to assert my position; it now needs to be demonstrated.

Neither Michelson nor any of the other experimental or theoretical principals seems to have paid any attention to the theory-dependent character of the measurement of interferometer-arm length when using their measurements to provide input values for the calculation of fringe shift. This uncritical attitude extends, or so it appears, even more broadly than this. Consider, for example, the fact that the Michelson--Morley interferometer (in virtually all its versions) consisted of a rotating platform on which were mounted sixteen or more mirrors aligned so as to cause the light beams (once separated) to travel back and forth across the platform several times before realignment at the sighting telescope[7] Overall path length was calculated by simply measuring the platform diameter or diagonal and then multiplying by the number of ray crossings. No attempt was made to take into account the small but measurable displacements of the mirrors from orthogonality needed to insure multiple non-coincident reflections. Michelson's 1887 value for D of 'about eleven meters' is not a determination of great precision, but it does properly reflect the simple measurement procedures employed.[8] This lack of precision needs to be explained given that the Michelson–Morley experiment was designed to measure a second-order effect.

An important clue is provided in Silberstein (1914).[9] He gives a variant of Michelson's simplified analysis, but one that explicitly retains the possibility that the interferometer arm lengths be different. If D_h is the length of the interferometer arm initially parallel to the direction of motion, and D_v is the length of the interferometer arm initially orthogonal, then (ignoring terms of order higher than the second) for a 90° rotation of the interferometer, the anticipated path length change is $\beta^2(D_h + D_v)$, where β is as usual v^2/c^2. So the sought-after effect is a function of the sum (and not the equality or near equality) of the interferometer arm lengths. This means that the equality of the interferometer arms is *not* crucial for the experiment. And since the effect varies as the simple sum, small errors in length determination will have little relative effect on the calculated effect. All of this means that while it is true that the actual length of the interferometer arms is a varying function of the very theories to be tested, when measuring that length the *computational* effect of assuming different theories from among the set to be tested is inconsequential. Hence, the phenomena (of fringe shift in rotating equal-arm interferometers) can be determined with an accuracy sufficient for testing the relevant theories *regardless* of which of the competing theories is chosen to specify the measurement procedures to be used to determine the experimental initial condition of length. The contraction hypothesis, for example, can be included in what we earlier referred to as the means of measurement production, and measurement values so obtained can then be used in a calculation that is based on the *non-existence* of contraction. This sounds paradoxical given that the Lorentz contraction was introduced precisely to counteract aethereal motion and yield a null shift. And so it does, but only if it is assumed at the beginning of the derivation of the theoretical prediction. If we were to apply contraction to an ordinary use of rulers and then use this corrected input value in the bottom line of the original Michelson calculation of fringe shift, there is, ignoring terms higher than second order, no change in the predicted effect. This shows the sensitivity of theoretical calculations to the order and location at which approximations are applied.[10] An example should nail down the point. Say the use of an ordinary ruler yields a value of 11 meters for both D_h and D_v (where we are here ignoring ordinary errors of measurement). Before outputting this value from the means of measurement production, assume that we correct D_h for Lorentz-contraction, obtaining $11(1 + v^2/c^2)^{-1/2}$, or ignoring higher-order terms, $11(1 + v^2/2c^2)^{-1}$. Passing this value for D_h (along with $D_v = 11$) to Michelson's Fresnel-aether-theory-based calculation yields $2[11 + 11(1 + v^2/2c^2)^{-1}]\beta^2$. But expanding and ignoring terms of higher order yields $2(11 + 11)\beta^2$, which is exactly as if no contraction were assumed in the means of measurement production.

Therefore, Michelson *could have* assumed contraction in his determination of interferometer arm length and then used this value in the calculation of the fringe shift to be expected given aether theory *without* contraction. Of course, he did not do this; such a procedure would have added complication without compensating advantage. However, during the years 1902–1904, Morley and Miller conducted a series of interferometer experiments the aim of which was explicitly to test the contraction hypothesis. (Miller and Morley, 1904, 1905a; Miller, 1933, pp. 208–209.) The idea was to construct interferometers of different materials where it was assumed that different materials would suffer different degrees of contraction when moved through the aether. In the analysis of the interferometers used in those experiments, Morley and Miller determined interferometer arm length on the assumption of no contraction. Therefore, they *in fact* violated what I have called the Fair

Play Principle (so there is historical actuality and not mere possibility). But here there was good reason to violate the principle, since violation brought about a simplification of measurement procedures without any real cost in predictive accuracy.[11] It is also to be noted that Miller insisted (as late as 1933) that the interferometer be seen as being neutral between competing theories, and this despite the fact that he assumed (without explicit comment) the absence of contraction when measuring the length of his interferometer paths (Miller, 1933, page 206).

My analysis of the Michelson–Morley experiment shows the Independent Evidence Thesis to be false, since the determination of some of the initial conditions of the Michelson–Morley experiment were dependent on the very theory to be tested or on an alternative.[12] The analysis also shows that the Fair Play Principle was violated, without ill effect. These refutations have been based on an acceptance of Silberstein's modification of Michelson's 1887 single-ray analysis of the interferometer and the (implicit) assumption that the sorts of error dependencies discussed above were so obvious as not to warrant explicit comment by the historical participants. But these refutations are somewhat questionable because Michelson's single-ray analysis came under attack (beginning in 1902) for being excessively idealized to the point of misrepresenting the phenomenon.[13] So the legitimacy of these attacks will have to be considered, as well as their relevance for my analysis of interferometer arm-length measurements. There is, in addition, this puzzle that actual historical practice poses for my analysis: if the effect of fringe shift is a function of the sum of the interferometer arm lengths, why did Michelson, for example, and Miller after him, persist in the use of equal-arm interferometers and in using a claimed equality of interferometer arms in their calculations of anticipated effect? Oddly there is no explicit explanation or justification of this persistence in any of the published experimental reports.[14] What did occur was an attack on the simplified *single* ray analysis used by Michelson.

It is easy to see *a priori* that supplementation or replacement of Michelson's single-ray account is necessary. Say the interferometer arms were exactly equal and the mirrors exactly orthogonal to the light paths. In such a case there should, according to the Fresnel aether theory, be no production of interference fringe. Now all versions of the Michelson––Morley experiment yielded interference fringes. Therefore (on the assumption of perfect alignment) the interferometer did not have to be rotated in order to refute the Fresnel aether theory. Of course, no one in fact reasoned this way, which indicates there was more to the historical understanding of the experiment than revealed in the single-ray account. In fact, the interferometer was not deemed ready for use until fringes were produced by adjustment of the mirrors. That is, the existence of fringes was a *precondition* for the suitability of the interferometer for experimental use. But exactly what fringes are indicative of is, as already noted, a function of the theories to be tested. Furthermore, the existence of fringes implies, given the Fresnel aether theory, that the optical paths are not equal. But given this it follows that Michelson's single-ray account is formally *inconsistent* with the actual experimental procedure.

In his 1882 paper, Michelson published an analysis of interferometer fringe production in terms of ordinary optical ray theory, that is, he ignored the complication of motion through the aether. His argument (not to be analyzed here) was that the interferometer could be analyzed *as if* there were thin film interference. That is, one could mathematically superimpose the two arms of the interferometer since the half-silvered mirror reunited the separated rays. Ordinary ray optics was then used to show how observable fringes would be produced if one of the mirrors were displaced from orthogonality (thus providing the thin film analog). Now Michelson's thinking seems to have been that his single-ray analysis (originally developed around 1881) of the Michelson–Morley experiment could be vectorially superimposed on his thin-film account in the sense that each ray (of the thin-film account) would be displaced by the amount $2D(v^2/c^2)$ during the rotation. If this was Michelson's thinking circa 1882, an objection made by Lorentz (in his 1887, a paper that Michelson read) should have convinced him that additional reasoning is required to support the accuracy of his estimate of anticipated fringe shift. Lorentz notes that if the paths of the rays (in the single-ray analysis) are calculated with due care, then when reunited they will be displaced from one another by a second-order difference, that is, of the *same* order of magnitude as that to be detected by the experiment. One had to show therefore that this displacement would not counteract the difference in arrival times with respect to production of interference fringes. Lorentz gave a proof (to be discussed below) showing that the result of this second-order displacement would result in at most a fourth-order modification of the value of fringe shift calculated on the basis of the single-ray account. Therefore one could safely ignore, according to Lorentz, the complication of displaced rays.

In his 1887 paper, Michelson somewhat surprisingly does not refer to his 1882 thin-film analysis. He does note that in his earlier 1881 treatment of the experiment (Michelson 1881), he neglected the effect of aethereal motion on the light beam that moves orthogonally to the relative aethereal velocity. And while he credits Lorentz for his 'searching analysis' of 'this oversight and of the entire experiment', his calculation of the path change of the orthogonal ray *assumes* equality of incidence and reflection at the end mirror. (See Michelson's analysis as reproduced above.) With possible reference to Lorentz in mind, Michelson asserts without argument that:

It may be remarked that the rays ba_1 and ca_1 do not now meet exactly in the same point a_1, though the difference is of the second order; this does not affect the validity of the reasoning (1887, p. 452).

Lorentz begins his proof that the validity of the reasoning is not affected with the following construction.[15] Assume that P is a point on a wave front at time t. According to Huygens' principle, this point (along with all other points on the wave front) is to be construed as a center of a vibration for the purposes of calculating the progression of the original wave front. Letting c be the velocity of wave propagation, in time dt the secondary wave will have progressed a distance cdt. Assume also the aether is moving at velocity w with respect to the physical system. Therefore in this interval of time dt, the center P will be displaced an amount wdt to "another point Q, namely, at the point that is reached at the time $t + dt$ by a particle of the ether which had the position P at the time t". The perceived velocity of a ray of light therefore will be vdt. (See Figure 3.)

Therefore, $c^2 = v^2 + w^2 - 2wv \cos \theta$. If one ignores second-order quantities this expression simplifies to: $v = c + wv \cos \theta$. It can then be shown, using Fermat's principle and the calculus of variations, that the minimization $\int ds/v$ (where the integration is along optical path length) is the same for both moving and stationary systems. This means that no first-order effect can be detected. But since this 'beautiful theorem' does not extend to quantities of the second order, experiments, such as the Michelson–Morley, designed to measure second-order effects will be possible from the point of view of the Fresnel aether theory. The above construction can be utilized to analyze such experiments if, as Lorentz notes, second-order terms are included in the equation for v. Including such terms one gets,

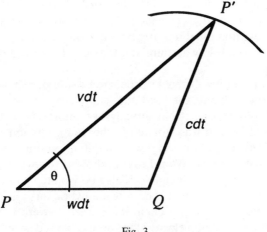

Fig. 3

$$v = c(1 + \frac{w}{c}\cos\theta - \frac{w^2}{2c^2}\sin^2\theta).$$

For each of the two paths of the interferometer rays, it is required therefore that the following integral be a minimum:

$$\int \frac{ds}{v} = \frac{1}{c}\int ds - \frac{1}{c^2}\int w\cos\theta\, ds + \frac{1}{c^3}\int w^2\,(\cos^2\theta + \tfrac{1}{2}\sin^2\theta)\, ds$$

But as earlier shown (the 'beautiful [first-order] theorem') only the third term in this integration need be taken into account. For the ray parallel to the aether velocity, one has $\cos^2\theta = 1$, while for the ray orthogonal to the aether velocity, one has $\cos^2\theta = 0$, *assuming* that the paths are as specified in the simplified single-ray account. The expected phase difference therefore will be $D(w^2/c^2)$, just as in the simplified single-ray account. Rotating the interferometer therefore will double the size of this difference.

Lorentz points out a slight defect in this analysis, namely, that the path over which $\int ds/v$ was calculated is not in fact the actual path followed by the ray orthogonal to the aether velocity. The calculated path was assumed to be symmetrical with respect to reflection at the end mirror of the orthogonal interferometer arm. But this symmetry will be disturbed because of motion with respect to the aether. This disturbance, though, because of the first-order theorem, can only be of the second order. And given that the actual path is a minimization of $\int ds/v$, it follows that the difference between the integral calculated over the path of the single-ray account and the actual path will be at most of the fourth order.

Therefore, assuming Lorentz's proofs to be sound, one need not have been concerned about the extreme simplifications of Michelson's elementary ray analysis. It has already been noted that the Fair Play Principle could be and was violated in the determination of values for the length of the interferometer arms, an initial condition of the calculation of anticipated effect. What Lorentz showed was what I have called the Extended Fair Play Principle can also be violated without ill consequences. One could employ, for example, a hybrid analysis that assumed that the ordinary laws of reflection and refraction were independent of aether movement but that velocity along optical paths determined by those laws is a function of aethereal velocity. And in fact Michelson's single-ray analysis is hybrid in exactly this way and as such violates the Extended Fair Play Principle. But just as in the case of the violation of fair play with respect to interferometer length determination, Michelson's violation in the single-ray account was not *computationally* significant. This is shown by Lorentz's argument.

That argument though was either unnoticed or found unpersuasive by at least some scientists interested in the experiment, and there were several attempts to analyze *more realistically* the Michelson–Morley experiment. These analyses all claimed to show that the positive fringe shift predicted by Michelson's single-ray account is not a consequence of the Fresnel aether theory when more realistic treatments of the interferometer are employed. Proponents of these analyses had in mind the need to take into account the finite size of the wave fronts, the variation of the mirrors from exact orthogonality, variations in the law of reflection, and variations in the frequency of light. Exactly what was to be expected on the basis of the Fresnel aether theory varied among the analyses.[16] Hicks analysis (1902) as modified by Miller and Morley (1905b) indicated fringe shifts of both half-turn and full-turn periodicity. If this is correct, then the averaging techniques employed by Michelson (in his data analysis) would not be sufficient to reveal the predicted effect. Miller therefore employed in his data analysis a mechanical differential analyzer in order to decompose the fringe shift data into their harmonic components. It is interesting to note that none of these analyses attacked or questioned Lorentz's proof for its lack of mathematical rigor (even by then contemporary standards). The strategy of the creators of these analyses was essentially just to apply Galilean transformations to elementary ray optics. Lorentz's proof and its utilization of the calculus of variations was just ignored.

The proliferation of competing analyses of the Michelson–Morley experiment became so severe that Lorentz felt obliged to repeat his 1887 proof in a paper published in *Nature* (Lorentz, 1921). But Lorentz's discussion, like the 1887 version, is somewhat sketchy and does not contain references to rigorous proofs of the theorems necessary for his proof. Lorentz was to repeat his argument for the adequacy of the simplified ray account one more time at the 1927 Conference on the Michelson–Morley Experiment held at the Mount Wilson Observatory.[17] Among the conference participants were Michelson, Dayton Miller and Roy Kennedy (who was later to collaborate in the Kennedy–Thorndike experimental test of time dilation).

Lorentz gave the second presentation (after Michelson) and essentially repeated, without significant change, his 1887 justification of the adequacy of the single-ray account of the Michelson–Morley experiment. E. R. Hedrick (of UCLA) in a later presentation questioned the adequacy of Lorentz's justification. Hedrick first computed, using ordinary ray optics and relative displacements between earth and aether, the second-order change in path experienced by motion orthogonal, or at an angle, with respect to the aether. He then computed the difference in path lengths for each of the interferometer rays and verified the correctness of the ordinary single-ray account. All of this was so far in accordance and as predicted by Lorentz's more abstract considerations. But Hedrick then went on to repeat with approval an argument given by Righi.

[These results were] obtained by Righi, who concluded from [them] that a rotation of the apparatus . . . through 90° would produce absolutely no effect, since, although the distances traversed by the two rays are exchanged, yet at the same time their positions are also exchanged; that is, the ray having the longer path occupies the same relative position with respect to that one having the shorter path, after the rotation as before. It follows that the pattern of the interference fringes after the rotation cannot be distinguished from that before the rotation (Michelson *et al.*, 1928, p. 379).

Hedrick was unable to see how Lorentz's constructions and proof were to be applied to the explanation of interference. There was this additional problem. Hedrick's calculations were based on 'ideal' mirror settings: end mirrors perfectly orthogonal, the beam splitter at exactly 45°. But since observable fringes required variation from orthogonality, actual fringe shift (according to the Fresnel aether theory) would be caused by *initial* deviations of mirrors from perfect orthogonality as well as changes in that orthogonality induced by motion through the aether. Hedrick went on to show what these functional dependencies would be (where the calculation was again based on elementary ray optic constructions). But since the mirror variations were *not* controlled (or measured) in the Michelson–Morley experiment,

we must suppose that the maximum and minimum positions for a series of experiments will have an entirely random distribution. It will not be legitimate, therefore, simply to average the readings of a series of observations, as was done in the Michelson–Morley experiment. In fact, there would seem to be a high degree of probability that this procedure would lead to a quite small result in case it is applied to a large number of observations (Michelson *et al.*, 1928, p. 382).

Lorentz (who was now 74 years old) was either unable or unwilling to show how his constructions could be made to apply to an analysis of interference caused by non-orthogonal mirrors. His response seems best interpreted as an attempt to be graciously polite:

I feel somewhat guilty in regard to the work of Righi. It was a long time ago that I read his papers, and I do not remember their contents very well, as I have been busy with quite different things these last years. I should have read them again of course for this meeting. But this good intention could not be materialized because of my being entertained so much by the people of Pasadena. . . . Until today I felt myself quite satisfied with the considerations which are based on Fermat's principle. After Mr. Hedrick's report, however, I shall have to reconsider these questions carefully. . . . My procedure seems to me still to be the easiest and most straightforward one. Still it must be found out where the discrepancy between the two ways lies. In case a method other than Fermat's is chosen, one has to do considerable work. One must distinguish, for instance, very carefully between the rays of light and the normals to the wave trains. Another difficult point is involved in the treatment of the reflection from moving mirrors. Fermat's principle, of which I have made use, gives in any case a much simpler treatment. But as there exists a discrepancy between the results obtained by the two methods, I intend to go through the detailed calculations as soon as possible. In the meantime. I still hope, of course, that my general considerations are right (Michelson *et al.*, 1928, pp. 389–390).

It was not until Kennedy (1935) that a detailed, direct constructional account was published that was fully in accordance with the Lorentz theorem and that thereby justified the use of Michelson's single-ray analysis for the analysis of second-order effects.

What is the relevance of this dispute about the adequacy of Michelson's single-ray analysis for the Independent Evidence Thesis and the Fair Play Principles? First of all, it shows that whether the Fair Play Principles can legitimately be violated depends on calculations that make essential use of idealizations and approximations. One needs to be able to show that their use does not itself bias the argument for or against the Fair Play Principles. Lorentz was able to show this in a powerful and elegant way. Second, my discussion of the adequacy of Michelson's single-ray analysis helps to explain why there did not develop in the scientific community any real analysis (as opposed to philosophical abstraction or textbook presentation) of the Michelson–Morley experiment in terms of measurements that were neutral between the competing theories. An analysis in these terms was not required because of the error dependencies revealed by Michelson's account. Finally, my discussion shows that scientists could legitimately violate the Fair Play Principles without prejudice to any of the historically considered competing theories.

I shall now consider a question introduced earlier: if one can justifiably employ the single-ray account, and in particular Silberstein's modification, why did Michelson and his followers persist in the use of equal-arm interferometers? Why were the interferometer arms historically made equal if the fringe shift is a function of the sum of the interferometer arm lengths? To understand this historical practice we need to remind ourselves that the experiment required the *continuous* observation of the fringes as the interferometer is rotated. Cinematographic techniques, however, were not sufficiently developed so as to allow Michelson the luxury of automated data collection as the interferometer was rotated. Therefore the fringe shift data were continuously collected by human observers. But given the frailties of human observers, it is necessary to allow for some breaks in observation. To make such breaks possible there must be a clearly *reidentifiable* fringe that can serve as a standard or zero point. If monochromatic light is used, all of the fringes will be of equal character, i.e., none will be distinguished. If, however, white light is used, colored fringes with a central dark, and hence reidentifiable, band can be produced. But white light will produce fringes *only if* the optical paths are equal to within a fraction of a wavelength. So historically the arms were made equal because of shortcomings of available experimental technique.[18]

4. CONCLUDING REMARKS

I want to depart now somewhat from the established format of these proceedings and consider the more general significance of the refutation of the Fair Play Principles. I have argued that, in the case considered, scientists were primarily interested in the computational consequences of their assumptions *regardless* of the consistency of those assumptions when viewed as sentences of some logical system. This attribution of interest should appear very counter-intuitive *if* one thinks of scientific theories, as the positivists did, as being sets of sentences written in first or possibly second-order predicate calculus. My hypothesis is that this central tenet of logical positivism is retained by the historicists though not, of course, in the precise self-conscious positivist version. Since it would be out of place here to develop a serious case for the truth of this hypothesis, I shall only briefly indicate some supporting evidence. Consider, for example, (GA1.1–1.5), (GA2.1–2.7), (GA3.1), (GA4.1–4.4), all of which are expressed in terms of sets of guiding assumptions. All of this is very natural if one construes theories as being sets of sentences whose consistency is defined or based on some sort of quantificational logic. My hypothesis, if correct, would also explain why it is that the historicists, *despite* their interest in real science, consistently missed the sorts of computational issues discussed above. For as Suppes has forcefully argued, logic is not a natural or revealing means for representing mathematical calculations. If Suppes is right, then it should not be surprising that the historicists did not develop an account of the philosophical importance of the use of idealizations and approximations in science. Notions such as closeness of fit and adequacy of approximation do not have natural expression in terms of a bivalent logic.[19]

It might be objected at this point that my illustrations of the violation of the Fair Play Principles are somehow atypical or special in the history of science. Such an attitude would be a mistake, I believe. Consider, for example, the standard relativistic analysis of the starlight-bending experiment. In this analysis the Schwarzschild metric is combined with a Newtonian analysis of the telescope and photographic plates. This combination of inconsistent theories can be analyzed exactly as above and shown to violate the Fair Play Principle.[20]

What is required therefore is a concept of scientific theories that allows for easy and revealing consideration of computational concerns and that allows for the coherent use within a calculation of inconsistent theories. The semantic view of theories as originally urged by Suppes and later by van Fraassen is a promising approach since it frees one from linguistic representations and allows for a natural use of mathematical structures. Since demonstrating this promise would be wildly off the mark for these proceedings, I shall contain my enthusiasm for the semantic approach and bring this paper to a halt.[21]

NOTES

[1]While it is obvious that a finer-grain set of distinctions is possible, that given here is sufficient for the purposes of this paper. See, e.g., Suppes (1962) for an example of the sort of finer-grained analysis I have in mind.

[2]See, e.g., Feyerabend (1970), pp. 219–222.

[3]Feyerabend's (1965) analysis of the use of Brownian motion as a refutation of thermodynamics is best understood as an examination of this thesis, at least according to Laymon (1977).

[4]This calculated effect is only the *maximum* to be expected since the interferometer might not be oriented in the favorable position assumed in the analysis. This complication creates some difficult problems for the experiment since the effect had to be given a fair chance to appear. However, I shall not discuss these sorts of problems here.

[5]See, e.g., Miller's (1933) review article (p. 210) where he says that "the distances between them [the mirrors] . . . are compared by means of light wooden rods and the mirrors are adjusted so that the two light-paths . . . are approximately equal."

[6]See, for example, Reichenbach's classic (1957), pp. 19–37.

[7]See, for example, Figure 4 in Michelson (1887). The use of the expression 'interferometer arm' was somewhat metaphoric and reflected Michelson's earliest and unsuccessful 1881 version of the experiment (Michelson, 1881). For a discussion of the difficulties one has in defining exactly what is meant by the Michelson–Morley *type* of experiment, and the relevance of these difficulties for the question of the ad hocness of the Lorentz–Fitzgerald contraction hypothesis, see Laymon (1980).

[8]This is in fact the value used in Michelson's calculation; it is not merely the value used in a summary description of the experiment!

[9]See also Sommerfeld's (1934) lecture notes as reflected in his (1964).

[10]See Cartwright (1983), pp. 119–126, for a very interesting discussion of how the presence of the Lamb shift depended on the order in which certain key approximations were employed in the quantum calculations. One should also compare Silberstein (1914), pp. 72–77, and Sommerfeld (1964), p. 78.

[11]The Trouton–Rankine (1908) test of contraction can also be analyzed as violating the Fair Play Principle.

[12]This refutation of the Independent Evidence Thesis suggests two requirements for a good scientific experiment. First, that the calculation of experimental effect be insensitive to errors in the determination of initial conditions. And second, as a sort of corollary, that the calculation of experimental effect be insensitive to errors or differences caused by the use of competing theories in the determination of initial conditions. But it must be noted that it need not always be the case that the computational effect of assuming different theories in the determination of initial conditions be negligible. In Laymon (1980), I show that the Kennedy–Thorndike experiment would not have been feasible when viewed from the perspective of the Fresnel aether theory unmodified by the contraction hypothesis. For an abstract simplified example and discussion see Laymon (1985).

[13]If both the Independent Evidence Thesis and the Fair Play Principle are interpreted descriptively, then the attacks on Michelson's analysis are irrelevant for my refutations, since I have shown historical violation. If, however, those theses are interpreted normatively, then Michelson's analysis (or some functional equivalent) must be preserved in order to provide the justification for the historical practice.

[14]Nor is this explained in Swenson's otherwise historically useful (1972).

[15]This proof is repeated, though without some of the mathematical detail, in chapter 5 of Lorentz's (1906) Theory of Electrons (Lorentz, 1952), as well as in Lorentz (1921), (1927), and Michelson et al. (1928). My presentation here is a simplified amalgam of these various presentations.

[16]For references and discussion of these accounts see Kennedy (1935) and Michelson et al. (1928), p. 374.

[17]Michelson et al. (1928). The proceedings also contain a transcript ("reviewed by the authors") of the general discussion that followed the paper presentations.

[18]Kennedy was able to construct a feasible monochromatic version of the experiment where orthogonal mirrors were used to produce wide circular interference fringes. Fringe shifts were measured by comparing intensity changes of the two rays. For details see Kennedy (1926) and Illingworth (1927). For the introduction of photographic techniques see Kennedy and Thorndike (1932).

[19]For discussion of this, see Laymon (1985).

[20]The starlight- bending experiment is described in similar terms in Laymon (1984). In Laymon (1985), I discuss Stokes' use of inconsistent assumptions in his analysis of pendulum motion.

[21]For an excellent review of Suppes' program see Moulines and Sneed (1979). Van Fraassen's version of the semantic program is explained in his (1970) and (1980). For a case study of the sort of improvement the semantic approach brings, see van Fraassen's (1983) semantic modification of Glymour's bootstrapping theory of confirmation. See also Glymour's (1983) adoption. In Laymon (1987) I give a semantic account (using continuous lattice theory) of the scientific use of idealizations and approximations.

REFERENCES

Cartwright, N. (1983), *How the Laws of Physics Lie*, Oxford: Clarendon Press.

Feyerabend, P. K. (1965), 'Problems of Empiricism', in *Beyond the Edge of Certainty*. (ed. R. Colodny), Pittsburgh: University of Pittsburgh Press, pp. 145–260.

Feyerabend, P. K. (1970), 'Consolations for the Specialist', in *Criticism and the Growth of Knowledge*, (eds. I. Lakatos and A. Musgrave), Cambridge: Cambridge University Press, pp. 197–230.

van Fraassen, Bas (1970), 'On the Extension of Beth's Semantics of Physical Theories', *Philosophy of Science* **37** 325–339.

van Fraassen, Bas (1980), *The Scientific Image*, Oxford: Clarendon Press.

van Fraassen, Bas (1983), 'Theory Comparison and Relevant Evidence', in *Minnesota Studies in the Philosophy of Science X: Testing Scientific Theories* (ed. J. Earman), Minneapolis: University of Minnesota Press, pp. 27–42.

Glymour, C. (1983), 'On Testing and Evidence', *Minnesota Studies in the Philosophy of Science X: Testing Scientific Theories* (ed. J. Earman), Minneapolis: University of Minnesota Press, pp. 3–26.

Hicks, W. M. (1902), 'On the Michelson–Morley Experiment Relating to the Drift of the Ether', *Philosophical Magazine*, 6th ser. **3**: 9–36.

Illingworth, K. K. (1927), 'A Repetition of the Michelson–Morley Experiment Using Kennedy's Refinement', *Physical Review* **30**, 692–696.

Kennedy, R. J. (1926), 'A Refinement of the Michelson–Morley Experiment', *Proceedings of the National Academy of Sciences* **12**, 621–629.

Kennedy, R. J. (1935), 'Simplified Theory of the Michelson–Morley Experiment', *Physical Review* **47**, 965–968.

Kennedy, R. J. and Thorndike, E. (1932), 'Experimental Establishment of the Relativity of Time', *Physical Review* **42**, 400–418.

Laymon, R. (1978), 'Feyerabend, Brownian Motion, and the Hiddenness of Refuting Facts', *Philosophy of Science* **44**, 225–247.

Laymon, R. (1980), 'Independent Testability: The Michelson-Morley and Kennedy-Thorndike Experiments', *Philosophy of Science* **47**, 1–37.

Laymon, R. (1984), 'The Path from Data to Theory', in *Scientific Realism* (ed. J. Leplin), Berkeley: University of California Press, pp. 108–123.

Laymon, R. (1985), 'Idealizations and the Testing of Theories by Experimentation', in *Experiment and Observation in Modern Science*, (ed. Peter Achinstein and Owen Hannaway), Boston: MIT Press-Bradford Books, pp. 147–173.

Laymon, R. (1987), 'Using Scott Domains to Explicate the Notions of Approximate and Idealized Data', *Philosophy of Science* **54**, 194–221.

Lorentz, H. A. (1887), 'De l'influence du mouvement de la terre sur les phénomènes lumineux', *Archives Néerlandaises* **21**, 103–172; reprinted in H. A. Lorentz, *Abhandlungen über Theoretische Physik*, vol. 1: Leipzig: Druck and Teubner, pp. 341–390.

Lorentz, H. A. (1921), 'The Michelson–Morley Experiment and the Dimensions of Moving Bodies', *Nature* **106**, 356–362.

Lorentz, H. A. (1927), *Lectures on Theoretical Physics Delivered at the University of Leiden*, trans. I. Silberstein and A. Trivelli, vol. 1, London: Macmillan.

Lorentz, H. A. (1952), *The Theory of Electrons*, New York: Dover.

Miller, D. C. (1933), 'The Ether-Drift Experiment and the Determination of the Absolute Motion of the Earth', *Rev. Modern Physics* **5**, 204–242.

Michelson, A. A. (1881), 'The Relative Motion of the Earth and the Luminiferous Ether', *Amer. J. Science*, 3rd ser. **22**, 120–129.

Michelson, A. A. (1882), 'Interference Phenomena in a New Form of Refractometer', *Amer. J. Science*, 3rd ser. **23**, 395–400.

Michelson, A. A. and Morley, E. W. (1887), 'On the Relative Motion of the Earth and the Luminiferous Ether', *Amer. J. Science*, 3rd ser. **34**, 333–345; also in *Philosophical Magazine*, 5th ser. **24**, 449–463.

Michelson, A. A. and Lorentz, H. A. *et al.* (1928), 'Conference on the Michelson–Morley Experiment', *Astrophysical J.* **68**, 341–402.

Morley, E. W. and Miller, D. C. (1904), 'Extract from a Letter dated Cleveland, Ohio, August 5th, 1904', *Philosophical Magazine*, 6th ser. **8**, 753–754.

Morley, E. W. and Miller, D. C. (1905a), 'Report of an Experiment to Detect the Fitzgerald-Lorentz Effect', *Proc. Amer. Academy of Arts and Sciences* **41**, 321–328.

Morley, E. W and Miller, D. C. (1905b), 'On the Theory of Experiments to Detect Aberrations of the Second Degree', *Philosophical Magazine*, 6th ser. **9**, 669–680.

Moulines C. and Sneed J. (1979), 'Suppes' Philosophy of Science', in *Patrick Suppes*, (ed. R. Bogdan), Dordrecht: D. Reidel, pp. 59–91.

Reichenbach, H. (1958), *The Philosophy of Space and Time*, trans. M. Reichenbach and J. Freund, New York: Dover.

Silberstein, L. (1914), *The Theory of Relativity*, London: Macmillan.

Sommerfeld, A. (1964), *Optics*, New York: Academic Press.

Suppes, P. (1962), 'Models of Data', in *Logic, Methodology and Philosophy of Science*, (ed. E. Nagel, P. Suppes, and A. Tarski), Stanford: Stanford University Press. pp. 252–261.

Swenson, L. S. (1972), *The Ethereal Aether: A History of the Michelson-Morley-Miller Aether-Drift Experiments, 1880–1930*, Austin: University of Texas Press.

Trouton, F. T. and Rankine, A. O. (1908), 'On the Electrical Resistance of Moving Matter', *Proc. Royal Society of London*, ser. A, **80**, 420–435.

4. RECENT GEOLOGICAL THEORY

HENRY FRANKEL*

PLATE TECTONICS AND INTER-THEORY RELATIONS

1. INTRODUCTION

The aim of this paper is to test (T1.1), (T2.1), (T2.2), and (T1.2) concerning the sorts of specific scientific theories scientists prefer and how they appraise specific theories by drawing upon the sixty year controversy over the reality of continental drift.

This controversy essentially began with the development of the first detailed theory of continental drift by Alfred Wegener in the early teens of this century and ended with the acceptance of sea-floor spreading and development of plate tectonics, the modern version of continental drift, in the late 1960s.

Before discussing the relevant aspects of the controversy over continental drift, I should like to make several introductory remarks about the methodology and nature of testing the particular theses under investigation. First, in testing these theses one not only needs to examine the sorts of problems that the specific theories in question solved, one should also examine the way in which theorists defend their theories and attack rival ones, if one is to gain a more complete understanding of the kinds of theories scientists prefer. Suppose, for example, that a scientist proposes and prefers a theory which contains a solution to a problem that none of the rival theories, despite attempts by their proponents, have been able to solve. Such an occurrence constitutes a positive instance of (T1.1), for there is a theory preferred by a scientist which solves at least one problem which proponents have failed to solve. However, it clearly does not follow that the scientist prefers the theory wholly or partly because it possesses such a problem-solution. But, if in defending his theory, he argues that a reason for preferring his theory is its possession of such a solution, then one has a stronger case in support of (T1.1). In addition, if proponents of the rival theories either attempt to show that the preferred solution is inadequate or continue attempting to develop their own solution to the problem, we have even better grounds for maintaining that we have strong evidence for (T1.1).

Second, it is worth examining whether confirmatory instances of one

269

thesis confirm other theses. For example, one might find that a solution to a problem was valued by proponents of a theory because (a) only that theory possessed a solution to the problem, (b) neither rival nor predecessor theories solved the problem, and (c) the theory was not designed to solve problem. In such circumstances, the situation would offer confirmatory instance of (T1.1), (T2.2) and (T1.2). Because most of the confirmatory instances of (T1.1), (T2.1), (T2.2), and (T1.2) in this controversy from the earth sciences confirm more than one thesis, I shall, besides isolating confirmatory (disconfirmatory) instances of each thesis, indicate whether a confirmatory instance of one thesis confirms other relevant theses and look at the question of whether any interesting correlational patterns emerge.

Third, all theses under consideration deal directly with why scientists prefer and how scientists appraise *individual scientific theories* rather than how scientists evaluate sets of *guiding assumptions*. Thus, I shall be primarily concerned with individual theories such as Alfred Wegener's theory of continental drift or Maurice Ewing's fixist theory, rather than the set of guiding assumptions shared by most theories of continental drift or fixism. But I shall classify particular theories in terms of being drift or fixist theories and sometimes refer to a particular solution or attack upon a solution as a drift or fixist solution or attack, since most of the scientists involved in the controversy over the reality of continental drift considered themselves to be either drifters or fixists. Drifters took as guiding assumptions the substantive claims that major landmasses had shifted their positions relative to one another throughout geological time by processes which did not involve *differential* vertical changes of one region of the earth's outer layer relative to another. Fixists were united in their *denial* of the claim that major horizontal shifts in the relative position of landmasses or ocean basins had occurred throughout geological time. They attributed major geological changes to differential vertical rather than horizontal movements. *An* additional reason why I shall pay some attention to the idea of guiding assumptions and distinguish between drift and fixist theories is that (T1.1), (T1.2), and (T1.3), at least given my reading, presuppose grouping individual theories in competing schools (be they Lakatosian research programmes or Laudanian research traditions) having different sets of guiding assumptions. Thesis (T1.1) addresses the question of how well an individual theory having one set of guiding assumptions compares with individual theories possessing a different set of guiding assumptions. Theses (T1.2) and

(T1.3) concern the question of how well an individual theory compares with previous theories having the same or a similar set of guiding assumptions.

Fourth, the issue of whether the theses in question find support in this or any case study depends, in part, upon what counts as a scientist preferring a theory. Does it require that a scientist accept the theory or does it have a broader interpretation and include instances where the scientist only pursues or just thinks the theory is interesting but is not in a position to pursue it? I shall interpret the notion of a scientist preferring a theory broadly.

Fifth, my account of the overall controversy is not meant to be anything approaching an exhaustive summary. Rather, I shall accent only some of those aspects of the controversy which have a direct bearing on the theses in question. Of course, this leaves me open to the charge that I have cooked the data. In defense, I shall simply refer the reader to detailed accounts of various aspects of the controversy offered by myself and others.

2. HIGHLIGHTS OF THE CONTROVERSY[1]

Alfred Wegener defended his theory of continental drift by arguing that it solved a number of empirical problems within the earth sciences. Some of the more important ones were as follows: (1) Why did the contours of the coastlines of eastern South American and Western Africa fit together so well, and why were there many similarities between the respective coastlines of North American and Europe? Wegener posited that the continents originally had been united and subsequently broke apart. (2) Why were there many examples of past and present-day life forms having a geographically disjunctive distribution?[2] Wegener argued that life forms had become disjunctively distributed through the separation of land areas with the breakup of the super continent, Pangea. He also brought up an objection to the fixists' landbridge solution to the problem. This solution solved the problem by positing the former existence of landbridges between continents having similar life forms. The landbridges served as migratory routes for the life forms. Once a landbridge had served as a migratory route, it disappeared by sinking into the ocean floor. Wegener argued that to propose that landbridges, which were supposed to be made up of continental material, could sink into the ocean floor was inconsistent with the well-founded principle of isostasy, since continental material was less dense than oceanic material. (3) Why were there Permo-Carboniferous glacial deposits found in South Africa, Argentina, southern Brazil, India and in western, central and eastern Australia during the Permian and Carboniferous periods?[3] Wegener's solution was to suppose that the respective continents had been united during the Permo-Carboniferous with the south pole located in southern Africa. (4) How do mountain ranges form? Wegener argued that mountain ranges form on the leading edge of drifting continents, and when moving continents collide. As the continent plowed its way through the sea floor, its leading edge would crumble and fold because of the resistence of the sea floor. His best applications were the formation of the mountain ranges extending up the western coast of the Americas and the Himalayas. Wegener also proposed several solutions to the problem of what forces could propel the continents through the sea floor. Wegener was never too sanguine about his solution to this theoretical problem, admitting that the Newton of drift theory had not yet arrived.

Wegener's theory was the first detailed specific theory of continental drift – the first of several *drift* theories to have as a guiding assumption the postulation of large-scale (horizontal) displacement of continents relative to each other. When he proposed his theory all other important global theories in the earth sciences shared the guiding assumption that major changes in the earth's surface were caused by vertical movements and the continents do not undergo major shifts in the latitude or longitude during their lifetimes. These *fixist* theories differed in many ways. For example, some fixists believed the former continents and landbridges had sunk into the ocean floor (*landbridgers*) while others believed in the permanency of continents and oceans (*permanentists*), but they shared in common the rejection of major longitudinal or latitudinal changes of particular continental masses.

Soon after Wegener's theory was proposed, it attracted some interest; however, within a few years it was severely attacked. The most important objection, or set of objections, raised by the fixists against the theory was what became known as the 'mechanism' objection. It amounted to the following: The forces postulated by Wegener to drive the continents through the sea floor are many times too weak. Moreover, there simply were no known forces of sufficient strength to cause such lengthy migrations of the continents. And, even if there were, the continents would not be able to survived such a voyage. They would break apart since continental material is weaker than oceanic material. Wegener defended his theory by attempting to answer various fixist objections. Other drifters such as Emile Argand, Alex du Toit, Arthur Holmes and Beno Gutenberg constructed their own specific theories of continental drift. Fixists also developed and amended problem-solutions to several problems that drifters had solved. By the late 1940s drifters were few in number and the controversy came to a standstill.

After World War II the fields of paleomagnetism and marine geology underwent a tremendous amount of growth. Work in these fields had more to do with the eventual acceptance of the modern theory of continental drift, plate tectonics, than any of the arguments used to support continental drift prior to World War II.

During the second half of the 1950s a number of paleomagnetists developed a way to determine the previous position of the continents. Although they often differed about specifics, many paleomagnetists argued that their work supported various theories of continental drift and gave paleogeographical reconstructions of the continents roughly consistent with those offered by Wegener and other drifters. Although former drifters and paleomagnetists argued in favor of drift because of the paleomagnetic support, non-drifters who were not paleomagnetists remained unconvinced. For the most part they took a wait-and-see attitude toward paleomagnetism, arguing that it was still in its infancy, was based upon a few questionable assumptions, relied upon an incomplete and perhaps unreliable data base, and required extremely complicated field and laboratory work.

Marine geologists and geophysicists, utilizing the techniques of exploratory geophysics, turned the sea floor from an area of ignorance into one of backyard familiarity. By the late 1950s and early 1960s they had discovered, among other things, the following: (a) That a system of ephemeral mid-ocean ridges ran through every ocean basin. These ridges were characterized by a central rift valley, shallow earthquakes, high heat-flow data, lack of sediment, a strong positive magnetic anomaly, and were made up of segments offset from each other by fracture zones. (b) That the sediment upon the ocean floor was, geologically speaking, young, and that the depth of the sediment was much less than expected if ocean basins were ancient. (c) That ocean trenches were often located at the edge of ocean basins, and were characterized by deep earthquakes, negative gravity anomalies and low heat flow data.

(d) That the eastern Pacific Basin off the western United States and
Canada was associated with a zebra pattern of magnetic anomalies
trending in a north-south direction and cut through by east-west trending fracture zones. The zebra patterns of 'high' and 'low' magnetic
anomalies between adjacent sets of fracture zones were found to match,
if relative horizontal displacement of the sea floor between fracture
zones was assumed, and the fracture zones appeared to be seismically
inactive.

These discoveries gave rise to a number of unsolved empirical problems, and the one receiving the most attention around 1960 concerned
the origin of oceanic ridges. Two of the more important solutions were
developed by Harry Hess and Maurice Ewing. Hess's solution (1960),
called sea-floor spreading, postulated that new sea floor is created along
ridge axes by material forced up from the mantle by rising convection
currents; that material spreads out perpendicularly from the axes creating new ocean basins; and that it subsequently sinks into the mantle at
oceanic trenches along the periphery of the basins.[4] His hypothesis
offered a solution to the problem of the origin of oceanic ridges, for it
could account for their median position, rift valley, high heat flow data,
shallow earthquakes, lack of sediments and ephemeral nature. Moreover, it explained the formation of trenches as well as the youth and
paucity of sediment upon the sea floor. Although Hess had not been a
proponent of any theory of continental drift prior to 1960, he grafted his
solution onto the basic idea of continental displacement and argued that
his theory of continental drift solved the mechanism difficulty which had
plagued other drift theories. According to Hess, the Americas separated
from Europe and Africa as the Atlantic Basin began to form. They did
not plow through the sea floor but moved passively atop convection
currents. Although Hess gave no paleographic reconstruction of the
continents, he appealed to paleomagnetic studies supportive of drifting
continents.

Maurice Ewing's rival theory of 1959 for the formation of ocean
ridges also employed convection currents. But, unlike Hess's theory,
Ewing's account restricted convection to the earth's mantle, did not
invoke the creation of new ocean basins, viewed ocean basins as ancient
and the continents as basically fixed. According to Ewing, rising convection currents caused the formation of the rift valley and shallow earthquakes, and the material which upwells through the rift covered any
surrounding sediment and helped to form the ridge. Ewing's fixist

theory accounted for the lack of sediment atop ridges, shallow earthquakes, rift valley and high heat-flow data. It did not account for the general paucity of sediments, but others explained this away by suggesting that the sediment had compacted and even metamorphized. Both Hess's drift theory and Ewing's fixist theory sparked the interest of researchers working on problems of marine geology. It is not that either of them were accepted by other earth scientists working in marine geology. But their hypotheses were discussed in the literature, and occasionally someone would relate his own work to one of the hypotheses.

During the next few years two important corollaries of Hess's sea floor spreading were proposed. One, developed in 1963 by F. Vine and D. Matthews and, independently by L. W. Morley, the Vine–Matthews–Morley hypothesis (hereafter, VMM) related sea-floor spreading and the idea of repeated reversals in polarity of the geomagnetic field.[5] According to VMM, new sea floor becomes magnetized in the direction of the existing geomagnetic field. Thus, given repeated reversals of the field and sea floor spreading, the sea floor will possess a zebra pattern of magnetic anomalies of alternating strips of normally magnetized (material magnetized in the direction of the present field) and reversely magnetized material, and the magnetic anomalies surround ridges and display a bilateral symmetry on both sides of the ridge. VMM solved the problem of why oceanic ridges were associated with a positive magnetic anomaly and also began to make sense out of the zebra pattern of magnetic anomalies in the Pacific Basin. The other corollary of sea-floor spreading was J. Tuzo Wilson's (1965) transform-fault hypothesis. Without going into detail, Wilson, who became a drifter in the early sixties, hypothesized the existence of a new kind of fault, the transform fault.[6] He argued that its existence should be expected if sea floor spreading occurred, but not if hypotheses like Ewing's fixist theory were correct. Wilson pointed out that if he were correct, the shallow earthquakes associated with the fracture zones which offset adjacent ridge segments should be confined to the portion of the fracture zone between ridge offsets, while if his analysis were incorrect, the earthquakes would not be so confined. He also provided a number of examples and suggested that his hypothesis explained why the lengthy fracture zones in the eastern Pacific were not associated with earthquakes.

Ewing's theory was also expanded to account for the large positive magnetic anomaly associated with oceanic ridges. Several workers at

Lamont–Doherty Geological Observatory (M. Ewing was then director of the obsevatory) proposed (1965) that upwelling basalt, which according to Ewing formed ocean ridges, was more susceptible to magnetization than the surrounding sea floor. This hypothesis accounted for the possitive magnetic anomaly. The Lamont researchers who developed this hypothesis for the origin of marine magnetic anomalies argued against VMM on the grounds that, although magnetic anomalies had been found upon the flanks and area surrounding the Mid-Atlantic Ridge, the revealed pattern was unlike the one predicted by VMM.

The magnetic group at Lamont gave up its view in the early spring of 1966 in favor of VMM and Hess's sea floor spreading. Once they obtained textbook quality support for VMM most of them switched within a few months, although Ewing took until 1967 to give his reluctant endorsement of Hess's sea floor spreading and VMM. What changed their minds, and the minds of others who were working on many of the same problems, were the following: *First*, at the time when VMM was proposed it faced two serious empirical difficulties. Both concerned the combined presence of a ridge and the predicted pattern of magnetic anomalies. The one place where there were the alternating strips of magnetic anomalies had not been associated with a ridge. Conversely, when VMM was proposed, none of the existing magnetic surveys which cut across ridges displayed the predicted pattern. It was not that they failed to reveal a pattern that could be interpreted in terms of VMM. They did, but they were far from clear cut, and, as already pointed out, the marine magnetics people at Lamont who favored Ewing's hypothesis thought they were not consistent wth VMM. In 1965, Vine and Wilson realized that the East Pacific Rise extended into the area of magnetic anomalies. In early 1966, Lamont obtained profiles from two ridges (the East Pacific Rise and the Reykjanes Ridge) which gave textbook confirmation of VMM. The magnetic anomaly patterns surrounding both ridges were precisely the type predicted by VMM: the bilateral symmetry and the magnetic anomalies they displayed were so clearly defined that the anomaly pattern could be matched, anomaly for anomaly, from one ridge profile to the next. Indeed, they found that the ratio of the change in widths of successive magnetic anomalies from each profile was the same for every profile. The matchups were that good! Thus, not only were the empirical difficulties removed, but the confirmatory evidence gave a decided advantage to VMM over the Lamont hypothesis, for the tremendous symmetry, linearity and anomaly matching from profile to profile remained unexplained by the latter hypothesis. *Second*, VMM faced a conceptual difficulty when it was first proposed, for in 1963 the idea of repeated geopolarity reversals was still dubious. By 1966 it had been solidly substantiated by two sets of workers studying lava flows and another group who found an even better record of the reversals through examination of sedimentary cores taken from the sea floor. Thus by 1966 the tie between VMM and the idea of geopolarity reversals had changed from a conceptual difficulty into an advantage. *Third*, Vine, and those at Lamont, were able to determine sea-floor spreading rates for many oceans, and the predicted rates were similar to those determined by previous proponents of continental drift. *Fourth*, seismologists at Lamont, aware of the marine magnetic support for VMM and sea-floor spreading, tested Wilson's idea of transform faults and found that the epicenters of earthquakes associated with the fracture zones connecting segments of the Mid-Atlantic Ridge were, as predicted by Wilson, confined to those portions of the fracture zones between ridge offsets. By 1967 just about everyone who knew and understood what had happened accepted sea-floor spreading. Plate tectonics was developed in 1967.

3. PRESENTATION AND DESCRIPTION OF DATA RELEVANT TO THESES

Confirmation of (T1.1)
Much of Wegener's early defense of his theory (1915) involved his showing that it could solve empirical problems that no fixist theory had been able to solve, and doing as such, his theory was worthy of serious consideration. For example, neither the problems about the Permo-Carboniferous ice cap nor the one about the coastline similarities had been solved by fixist theories. (The ice cap problem-solution also confirms (T1.2), since Wegener's theory was the first drift theory to solve the problem.) He also argued that a weighty reason for preferring his solution to problems of the geographically disjunctive distribution of past and present day life forms was that it, unlike fixist solutions to the problem, explained the unusual combination of Australian fauna. (Wegener's solution to the peculiar combination of Australian fauna also is a confirmatory instance of (T1.2) since no drift theory had offered a solution to the problem prior to the development of Wegener's

solution.) Fixists later (*circa* 1930) attacked Wegener's general solution to the problem of disjunctively distributed life forms by arguing that it could not account for the disjunctive distribution of similar life forms on opposite sides of the Pacific. Fixists, who were landbridgers, added that they had little difficulty explaining the distribution of such life forms. Unlike the magnetic anomaly pattern developed at Lamont–Doherty Geological Observatory, the 1966 discovery that two oceanic ridges were surrounded by a magnetic anomaly pattern (which VMM could explain), produced an instant empirical difficulty for the Lamont alternative. (This instance also confirms (T2.1), (T2.2) and (T1.2).

The 1966–67 confirmation of Wilson's idea of transform faults through the analysis of earthquakes on the Mid-Atlantic Ridge created an empirical difficulty for Ewing's fixist view about the origin of oceanic ridges. (This analysis of the earthquakes on the Mid-Atlantic Ridge also confirms (T2.2) since Hess's hypothesis of sea-floor spreading, although not designed to solve the problem of the confinement of earthquakes along fracture zones to those portions between ridge offsets, solved the problem with the addition of Wilson's corollary. It also confirms (T1.2) since the combined view of sea-floor spreading and Wilson's idea of transform faults solved problems which previous theories of continental drift could not solve.)

Confirmation of (T2.1)

Fixists (*circa* 1930) developed a solution to the problem of the origin of the Permo-Carboniferous ice cap. This also confirms (T1.2) since previous fixist theories, at least according to some of the participants, had not been able to solve the problem. Fixists (*circa* 1940) also developed a solution to the problem of life forms having a geographically disjunctive distribution that included an explanation for the puzzling combination of the Australian fauna. This also confirms (T1.2) since previous fixist theories had not been able to solve this aspect of the problem. The discovery of the magnetic profiles from the East Pacific Rise and the Reykjanes Ridge as well as the realization of an ocean ridge in the northeastern part of the Pacific Basin removed the empirical difficulties from VMM. (As pointed out above, these discoveries also confirm (T1.1), (T2.2) and (T1.2).) Two other confirmatory instances of (T2.1) deserve mention. Fixists criticized Wegener's solution to the origin of the Permo-Carboniferous ice cap on the grounds that the existence of glacial deposits in North America during the Permo-Carboniferous presented a counter-example to his solution. Wegener argued that the evidence in support of the supposed glacial deposits was unreliable, and he strengthened his argument by pointing out that the presence of such glaciers was inconsistent with the well-supported view that North America was extremely hot and dry during the Permo-Carboniferous. Alex du Toit made a similar sort of move (*circa* 1940) in response to an early and repeated criticism that the discovery of a particular kind of flora in Russia destroyed one of the most successful drift solutions to the problem of life forms having a geographically-disjunctive distribution by happily reporting that the Russian finds had been incorrectly identified. In both cases the scientist, rather than developing a new solution, saved the original solution by showing that the relevant observations which supported the counter-example were unreliable or mistaken.

Before moving to the next thesis, I should like to mention two other examples of how fixists got rid of apparent counter-examples in ways not quite captured by (T2.1). In both cases, the fixists got rid of the apparent counter-examples, or at least took out much of their sting, by simply showing that the data giving rise to the apparent counter-example were unreliable *without* developing a new (or preserving an old) solution. Some fixists argued that they did not have to worry about trying to account for the similarity in shape between the facing coast-lines of the Americas and Europe and Africa since the degree of similarity had been greatly overestimated by Wegener and other drift-ers. During the late 1950s and early 1960s many non-drifters who were not paleomagnetists argued that they did not have to develop a solution explaining the paleomagnetic data supportive of drift theories since the data were unreliable and quite incomplete for providing paleogeographic reconstructions. In these two cases, fixists were in an unhappy situation from which they could not escape by giving their own solution. If the data from paleomagnetism were reliable and complete and there were no conceptual problems with the analysis of the data, every fixist theory would have been in serious trouble, since even the possibility of giving a fixist interpretation of the data seemed remote. Thus, the best strategy open to fixists was to argue that the data were unreliable and, therefore, that the paleomagnetic difficulty faced by fixist theories was a pseudo-difficulty.

Confirmation of (T2.2)
Two confirming instances already have been cited: (a) the magnificent

magnetic profiles of 1966 supporting VMM supporting (T1.1), (T2.2) and (T1.2) and (b) the 1967 confirmation of Wilson's idea of transform faults (also a positive instance for (T1.1) and (T1.2). Both provided Hess's continental drift theory of sea-floor spreading with solutions to problems it had not been designed to solve. Another example is found in the ability of drift theories to solve the problem of why paleomagnetic studies indicated that the continents had once been united. No drift theory developed prior to the early 1950s had been designed to solve the problem since nobody had seriously thought about using paleomagnetism to test drift until the middle 1950s. Another confirmatory instance of (T2.2) and (T1.1) is found in Wegener's excitement about the purported ability of his theory to solve the problem of why geodetic studies supported a westward movement of Greenland. It turned out that the geodetic data, as even the experimentalists admitted, were extremely unreliable. These four positive instances for (T2.2) exhaust the support for (T2.2) from this case study. Thus every positive instance of (T2.2) also supports (T1.1) and (T1.2), and one also supports (T2.1).

Confirmation of (T1.2)
Wegener's solution to the problem of disjunctive distribution of life forms constitutes a positive instance of (T1.2) since previous theories of continental drift had not addressed the problem. Other confirming instances of (T1.2), which also confirm additional theses, are as follows: Wegener's solution to the ice cap problem and the problem of the unusual nature of the Australian fauna, the fixists' solution to the ice cap problem and the question of the Australian fauna, Wegener's explanation for the westward drift of Greenland based upon geodetic measurements, the pro-drift paleomagnetic studies of the 1950s and, by 1967, Hess's sea-floor spreading when combined with VMM and Wilson's idea of transform faults.

Disconfirmation of (T1.3)
Scientists might like new theories to solve all the problems solved by their predecessors plus new problems, but scientists do not always get what they want. Moreover, if they thought that this condition had to be satisfied by new theories, there would not be any new theories. Hess's theory of sea-floor spreading, even when supplemented with VMM and Wilson's idea of transform faults, did not include a solution to the origin of mountains, and it did not address many other problems concerned

with continental geology. The victory enjoyed by Hess's sea-floor spreading and its two corollaries was a victory at sea. Of course, sea-floor spreading hac at hand solutions to the problems concerned with the coastline similarities, the Permo-Carboniferous ice cap and the origin of life forms having a geographically disjunctive distribution once it was grafted onto continental drift. However, it did not have a solution to mountain building at the time it was accepted. Hess could not utilize Wegenerian-type theories since the continents did not plow through the sea floor, and Hess hac not developed one of his own.[7] Specific fixist theories often dealt with different groups of problems. Even those addressing global problems usually had a large regional component. The theorist typically concentrated upon problems from his region (or area of specialization) rather than problems from other regions of world (or areas of the earth sciences).[8]

4. SUMMARY AND DISCUSSION

Table I summarizes the confirmatory instances of (T1.1) through (T2.2) and illustrates which positive instances confirm more than one of the theses.

The number of positive instances of (T1.1), (T2.1) and (T1.2) in Table I is striking. Even this brief summary of the controversy over continental drift has unearthed eight positive instances of (T1.1) and (T2.1) and nine positive instances of (T1.2). This large number of confirmatory instances suggests that scientists prefer theories which solve empirical difficulties confronting rival theories (T1.1), turn counter-examples into solved problems, or, at least remove counter-examples (T2.1), and solve problems not solved by their predecessors (T1.2). Why do scientists prefer such theories during a scientific controversy? And why do scientists exploit the fact that the theories they prefer possess one or more of these characteristics? I first consider why scientists desire theories which satisfy (T1.1) and (T1.2).

When a scientist presents a new theory, he often has individuals from three groups to convince: Those who are in the same camp (prefer theories having the same or a similar set of guiding assumptions), those who are in a different camp (prefer theories having a very different set of guiding assumptions) and those who are undecided about which camp offers the best, or promise of having the best, theory.[9] In order to

TABLE I.

Particular theses confirmed

Confirmatory instance	Thesis confirmed			
	(T1.1) not solved by rivals	(T2.1) remove counter-example	(T2.2) not designed to solve	(T1.2) predecessors have not solved
1. Wegener's solution to coastline similarities	x			
2. Wegener's solution to the origin of the Permo-Carboniferous ice cap	x			x
3. Wegener's solution to the general problem of disjunctively distributed life forms				x
4. Wegener's solution to origin of Australia's peculiar combination of fauna – particular aspect of disjunctive distribution problem	x			x
5. Fixists' solution to the circum-Pacific aspect of the general problem of disjunctively distributed life forms	x			
6. Fixists' solution to the ice-cap problem [developed after Wegener's solution]		x		x
7. Fixists' attack upon Wegener's solution to the problem of the similarity of coastlines		x		
8. Wegener's removal of apparent counter-example raised by fixists to his ice-cap solution		x		
9. Du Toit's removal of apparent counter-example raised by fixists against drift's solution to (3)		x		
10. Fixists' solution to the origin of Australia's peculiar combination of fauna		x		x
11. Paleomagnetic studies during 1950's supportive of continental drift	x		x	x
12. Geodetic studies initially supporting westward drift to Greenland	x		x	x
13. Fixists' attack upon paleomagnetic studies supportive of continental drift		x		
14. Fixists' argument that geodetic evidence about Greenland's drift is unreliable		x		
15. Discoveries which confirmed Hess's sea-floor spreading and VMM	x	x	x	x
16. Discoveries confirming sea-floor spreading and Wilson's transform faults	x		x	x

convince members of his own camp that his new theory or problem-solution is better than his predecessors', he can engage in a strategy captured by (T1.2), namely, he can show that his theory solves a problem that none of his predecessors' theories have been able to solve. This strategy was used by Wegener in instances (2), (3), (4) and (12) of Table I, various fixists used it in (6) and (10), paleomagnetists and committed drifters engaged in such a strategy with (11), and the confirmation of VMM and Wilson's idea of transform faults in (15) and (16) showed that Hess's theory, when supplemented with VMM and Wilson's idea of transform faults, could solve problems that none of its predecessors could solve. This strategy is also important in getting scientists who are neutral to reassess their evaluation of the merits of the new theories coming out of the rival camps. It might not result in their changing their minds, but it could lead to their re-evaluating their own work or designing an experiment to test the new theory or problem-solution. In order to convince members of different camps, the scientist must show that his theory has an advantage over their best theory. One strategy which can be used to make such a point is for the scientist to argue that his theory or problem-solution satisfies (T1.1). This is exactly what Wegener attempted to do with (1), (2), (4) and (12) in Table I; fixists accomplished with (5); paleomagnetists gave drifters with (11); and the discoveries confirmatory of VMM and Wilson's idea of transform faults (15) and (16) provided for Hess's continental drift theory of sea floor spreading. Of course, this strategy is a way for a scientist to convince those who are neutral that his theory is preferable to the best theory offered by rival camps. However, those who are neutral might agree that the scientist's theory is preferable, but nevertheless, remain neutral.

Table I illustrates a strong correlation between the positive instances of (T1.1) and (T1.2). Of the eight positive instances of (T1.1), six also confirm (T1.2); of the nine positive instances of (T1.2), six also confirm (T1.1). Why is there such a high correlation? Again the answer lies in the fact that a scientist typically has three different groups to convince. If a scientist can present a theory or a problem-solution that satisfies both (T1.1) and (T1.2), he can direct his arguments in favor of his theory to members belonging to any of the groups. This is the way Wegener argued in (2), (4) and (12). It is basically how the paleomagnetic studies of the 1950s were used. Those who were drifters before the rise of paleomagnetism argued that the results of the paleomagnetic studies favored drift theories over fixist ones, and the paleomagnetists involved in such studies argued that they had found a valuable tool that drifters and non-drifters should consider when addressing paleogeographical questions. Proponents of sea-floor spreading engaged in such a strategy with the discoveries of (15) and (16). They argued that the sea-floor spreading version of continental drift was better than its predecessors and better than fixist theories such as the one developed by Maurice Ewing at Lamont.

What about (T2.1)? It has eight confirmatory instances in Table I. This suggests that scientists think, or at least thought during the controversy over continental drift, that it is important to remove apparent counter-examples facing their preferred theories and to point out how rival theories faced apparent counter-examples. The large number of confirmatory instances may also indicate that it is not too difficult for a scientist to remove apparent counter-examples from his theory or show how the theories of his rivals are plagued by apparent counter-examples. Perhaps something can be learned from examining the correlational pattern exhibited by the confirmatory instances of (T2.1) in Table I.

Of the eight confirmatory instances of (T2.1), five of them confirm only (T2.1), three of them also confirm (T1.2), and of these three, one of them also confirms (T1.1) and (T2.2). Instances (7), (8), (9), (13) and (14) constitute those positive instances of (T2.1) which do not confirm other theses. Thus, the scientists in question were able to remove the apparent counter-example without solving a problem. How was this done? In (7), (13) and (14), an apparent counter-example was raised against fixist theories because various theories of continental drift explained a set of data which fixists were unable to explain. In other words, drifters had produced problem-solutions that satisfied thesis (T1.1). Fixists responded to the situation by arguing that the data in question – the data explained by the drift solution — were unreliable and/or incomplete. They argued that drifters had solved a pseudo-problem and, therefore, that the inability of fixist theories in explaining the data was not a detriment. In the other two instances, (8) and (9), Wegener and du Toit used the same technique as employed by the fixists in (7), (13) and (14) to get rid of the apparent counter-example, namely, they argued that the data giving rise to the counter-example was unreliable and incomplete. However, the examples differ somewhat because the context differs. In (8) and (9) the two prominent drifters were attempting to save one of their problem-solutions from an attack of apparent counter-examples. They had a solution to defend; they defended it by undermining the empirical support for the apparent counter-example. In (7), (13) and (14) fixists had no solution at hand and thought it very improbable that they could develop one since the data were at such odds with their own theories: If continental coastlines on opposite sides of the Atlantic realy matched perfectly then the fixists were in trouble. If the geodetic data supportive of a westward movement of Greenland were reliable, fixists would find it difficult to explain how Greenland could have moved without agreeing to the drifting of at least one fairly large landmass. Likewise, fixists attacked the paleomagnetic studies in support of continental drift by attempting to find problems with data, in part because they knew that it would be difficult to explain the paleomagnetic results without adopting continental drift. But despite the contextual differences between the two sorts of examples, the scientists were able to remove, either permanently or temporarily, the apparent counter-example without developing a new problem-solution.

Table I indicates that there are fewer positive instances of (T2.2) than for any of the other theses. This might suggest it is less important for a scientist to have a theory which solves problems for which it was not designed than it is to have a theory which satisfies one of the other conditions. However, lack of many confirmatory instances of (T2.2) might simply indicate that it is harder for a scientist to have a theory which solves problems for which it was not designed than to develop a theory which turns apparent counter-examples into solved problems or solves problems that neither rival theories nor predecessor theories have solved. Table I also displays a correlation among confirmatory instances of (T2.2), (T1.1) and (T1.2), for all of the five confirmatory instances of (T2.2) also confirm (T1.1) and (T1.4). Is this correlation simply an accident? I do not think it is. Dismissing the many subtle analyzes of 'novel fact' that have been offered in the literature, scientists have a theory which solves a problem for which it was not designed when (a) they realize that an existing theory can solve a problem in some area where it has not been applied, (b) they become acquainted with new data that was not available until after the theory was developed, or (c) they make a prediction on the basis of their theory which is later confirmed with the gathering of the relevant data. In (a) and (b) situations, it is not surprising that positive instances of (T2.2) also confirm (T1.2). In (c) situations, situations where a scientists makes a prediction which is later confirmed, it is quite likely that neither rival theorists nor theorists in the same camp will have developed a problem-solution which also makes the same prediction, since there would be no reason to bother until and unless the prediction is confirmed. Thus in (c) situations, confirming instances of T2.2 probably will be confirming instances of (T1.1) and (T1.2).

NOTES

*I wish to thank the National Endowment for the Humanities and the National Science Foundation for supporting the research upon which this article is based.

[1] This account of the history of the controversy over continental drift is extremely brief. More extensive accounts may be found in Frankel (1987), Glen (1982), Hallam (1973), Marvin (1973) and Menard (1986).

[2] The problem of disjunctively distributed life forms is detailed in Frankel (1981 and 1984a).

[3] The role of the problem of accounting for the Permo-Carboniferous Ice Cap in the controversy over continental drift is detailed in Frankel (1984b).

[4] For a detailed discussion of the development of Hess's ideas and his hypothesis of sea floor spreading, see Frankel (1980) and Menard (1986).

[5] For detailed discussions of the development of the Vine–Matthews–Morley hypothesis, its reception and eventual acceptance, see Frankel (1982) and Glen (1982).

[6] For the details of how Wilson came up with the idea of transform faults see Laudan (1980) and Menard (1986).

[7] Mention, however, should be made of R. S. Deitz's work. Dietz also developed an hypothesis of sea-floor spreading. His account was published in 1961 and Dietz began working on the problem of mountain building soon after developing his own idea of sea-floor spreading. Thus one could argue that sea-floor spreading presented a general solution to the problem of mountain building around the time when it was accepted. Nevertheless, its possession of a solution to the problem of mountain building had nothing to do with why it was initially accepted. In addition, Arthur Holmes (*circa* 1944) also developed a solution to the mechanism problem facing continental drift which was quite similar to Hess's and Dietz's idea of sea-floor spreading which did offer an account of mountain building. For details, see Frankel (1978 and 1980) and Menard (1986).

[8] The regional component in the controversy over continental drift is discussed in Frankel (1984a) and Le Grand (1986). The importance of the specialization component in the drift controversy is discussed in Frankel (1976) and Hallam (1973).

[9]Of course, if a scientist is the first to propose a theory with a very different set of guiding assumptions than those theories with which it competes, then there are not any predecessor theories. The scientist, in such cases, will expend his efforts in converting those who are neutral and those who are committed to a rival theory having a very different set of guiding assumptions. This idea of how a scientist directs different sorts of arguments to scientists in the same camp and in different camps has been stressed by Nunan (1984).

REFERENCES

Frankel, H. (1976), 'Wegener and the Specialists', *Centaurus* **20**, 305–324.

Frankel, H. (1978), 'Arthur Holmes and Continental Drift', *British J. History of Science* **11**, 130–150.

Frankel, H. (1980), 'Hess's Sea-Floor Spreading Hypothesis', in T. Nickles (ed.), *Scientific Discovery: Case Studies*, Dordrecht: D. Reidel, pp. 345–366.

Frankel, H. (1981), 'The Paleobiogeographical Debate over the Problem of Disjunctively Distributed Life Forms', *Studies in History and Philosophy of Science* **12**, 211–259.

Frankel, H. (1982), 'The Development, Reception, and Acceptance of the Vine–Matthews–Morley Hypothesis', *Historical Studies in the Physical Sciences* **13**, 1–39.

Frankel, H. (1984a), 'Biogeography, Before and After the Rise of Sea-Floor Spreading', *Studies in History and Philosophy of Science* **15**, 141–168.

Frankel, H. (1984b), 'The Permo-Carboniferous Ice Cap and Continental Drift', in *Neuvième Congrès International de Stratigraphie et de Geologie du Carbonifere 1979, Compte Rendu, Vol. 1*, Southern Illinois University Press, Carbondale, Carbondale, Ill: Illinois University Press, Carbondale, Carbondale: Illinois University Press pp. 113–120.

Frankel, H. (1987), 'The Continental Drift Debate', in H. Tristram Engelhardt, Jr., and Arthur L. Caplan (eds.), *Scientific Controversies: Case Studies in the Resolution and Closure of Disputes in Science and Technology*, N.Y.: Cambridge University Press, pp. 203–248.

Glen, W. (1982), *The Road to Jaramillo*, Palo Alto: Stanford University Press, California.

Hallam, A. (1973), *A Revolution in the Earth Sciences: From Continental Drift to Plate Tectonics*, Oxford: Clarendon Press.

Laudan, R. (1980), 'The method of multiple working hypotheses and the development of plate tectonic theory', in T. Nickles (ed.), *Scientific Discovery: Case Studies*, Dordrecht: D. Reidel, pp. 331–343.

Le Grand, H. E. (1986), 'Specialties, Problems, and Localism: The Reception of Continental Drift in Australia 1920–1940', *Earth Sciences History* **5**, 84–94.

Marvin, U. B. (1973), *Continental Drift: The Evolution of a Concept*, Washington, D.C.: Smithsonian Institution Press.

Menard, H. W. (1986), *The Ocean of Truth: A personal history of global tectonics*, Princeton, NJ: Princeton University Press.

Nunan, R. (1984), 'Novel Facts, Bayesian Rationality, and the History of Continental Drift', *Studies in History and Philosophy of Science* **15**, 267–307.

RICHARD NUNAN

THE THEORY OF AN EXPANDING EARTH AND THE ACCEPTABILITY OF GUIDING ASSUMPTIONS[1]

Most scientists read as little as they can get away with anyway, and they do not like new theories in particular. New theories are hard work, and they are dangerous – it is dangerous to support them (might be wrong) and dangerous to oppose them (might be right). The best course is to ignore them until forced to face them. Even then, respect for the brevity of life and professional caution lead most scientists to wait until someone they trust, admire, or fear supports or opposes the theory. Then they get two for one – they can come out for or against without having to actually read it, and can do so in a crowd either way. This, in a nutshell, is how the plate-tectonics "revolution" took place. (Greene, 1984, p. 753)

1. INTRODUCTION

For over two decades plate tectonics has enjoyed a dominant position in the earth sciences. The intellectual grandchild of Alfred Wegener's theory of continental drift, plate tectonics is the now familiar explanation of continental displacement as the product of lateral movement of rigid lithospheric plates in which continental blocks are embedded. This theory is the direct descendant of Harry Hess's theory of sea-floor spreading (Hess, 1962), according to which mid-ocean ridges are a source of new crustal material upwelling from the earth's interior. In Hess's theory, this new oceanic crust is gradually transported away along crustal 'conveyor belts' driven by mantle convection currents, and ultimately dragged back down into the earth's mantle (subducted) at the trailing edges of these convection currents. Lateral drift of continents occurs as they ride piggy-back along these conveyor belts (see Figure 1).

Plate tectonics and Hess's sea-floor spreading theory emerged in response to a flood of new discoveries about ocean floors and remanent magnetism compiled in the 1950s and early 1960s. Collectively, this information ended the hegemony of the fixist and contractionist research traditions which had prevailed until that time. During those same years however, an alternative explanation of the evidence for continental displacement was also being developed: the hypothesis of global expansion (see e.g., Egyed, 1957; Carey, 1958; Heezen, 1960; Jordan, 1966).

289

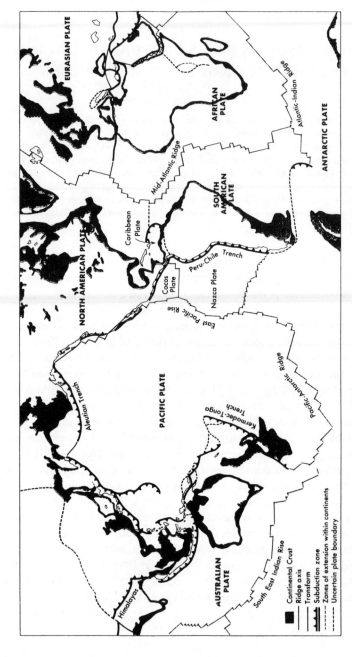

Fig. 1. The division of the earth into lithospheric plates. The direction of plate motion is away from mid-ocean ridges and rises, and towards roughly parallel oceanic trenches, e.g., the South American plate travels in a westerly direction from the mid-Atlantic ridge to the Peru-Chile trench. After Turekian (1976); by courtesy of *Scientific American* (Dewey, 1972).

Instead of treating continental displacement as the product of lateral movement across the earth's surface, expansionists think continental displacement is best explained, or at least partly explained, in terms of vertical movement. Expansion models treat the earth's crust as a contiguous series of rigid wedges extending into the earth's mantle. These blocks are being augmented at their boundaries by new crustal material emerging from the mantle at the mid-ocean ridges. On this point, expansionists and plate theorists are in agreement. But advocates of plate tectonics assume that oceanic trenches mark the locations of corresponding zones of crustal destruction, and that subduction at these sites proceeds at a rate equal to the rate of crustal creation at the oceanic ridges. Expansionists, on the other hand, subscribe to the view that subduction zones either do not exist at all (Heezen, 1960; Carey, 1976), or else the rate of subduction is significantly less than the rate of crustal creation (Owen, 1976; Steiner 1977). Consequently, the earth must be expanding at a rate sufficient to accommodate the net increase in new crustal material covering its surface. Continents then become displaced relative to one another because they are not also growing as the sea floor grows.[2] For those researchers who offer expansion as the primary explanation of continental displacement, the rate of expansion must be something in the neighborhood of a 20 to 25% increase in diameter over the past 200 million years.

I want to argue that this rivalry between plate tectonics and earth expansion undermines a particular account of rational theory choice, at least in a special class of cases. The account I have in mind is Larry Laudan's principal criterion of theory choice, the notion of 'problem-solving effectiveness' (Laudan, 1977, p. 119). On this view, rival theories are evaluated on the principle that "scientists prefer a theory which can solve the largest number of important empirical problems while generating the fewest important anomalies and conceptual difficulties" (T2.11). Laudan has proposed this conjunctive criterion as the chief standard for evaluating not only individual theories, but also the 'research traditions' (sets of guiding assumptions) of which they are members: "The *acceptability* of a research tradition is determined by the problem-solving effectiveness of its latest theories" (Laudan, 1977, p. 119).

This standard for assessing sets of guiding assumptions corresponds to (GA1.2), a fact which illustrates an important relationship between two of the thesis clusters under test (one not readily apparent from the manner in which the theses are organized). A number of the theses governing the appraisal of individual theories (Clusters (T1) and (T2)) have corresponding analogues among the theses governing appraisal of sets of guiding assumptions (Cluster (GA1)). In addition to the pair just noted, compare (T2.2) and (GA1.4), or (T1.2) and (GA1.3) (at least on some interpretations of the meaning of 'novel predictions'). These correlations emerge because, to some extent at least, the evaluation of research traditions is dependent on the relative success of their member theories. Consequently, (T1.1), (T1.2), (T2.1), (T2.2) have direct analogues in the evaluation of research traditions. Although they are not all explicitly listed in the first guiding assumption cluster, we can take them to be implicit in (GA1.2). For a theory's ability to solve empirical difficulties confronting its rivals (T1.1), to solve problems not solved by its predecessors (T1.2), to transform apparent counterexamples into solved problems (T2.1), and to solve problems it was not intended to solve (T2.2) all contribute to the overall problem-solving effectiveness of the theory referred to in (T2.11).[3] And the problem-solving effectiveness of rival research traditions (GA1.2) is then assessed in terms of a comparative evaluation of their latest member theories under (T2.11).

Now I think a number of these correlations between appraisal of individual theories and appraisal of rival research traditions are perfectly legitimate. But the evaluation of research traditions is a distinct enterprise, and the temptation to draw parallels with theory appraisal should be treated with some caution. In particular, I believe Laudan's criterion of problem-solving effectiveness works quite well as a model of rational theory choice within the confines of a single research tradition, but it does not function as an effective heuristic guide for choosing between rival research traditions. Within a particular research tradition, there is sufficient continuity among successive theories to generate a coherent sense of what would count as increased problem-solving effectiveness from one theory to the next. But for choices between research traditions, the determination of what are to count as *important* empirical problems, anomalies, and conceptual difficulties will often depend on who is doing the counting. Proponents of one research tradition will frequently weight problems differently from proponents of a rival research tradition. This problem is nicely illustrated by the debate over expansion and plate tectonics, two theories which I regard as representatives of distinct research traditions, and not merely as member theories within the same research tradition.[4] Expansionists regard certain problems afflicting plate tectonics as glaring anomalies or serious conceptual difficulties, while the opposing camp tends to trivialize the same problems. And of course this phenomenon works in reverse, too.

2. PROBLEM SOLVING EFFECTIVENESS AND EXPANSION

The story about the problem-solving effectiveness of plate tectonics is a fairly familiar one now, worked out in some detail by Henry Frankel elsewhere in this volume, where the analysis is conducted in terms of the theses governing theory appraisal discussed above. But how would expansionists do the counting differently? Before going into the details of this particular case, it is necessary to get a more detailed grasp of the measuring tool which Laudan is offering.

In Laudan's account of problem-solving effectiveness, a theory's ability to solve important empirical problems is weighed against two kinds of debits: empirical problems which remain unresolved (anomalies), and conceptual problems which the theory generates for itself. The distinction between empirical and conceptual problems is roughly this: empirical problems arise when a theory has to confront what we treat as 'facts about the world' (with the understanding that such locutions refer to observed regularities which are themselves 'theory-laden' [see Laudan, 1977, pp. 14–17]); conceptual problems, on the other hand, are characteristic of theories themselves. Were it not for the formulation of the theories, the conceptual problems would not exist, while empirical problems demand attention independently of the particular theories which attempt (or fail) to address them. For conceptual problems concern issues like internal coherence or clarity of a theory ('internal conceptual problems'), or questions about the theory's compatibility with other theories, already well-established in the scientific community ('external conceptual problems') (see, Laudan, 1977, pp. 48–54).

Now to the case. First of all, earth expansion, like plate tectonics, solves the major external conceptual problem associated with Alfred Wegener's original account of continental displacement: namely, how could the continents have ploughed through the sea floor without disintegrating as they crashed against the much denser rock which comprises the oceanic crust? Where plate tectonics circumvents this problem by embedding the continents inside of much larger crustal plates which are themselves being transported along convection currents in the earth's mantle, expansion deals with the problem by driving the continents apart through vertical expansion instead of lateral movement. Thinking of Wegener's theory as a common ancestor of the research traditions represented by earth expansion and plate tectonics, both theories could be credited with the ability to solve problems not solved by their predecessor (T1.2).

Another instantiation of (T1.2), but one which applies exclusively to earth expansion, emerged when S. Warren Carey attempted to reconstruct Wegener's two-dimensional model of Pangaea on a three-dimensional spherical surface. He found that, regardless of where on the globe he located the center of his maps, the extremities of his cartographic reconstructions were marred by ineliminable triangular gores (Carey, 1958; see Figure 2). The largest of these gores constitutes what has come to be known as the Tethys Ocean, a wedge of pre-Jurassic oceanic crust inferred to exist between the Eurasian landmass and Gondwana (the southern portion of Wegener's supercontinent, consisting of South America, Africa, Antarctica, Australia, and the Arabian and Indian sub-continents). Carey pointed out that these gores could be eliminated if Pangaea were reconstructed on a smaller globe, thus achieving a better fit among the continental shelves than Wegener could have hoped to provide (see Figure 3).

At this level of explanation, the issue of the triangular gores is an internal conceptual problem about 'goodness of fit' of the continental margins under mobilist hypotheses about the structure of the earth's crust. In making the claim that there is a need to explain the apparent similarities in the shapes of opposing continental margins, Wegener created along with his theory the hitherto nonexistent problem of determining how good a fit was necessary to warrant acceptance of particular reconstructions of continental movements. Because this is an internal conceptual problem, the evaluation of its severity is going to be at least somewhat subjective. Harold Jeffreys complained for years that there was a 15-degree misfit between South America and Africa under standard reconstructions of Pangaea. (See, e.g., Jeffreys (1929), p. 322, Jeffreys (1962), and the triangular gap between Africa and South America in Figure 2. From its apex to its terminus at the southern tip of South America, that gap is about 15 degrees long.) To Jeffreys this was a serious objection, but to confirmed drifters it was just so much carping

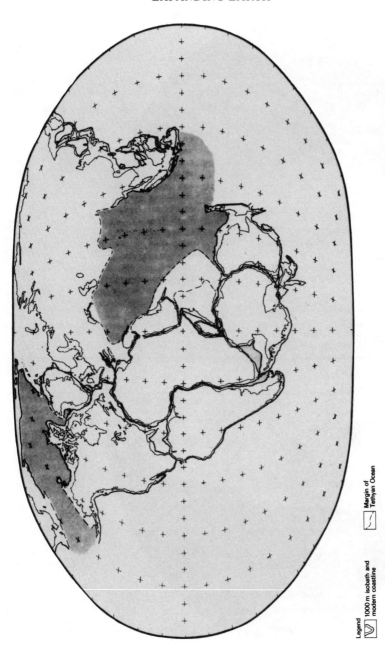

Fig. 2. Standard reconstruction of Pangaea about 180 million years ago, assuming an earth of modern dimensions. Oceanic crust required to be present in the reconstruction, but of which no evidence exists (Paleozoic Arctic and Tethys oceans, and the triangular gaps between Africa and South America, Africa and Antarctica), are shaded. By courtesy of the British Museum (Natural History) (Owen, 1981).

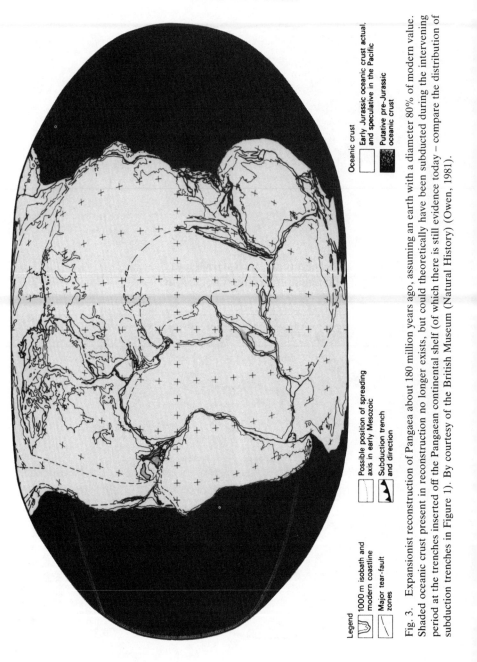

Legend

| 1000 m isobath and modern coastline | Possible position of spreading axis in early Mesozoic |
| Major tear-fault zones | Subduction trench and direction |

Oceanic crust

Early Jurassic oceanic crust actual, and speculative in the Pacific

Putative pre-Jurassic oceanic crust

Fig. 3.   Expansionist reconstruction of Pangaea about 180 million years ago, assuming an earth with a diameter 80% of modern value. Shaded oceanic crust present in reconstruction no longer exists, but could theoretically have been subducted during the intervening period at the trenches inserted off the Pangaean continental shelf (of which there is still evidence today – compare the distribution of subduction trenches in Figure 1). By courtesy of the British Museum (Natural History) (Owen, 1981).

on Jeffreys' part. By offering the model of an expanding earth as a substitute for lateral drift, Carey was able to provide a solution to this conceptual dispute. (See Figure 3, where the gap has disappeared entirely.) Because the conceptual problem afflicted a predecessor of earth expansion (Wegener's theory), Carey's achievement represents an instantiation of (T1.2).

Since the time Carey first demonstrated the topological virtues of earth expansion, the data collected about the ocean floor in the 1950s and 1960s has transformed the 'goodness of fit' issue from a conceptual problem into an empirical one. For in the mid-1970s Hugh Owen, a British paleontologist, argued that there is no evidence for the ancient oceanic crust which must have existed in these triangular gores if the earth's dimensions have remained constant. Consider, for example, Jeffreys' small triangular gap which remains between the South African and Argentinian plateaus on all constant dimensions reconstructions of Pangaea. Owen reasoned that, since Africa and South America started to break apart at the beginning of the Cretaceous, the gap, if it existed, would have had to consist of crustal material older than the Cretaceous (as contrasted with the younger oceanic crust we would expect to find in the newly created South Atlantic). But the oceanic crust just off the continental margins of South America and Africa is no more than 135 million years old, while the crust necessary to fill that triangular gap would have to be at least 200 million years old. Moreover, there are no subduction zones in the South Atlantic, so the missing crust could not have been swallowed up at some intervening date. It would have to still be present today (Owen, 1981, pp. 181–182). The fact that it is not creates a new anomaly for plate tectonics, but credits the expansionist research tradition with a new solved problem.

In other words, what was an example of (T1.2), earth expansion's solution of a conceptual problem confronting its intellectual ancestor, has since become an example of (T1.1): earth expansion's ability to solve an empirical difficulty which still confronts a contemporary rival. Moreover, as Henry Frankel points out in his contribution to this volume, positive instances of one of the theses under study may also be positive instances of another. That is the case with the empirical 'goodness of fit' problem, which instantiates (T2.2) as well as (T1.1). For Carey did not have access to any of the new oceanographic data when he proposed earth expansion as a solution to the problem of the triangular gores. Since Carey's focus was on the conceptual problem rather than the empirical one, the missing oceanic crust credits his theory with the triumph of solving a problem which it was not originally intended to solve.

Returning to conceptual issues, one thing that is statistically puzzling under plate tectonics and all earlier lateral drift theories is the fact that all the continents had to amass on one side of the globe 200 million years ago in order to produce Pangaea. But expansionists have a way to address this problem. Otto Hilgenberg, one of the original proponents of expansion (Hilgenberg, 1933), subscribed to the thesis that, prior to expansion, the Earth was entirely covered by a sialic shell which the earth's expanding interior has since split apart into the existing continental fragments. As our planet cooled and stabilized, the lighter sialic material rose to the surface and was evenly distributed across the earth's face. Subsequently, as this continental shell was rifted apart by expansionary forces within the earth (initially in the area that has become the Pacific Ocean), a morphologically distinct layer of crustal material (oceanic crust) was created to fill the expanding rifts. This hypothesis provides a statistically plausible account of past distributions of continental blocks. If the initial effect of expansion was to split the primitive continental shell along the boundary of what has since become the Pacific Ocean, that would explain how a hugh Triassic continent (Alfred Wegener's Pangaea) could "accrete" on one side of the globe. It didn't really accrete at all: the emergence of the Pacific *created* the other side of the globe, and Pangaea was the more or less intact residue of the original sialic shell.[5] Here again earth expansion represents a positive instance of (T1.2) relative to Wegener's theory.

To cite one final example of this research tradition's problem-solving capacities, consider the expansionists' argument that available evidence suggests that continental dispersion is a universal phenomenon. Advocates of plate tectonics believe that the fragmentation of Pangaea necessitated the simultaneous dispersal of the continental blocks created by that event. But these scientists would also insist that there must be points of convergence elsewhere (e.g., the Pacific Ocean). Expansionists, on the other hand, do not believe there are any points of convergence. The African plate, for example, is almost completely surrounded by oceanic rift zones, with no corresponding zones of subduction to swallow an equivalent amount of old crustal material. (see Figure 1). Plate theorists have dealt with this problem by holding the African plate fixed, and assuming that all other plates are diverging from that one. But the very same problem applies to the Antarctic plate, which is also surrounded by zones of crustal creation. Holding two distinct plates fixed on a spherical surface appears to be a topological impossibility. Moreover, Carey (1958) and Meservy (1969) argue that there is tectonic and paleobiographic evidence that the continental platforms bordering the Pacific have been simultaneously "drifting" apart from each other since the breakup of Pangaea (see Figure 4). In other words, the Pacific, too, has been expanding since the Paleozoic, in direct contradiction of standard models of lithospheric plate movements, which require contraction of the Pacific in order to accommodate the dispersal of other plates away from Africa. Meservy observes that the area marked off by the Pacific perimeter currently occupies less of the planet's surface than the area bounded by a great circle (i.e., half the surface of the globe). Consequently, the apparent expansion of the Pacific perimeter cannot be explained by arguing that the drift of the

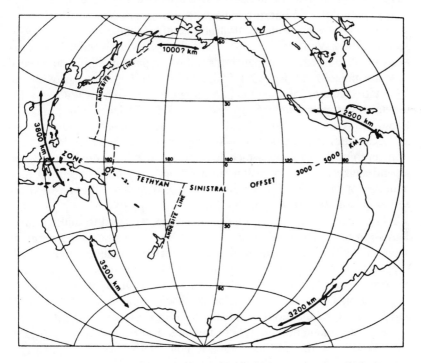

Fig. 4.   The present separation of points which, according to paleobiographic data and Carey's tectonic analysis, were formerly contiguous. Courtesy of Carey (1983).

continents toward the Pacific is just now passing over a great circle. Meservy also points out that this argument does not depend on any particular assumption about the rate of subduction. It is, in that sense, independent of theoretical assumptions associated exclusively with the expansion hypothesis. Consequently, if Carey and Meservy are right about the paleobiographic and tectonic evidence, then the evolution of the Pacific constitutes a genuine empirical problem confronting plate tectonics (T1.1), and not just a conceptual dispute between two opposing theories.

3. THE CASE AGAINST EXPANSION

Despite its claims of explanatory power, expansion was never a popular doctrine. By the late 1960s most earth scientists regarded it as a dead issue. Historians of plate tectonics who are themselves earth scientists working in the plate tectonic tradition (e.g., Hallam, 1973; Cox, 1973; Marvin, 1974; Van Andel, 1985) offer the following standard explanations for its demise: (1) expansion's failure to pass an empirical 'crucial test' formulated by Egyed (1960, 1961); (2) failure to provide an adequate account of the mechanism generating expansion; and (3) tectonic and paleobiographic evidence for a Paleozoic dispersal of Pangaea. The first and third of these represent claims that plate tectonics fares better than expansion with respect to (T1.1): expansion suffers from empirical problems which leave plate tectonics unaffected. The second suggests that expansion suffers from a serious conceptual flaw reminiscent of a similar problem afflicting Wegener's theory 50 years earlier (a counterinstance of (T1.2)).

To understand the first of these objections, we need to make a brief digression to explain paleomagnetism. This science studies the remnant magnetic field which exists in all sedimentary and igneous rocks which happen to contain ferro-magnetic particles. Igneous rocks, for example, are produced when molten material emerges from the earth's interior (e.g., through volcanic eruption or sea-floor spreading) and then cools and solidifies. As the molten material cools to a critical temperature, the Curie point, magnetic minerals in this material become magnetized in the direction of the earth's magnetic field at that location and time. Above that temperature the magnetic minerals have no magnetic properties, and below that temperature their magnetic properties remain unaltered. So long as the rocks have remained relatively undisturbed, this magnetic 'fingerprint' enables geologists to measure the direction and distance from those rocks to the Earth's magnetic poles as it was *at the time those rocks were formed.*

The relevance of paleomagnetism to earth expansion became apparent in 1960, when Egyed proposed a paleomagnetic test for expansion. The paleomagnetic inclination (the angle of the dip from horizontal which would be taken by a freely suspended magnetised needle) determines the paleolatitude of the rock exhibiting that remnant magnetic field. Assuming that continental displacement has occurred, the paleomagnetic inclination of a rock will not usually be identical to the modern magnetic inclination at the point in the earth's magnetic field where the rock is now located. That means the continental block in which the rock is found has changed its magnetic (and geographic) latitude over time. Consider now a *pair* of rocks (1) of the same age, (2) located at some distance from one another, and which (3) both happened to be created along the same paleomeridian, (4) on the same continental block, (5) in a region which has remained relatively free of tectonic disturbances ever since. If earth expansion has not occurred, the difference between the paleomagnetic inclinations of those two rocks should be approximately the same as the difference between the inclinations of the modern magnetic field at latitudes the same distance apart as the two rocks. But if the earth has been expanding, then the paleomagnetic fields of our pair of rocks were created the same distance apart on a smaller sphere. In other words, the rocks would have been created at paleolatitudes that were further apart. Consequently, the difference between the paleomagnetic inclinations of the two rocks would have been greater than the difference between modern magnetic inclinations at latitudes located as far apart as the distance between the two rocks. By combining the information about the extent of the difference with the age of the two rocks, a rate of expansion could be calculated.

Egyed's test, which has come to be known as the 'common meridian method', was first applied by Cox and Doell (1961) to rock pairs in eastern Siberia and western Europe. (It's important that the pairs be very far apart in order to minimize the effect of the large margin of error associated with paleomagnetic measurements.) They concluded that there was no evidence of significant earth expansion. A few months later Carey (1961) replied that their data was compromised by failure to meet one of the conditions of the test: that the region between the sites should have been relatively free of tectonic disturbances since the geologic period during which the rock samples were created. For according to Carey's own analysis the Asian block had been subject to elongation since the Paleozoic which would, of course, alter the original distance between the sample pair. Carey's final assessment of such tests is that it is extremely unlikely that *any* sample pairs sufficiently far apart on a single continental block will satisfy the requirement of no significant crustal disturbance (Carey, 1983, p. 388).[6]

More generally, Carey describes the current state of paleomagnetic technology as a "blunt tool", very useful for making certain gross measurements (e.g., rotation of continental blocks relative to other continental blocks, and large latitudinal changes of continental blocks), but not useful for the kind of refined measurements necessary to establish paleoradii (Carey, 1983, p. 387). He is not alone in this assessment. Ernst Deutsch, one of the pioneers of paleomagnetic research in the 1950s, said that none of the tests performed up to the mid-1960s (Cox and Doell, 1961; Ward, 1963; van Hilten, 1963) could be regarded as conclusive (Deutsch, 1965, p. 47). Nonetheless, the accumulated weight of subsequent applications of Egyed's test have helped to convince a number of earth scientists of expansion's empirical inadequacy (see, e.g., McElhinney, 1978).

But it is the second major objection to expansion, the one about the absence of a mechanism, which has probably been most influential within the scientific community. Even though advocates of plate tectonics acknowledge that a fully adequate account of the mechanism underlying lithospheric plate movements has yet to surface, there is general agreement among them that mantle convection currents are the source of plate motion. Despite the absence of any quantitatively precise model of the actual nature and effects of these convection currents, that general model constitutes a far less radical (i.e., more physically plausible) mechanism for explaining continental displacement than expansion on the scale necessary to account for that phenomenon. In Laudan's terminology, the absence of a plausible mechanism for expansion constitutes a serious *external conceptual problem*: the hypothesis conflicts with a number of well-entrenched theories in other scientific disciplines.

The third major objection to earth expansion, the evidence for the existence of a dismantled Pangaea during the Paleozoic, probably now carries almost as much weight among plate theorists as the mechanism objection. In the plate tectonics tradition, orogenesis is primarily a compressional phenomenon. Mountain ranges are normally created when the flat layers of sediment laid down at continental margins become crumpled once the continent has become aligned with the leading edge of a moving plate. Compressional crumpling occurs there for one of two reasons. If there is a subduction trench marking the plate boundary, the continental rock will be too light to be sucked down into the trench. Consequently, the lateral movement of the plate on which the continent rests will force those marginal sediments to pile up on themselves instead. The Andes along the western margin of South America are supposed to have been produced by this mechanism. The other type of compressional mountain building considered in plate tectonics is the product of continent to continent collisions, usually called 'suture zones', because they represent the merger of two continental blocks into a single unit. The Himalayas and the Tibetan Plateau, for example, are presumed to be the consequence of the northward drift of the Indian subcontinent along its plate, leading ultimately to a collision with Asia. (Note that the Tethys ocean separates the Indian subcontinent from the main Asian block in standard reconstructions of Pangaea [see Figure 2].)

Advocates of plate tectonics generalize on their compressional model of orogenesis, and use it as a tool for reconstructing Paleozoic plate movements. That is, they regard pre-Triassic mountain belts, such as the Urals and the Appalachians, as evidence of plate movements prior to the assembly of Pangaea. Since the Urals and the Appalachians slice across Laurasia (the northern hemisphere half of Pangaea), compressional orogenesis entails that Laurasia was previously subdivided into different continental blocks which collided to produce Laurasia and these particular mountain ranges. More generally, the entire set of Paleozoic orogenic belts get construed as a record of the Paleozoic continental collisions which ultimately produced Pangaea.

Morever, corroboratory evidence has been offered for this hypothesis. The most famous example is probably the disparate Ordovician faunal regions on opposite sides of the present day Atlantic. Distinctive species of marine fossils from that period have been found on opposite sides of the Atlantic, generally speaking. But some species found primarily on the North American side (known as the 'Pacific Province' faunal realm) are also found in Norway, Scotland, and the northern half of Ireland, while other species found mainly on the European/African side (known as 'Atlantic Province') are also found in southern

Newfoundland, Nova Scotia, coastal New England, and southern Florida. J. Tuzo Wilson (1966) suggested that this anomaly would be explained by the existence of a proto-Atlantic Ocean which separated these faunal realms during the Ordovician, subsequently closed up during the Devonian (the continental collision which produced the Appalachians), and then reopened as the modern Atlantic during the early Jurassic, but with bits of the older continents now left on opposite shores. There is similar faunal evidence for a Paleozoic ocean separating the Asiatic and European sides of the Urals during the Devonian (Hamilton, 1970).

Earth expansion, on the other hand, traditionally precludes direct collisions between crustal blocks: they are being simultaneously driven apart from one another. (Note that India is not detached from the rest of Asia in the expansionist reconstruction of Pangaea depicted in Figure 3. In the history of the subsequent break-up of Pangaea, India is assumed to remain attached to Asia, which splits apart from Australia, Antarctica, and Africa.) Consequently, expansionists need a very different explanation for the Paleozoic mountain building events which produced the Appalachians and the Urals. They argue that mountain building is primarily a diapiric rather than a compressional phenomenon, in which the upper layers of crustal material are rifted apart and pushed upwards by the intrusion of underlying material driven upwards by the expansionary forces within the earth (see Carey, 1958, pp. 322–339; Carey, 1976, pp. 54, 63–79). Thus, instead of regarding the Appalachians as the product of a suture zone between two Paleozoic continental blocks, Carey assumes that mountain belt is the site of an ancient rift valley (Carey, 1958, p. 338). Evidence of compressional folding, such as we find on a very large scale in the Alps and the Himalayas, can be attributed to the rotation of crustal blocks in response to the opening out of triangular wedges of oceanic crust produced by expansionary forces within the Earth (Carey, 1958, pp. 257–268, 333).

Expansionists have also attempted to deal with the paleobiographic evidence for a Paleozoic dispersal of Pangaea (e.g., the Atlantic and Pacific faunal provinces), but not very convincingly. Carey, for example, has argued that free faunal interchange should not automatically be expected to occur on an intact Paleozoic Pangaea:

The early Proterozoic lithosphere resembled a soccer football, broad stable epeiric basins separated by seams and welts where internal pressure strove to extrude, causing seismicity, high heat flux, vulcanism and rifts; some such rifts were little more than narrow plateau swells between adjacent basins, some were like rift valleys (Carey, 1983, p. 384).

Concerning VMM in particular, see Figure 5; note the symmetrical spreading pattern centered on the mid-Atlantic ridge.) While this series of events certainly aided the cause of mobilist accounts of the earth's structure, they are no help for choosing between plate tectonics and earth expansion. For both predictions are corollaries of the sea-floor *spreading* component of Hess's theory, the thesis that the sea-floor ridges and their environs were zones of crustal creation. Neither VMM nor the transform fault hypothesis depends on the existence of subduction zones, the opposite terminus of Hess's conveyor belt which distinguishes his model of continental displacement from earth expansion theories. Consequently, the confirmation of VMM and the transform fault hypothesis provide equally dramatic corroboration of earth expansion. They raised the stock of Hess's theory only with respect to fixist and contractionist alternatives (neither of which could explain the new data about the ocean floors), but not with respect to earth expansion.

Given this overall vision of the problem-solving effectiveness of expansion, can Laudan's criterion of theory appraisal explain why so many earth scientists opted instead for Hess's sea-floor spreading model, leading ultimately to their adherence to plate tectonics? I do not think so. For problem-solving effectiveness gets evaluated very differently from within the two traditions. Plate tectonics advocates might cite the truncated spreading pattern in the East Pacific as one example among a number of serious empirical problems for expansionists who reject subduction. (See Figure 5.) But Carey dismisses this objection as evidence only for the asymmetric growth of the Pacific. Expansionists take very seriously the triangular gores in Cenozoic maps of a constant dimensions earth, while plate tectonics advocates hardly even recognize their existence. The topological problem associated with the evidence for an expanding Pacific perimeter strikes expansionists as irrefutable evidence of an expanding earth, and the plate tectonics camp deals with this problem by ignoring it. On the other hand, the absence of a mechanism to explain the physical possibility of expansion has generated a response to that theory reminiscent of the opposition to Wegener's model of lateral drift 50 years earlier. But again, the weight attached to this problem will vary depending on who is making the assessment. Contemporary expansionists are inclined to put more weight on their empirical successes than on this conceptual failure. The proponents of lateral drift, now that they enjoy the luxury of having successfully formulated a mechanism (the convection current theory

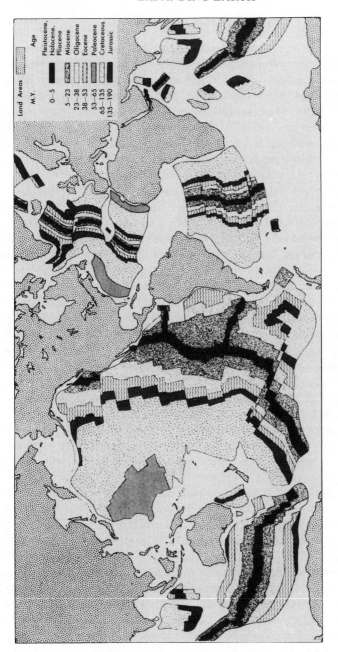

Fig. 5. Age distribution of oceanic crust based on magnetic anomaly patterns produced by the effect of polarity reversals in the earth's field, as new oceanic crust emerged at the mid-ocean ridges and was gradually displaced on either side of those locations. Compare with the distribution of ridges and trenches in Figure 1. (Blank areas indicate incomplete data.) After Turekian (1976); by courtesy of the Geological Society of America (Pitman, 1974).

associated with plate tectonics), are inclined to reverse the weightings.[7]

How then can scientists rationally choose between rival theories? When the focus is on comparisons between individual theories within a particular research tradition, (T2.11) is quite a plausible heuristic guide, because there is sufficient continuity within the tradition to generate a coherent sense of what would count as increased problem solving effectiveness from one theory to the next. In the lateral continental drift research tradition, for example, Arthur Holmes (1931) offered an account of a possible mechanism for drift which circumvented the most serious conceptual problem afflicting Wegener's version of drift: the fact that the continents didn't break apart as they 'ploughed through' the much harder basaltic rock of the sea floor. On Holmes' view, convection currents operating in the earth's mantle exert tensional force on the earth's crust, eventually pulling it apart where the convection currents ascend and sheer off laterally underneath the crust. Moreover, where the convection cells turn downward they exert metamorphic pressure on old oceanic rock, transforming it into heavy eclogite which is then subducted into the mantle. Thus the continents are no longer conceived as ploughing through the ocean floor. The stretching mechanism entails that the continents simply ride along inside a block of sea-floor rock, and the foundering eclogite provides room for the continuing advance of the continents.

Even Harold Jeffreys, whose physical geology text, *The Earth*, helped feed the negative reaction to Wegener's theory (he was responsible for the objection that the continents would break apart), grudgingly admitted that Holmes version of drift was at least possible (Jeffreys, 1931). Although Jeffreys did question the plausibility of such large-scale convection currents, here was a relatively hostile observer who clearly recognized an example of increased problem solving effectiveness in a research tradition which he did not accept. Similarly in the early 1960s, when Harry Hess offered his sea-floor spreading model of drift, earth scientists quickly recognized the increased problem-solving effectiveness of that model. For Holmes' model could not account for the ocean-floor age data emerging in the late 1950s and early 1960s. Sea-floor stretching (as distinct from sea-floor spreading) would not predict the symmetrical pattern of gradually increasing age of the oceanic crust on either side of oceanic ridge lines (see Figure 5). Where Holmes' drift theory was *conceptually* progressive relative to Wegener's, the new oceanographic and paleomagnetic discoveries rendered Hess's theory *empirically* progressive relative to Holmes' alternative.[8]

Here we have a series of three theories emerging from the same research tradition. The continuity between them is obvious, and consequently the locus of increased problem solving is obvious, too. Each theory preserves the major successes of its predecessors while alleviating some of the conceptual or empirical problems afflicting those earlier efforts. Consequently, (T2.11) does indeed seem to be a meaningful criterion for choosing between theories within a single research tradition. On some occasions problem-solving effectiveness is an effective method for distinguishing between acceptable and unacceptable research programs, too. But the instances where (GA1.2) appears to be confirmed are cases where the debate has already been settled, when problem-solving effectiveness has become so clearly weighted in favor of one alternative that the choice becomes easy for most participants in a given scientific discipline. This was surely true when earth scientists were confronted with mobilist and fixist alternatives in the early 1960s. But at that point fixist alternatives no longer counted as serious rivals to theories advocating continental displacement. The more difficult choices made between viable rivals, such as expansion and sea-floor spreading, tend to be based on rather subjective responses. The dimensions of theory appraisal discussed in this paper underdetermine the choice between research programs. If both rivals can reasonably claim to satisfy the criteria examined here, then the choice rests finally with each individual's assessments of the weight of such claims. Those assessments can be quite idiosyncratic.

The most interesting feature of this phenomenon, I think, is the fact that choices do get made. Researchers tend to commit their energy and resources to one research tradition, ignoring or denigrating serious rivals. But that fact still doesn't explain why the bulk of earth scientists linked their professional fortunes to plate tectonics rather than expansion. I suspect part of the explanation falls outside the scope of rational considerations, that mere familiarity breeds adherence: the familiarity of the *concept* of lateral drift for nearly half a century versus the relative novelty of expansion; the familiarity of compressional models of orogenesis, as contrasted with diapiric models; the familiarity of the leading scientists who chose to advocate lateral continental displacement (Arthur Holmes, Harry Hess, J. Tuzo Wilson, all very influential figures within the English-speaking scientific community), versus the relative obscurity (in that community) of Carey, Egyed, and certainly the earlier German expansionists like Hilgenberg, whose work in the 1930s was completely forgotten until Carey resurrected it. Geography may also have something to do with the explanation: many of the proponents of expansion work in regions relatively isolated from the professional community in North America and Western Europe (e.g., Australia and other Southern Hemisphere nations, the Soviet Union).

This is not to say that exclusive adherence to the plate tectonics research tradition is irrational. In terms of many dimensions of theory appraisal, plate tectonics had, and still has, a lot to offer (as Frankel here clearly demonstrates). If it did not, or if new evidence significantly reduced the problem-solving effectiveness of plate tectonics (so that, relative to expansion, it was comparable to fixist alternatives), most current advocates of plate tectonics would surely abandon it for the more progressive alternative research tradition. It's only when a good case can be made for the progressive problem solving capacities of both alternatives, so that adherence to either has some legitimate justification, that scientists will fall back on more blatantly subjective responses. Opting for the alternative which is in some sense more familiar is perhaps a *nonrational*, but surely not *irrational*, response to such circumstances.

I have been arguing that problem-solving effectiveness, Laudan's criterion of theory choice, is ineffective as a heuristic for choosing between theories belonging to rival research traditions. Laudan's ambition in offering this measure of theory appraisal was to circumvent Kuhn's charge of theory incommensurability between research traditions, and the theory choice anarchism which Lakatos expresses when he says "the pursuit of *any* research tradition – no matter how regressive it is – *can always be rational*" (Laudan, 1977, pp. 113–114). Laudan believes that even in a case like the debate between expansion and plate tectonics, where both research traditions apparently have a lot to offer, the choice between the two need not be made on nonrational grounds. The criterion of problem-solving effectiveness offers the promise of resolving the dispute by rational means.

But if my analysis of the reception of earth expansion is right, Kuhn's claims concerning incommensurability (Kuhn, 1970a, pp. 111–135, 144–159, 198–207; 1970b, pp. 259–277), and Lakatos's observations about the difficulty of determining precisely when it's rational to abandon a particular research tradition (Lakatos, 1970, pp. 154–159), reflect an ineliminable feature of theory choice. The kind of incommensurability revealed by this analysis is not Kuhn's meaning of incommensurability, but a kind of value incommensurability: scientists trained within the confines of a particular research tradition learn a system of weighting evidence and problems that may be irreconcilable with the weighting system associated with another research tradition. In my opinion that does *not* mean that the choice between rival research traditions is reduced to a purely irrational phenomenon, but it does mean that the precision implied by Laudan's criterion of theory appraisal is illusory.

NOTES

[1] I wish to thank the organizers and participants of this conference for helpful comments on an earlier draft of this paper. I am similarly indebted to my colleagues in the College of Charleston Philosophy Department, and to geologists Michael Bevis, Geoffrey Feiss, Kem Fronabarger, and Nancy West.

[2] There would be no evidence of displacement if all areas of the earth's surface were expanding uniformly. Think of drawing tiny 'continents' on the surface of a balloon and then blowing it up, in which case the regions representing continents would grow at the same rate as the regions representing sea floor, and no displacement would occur.

[3] Thesis (T1.3), that a 'good' theory solves *all* the problems solved by its predecessors (plus some new ones) would be difficult to prove, owing to the categorical nature of the claim. Because successive theories, even within the same research tradition, are not usually related by logical entailment (see Laudan, 1977, pp. 84–85), I suspect it is probably false in any case.

[4] One might object that, because plate tectonics and earth expansion both explain the earth's structure in terms of large scale and on going movements of the earth's crust, they are really member theories in a single 'mobilist' research tradition, as distinct from 'fixist' alternatives. (See Henry Frankel's paper in this volume for an explanation of the fixist research tradition.) But I have argued elsewhere (Nunan, 1984) that the concept of a research tradition is flexible, so that the partition of research traditions depends partially on the nature of the scientific debate at a particular point in time. If this were not the case, we might want to argue that fixism and mobilism were members of the same research program: uniformitarianism, as distinct from catastrophism (if *this* distinction was ever a meaningful way of partitioning geological research traditions; see Greene (1982) on this subject). While it is true that expansion and plate tectonics have a lot more in common with each other than either does with fixist rivals, there are nonetheless sufficiently significant differences between the two to warrant their classification as distinctive research traditions. Thus, for example, the thesis that the rate of subduction equals the rate of crustal creation is an axiom of the 'hard core' of plate tectonics (to borrow Lakatos'

terminology) which has been rejected by expansionists. Similarly, diapiric orogenesis models are characteristic of earth expansion, while plate tectonics regards orogenesis primarily as a compressional phenomenon. (See Section 3 below for a discussion of this issue.)

[5]H. G. Owen's expansionists reconstruction of the growth of the Pacific prior to the breakup of Pangaea (Figure 3) illustrates the appeal of the sialic shell hypothesis. The thin band of light shading dividing the eastern and western halves of the Pacific, and cutting across the northern and southern halves, represents a hypothetical location for the mid-ocean spreading ridge system which must have opened out the Pacific since the Proterozoic. Since new crust is being created there, the material just to either side of the spreading ridge would be early Jurassic, the geological period which this map is supposed to represent. The darker shading in the Pacific represents older crustal material, all of which would have been subducted during the last 200 million years, leaving no evidence of the pre-Mesozoic Pacific. That is the function of the subduction trenches Owen has inserted just off the Pangaean continental shelf at the Pacific's eastern and western perimeters. Prior to the existence of the Pacific spreading ridges (800 million years ago), the eastern and western extremities of Pangaea would have been joined on a globe 55% of the present size.

[6]Egyed (1961) subsequently developed a second test, known as the 'minimum scatter method', applications of which have usually yielded paleoradius estimates identical to the Earth's current radius. But Carey (1976), pp. 190–194, has argued that the standard method of applying this test is statistically biased in favor of such results.

[7]Ironically, this priority of weighting values was not always characteristic of the lateral drift movement. When Wegener and Du Toit were advocating continental drift in the Twenties and Thirties, the absence of a plausible mechanism mattered less to these scientists than what they perceived as the collective weight of empirical evidence favoring drift.

[8]For an account of the distinction between conceptually and empirically progressive research traditions, see Laudan (1977), pp. 14–8, 45–54, and 106–114. This distinction should not be confused with Lakatos's demarcation between theoretically and empirically progressive research programs (Lakatos, 1970, p. 118).

REFERENCES

Carey, S. W. (1958), 'A Tectonic Approach to Continental Drift', in S. W. Carey (ed.), *Continental Drift: A Symposium*, Hobart: University of Tasmania, pp. 177–355.

Carey, S. W. (1961), 'Palaeomagnetic Evidence Relevant to a Change in the Earth's Radius', *Nature* **190**, 36.

Carey, S. W. (1976), *The Expanding Earth* (Developments in Geotectonics 10), Amsterdam: Elsevier.

Carey, S. W. (1983), 'The Necessity for Earth Expansion', in S. W. Carey (ed.), *The Expanding Earth: A Symposium*, Hobart: University of Tasmania, pp. 377–396.

Cox, A. (1973), *Plate Tectonics and Geomagnetic Reversals*, San Francisco: W. H. Freeman.

Cox, A., and R. R. Doell (1961), 'Paleomagnetic Evidence Relevant to a Change in the Earth's Radius', *Nature* **189**, 45–47.

Deutsch, E. R. (1965), 'The Rock Magnetic Evidence for Continental Drift', in G. D. Garland (ed.), *Continental Drift*, Toronto: University of Toronto Press, pp. 28–52.

Dewey, J. O. (1972), 'Plate Tectonics', *Scientific American* **226** (May), 56–66.

DuToit, A. L. (1937), *Our Wandering Continents, An Hypothesis of Continental Drifting*, Edinburgh: Oliver and Boyd.

Egyed, L. (1957), 'A New Dynamic Conception of the Internal Constitution of the Earth', *Geologische Rundschau* **46**, 101–121.

Egyed, L. (1960), 'Some Remarks on Continental Drift', *Geofisica Pura e Applicata* **45**, 115–116.

Egyed, L. (1961), 'Paleomagnetism and the Ancient Radii of the Earth', *Nature* **190**, 1097–1098.

Greene, Mott T. (1982), *Geology in the Nineteenth Century: Changing Views of a Changing World*, Ithaca: Cornell University Press.

Greene, Mott T. (1984), 'Alfred Wegener', *Social Research* **51**, 739–761.

Hallam, A. (1973), *A Revolution in the Earth Sciences*, London: Oxford University Press.

Hamilton, W. (1970), 'The Uralides and the Motion of the Russian and Siberian Platforms', *Geological Society of America, Bulletin* **81**, 2553–2576.

Heezen, B. C. (1960), 'The Rift in the Ocean Floor', *Scientific American* **203** (October), 98–110.

Hess, H. H. (1962), 'History of the Ocean Basins', in A. E. J. Engel, *et. al. (Eds.)*, *Petrological Studies* – A Volume in Honor of A. F. Buddington, Geological Society of America, pp. 599–620.

Hilgenberg, O. C. (1933), *Vom Wachsenden Erdball*, Berlin, 1933.

Holmes, A. (1931), 'Problems of the Earth's Crust', *Geological Journal* **78**, 445 and 541.

Jeffreys, H. (1924), *The Earth*, London: Cambridge University Press.

Jeffreys, H. (1931), 'Problems of the Earth's Crust', *Geological Journal* **78**, 451.

Jeffreys, H. (1962), 'A Suggested Reconstruction of the Land Masses of the Earth as a Complete Crust – Comment', *Nature* **195**, 448.

Jordan, Pascual (1971), *The Expanding Earth: Some Consequences of Dirac's Gravitation Hypothesis*, Oxford: Pergamon Press. German edition originally published in 1966: *Die Expansion der Erde*, Braunschweig: Freidrich Vieweg.

Kuhn, T. S. (1970a), *The Structure of Scientific Revolutions*, 2nd edn., Chicago: University of Chicago Press.

Kuhn, T. S. (1970b), 'Reflections on my Critics', in I. Lakatos and A. Musgrave (eds.), *Criticism and the Growth of Knowledge*, London: Cambridge University Press, pp. 231–278.

Lakatos, I. (1970), 'Falsification and the Methodology of Scientific Research Programmes', in I. Lakatos and A. Musgrave (eds.), *Criticism and the Growth of Knowledge*, London: Cambridge University Press, pp. 91–196.

Laudan, L. (1977), *Progress and its Problems*, Berkeley: University of California Press.

Marvin, U. B. (1974), *Continental Drift: The Evolution of a Concept*, Washington: Smithsonian Press.

McElhinny, M. W., *et al.* (1978), 'Limits to the Expansion of Earth, Moon, Mars and Mercury and to Changes in the Gravitational Constant', *Nature* **271**, 316–321.

Meservy, R. (1969), Topological Inconsistency of Continental Drift on the Present-Sized Earth', *Science* **166**, 609–611.

Nunan, R. (1984), 'Novel Facts, Bayesian Rationality, and the History of Continental Drift', *Studies in History and Philosophy of Science* **15**, 267–307.

Owen, H. G. (1976), 'Continental Displacement and Expansion of the Earth During the Mesozoic and Cenozoic', *Proceedings and Philosophical Transactions of the Royal Society, Series A* **281**, 223–291.

Owen, H. G. (1981), 'Constant Dimensions or an Expanding Earth?', in L. R. M. Cocks (ed.), *The Evolving Earth*, London: British Museum and Cambridge University Press, pp. 179–192.

Pitman, W. C., *et al.* (1974), *The Age of the Ocean Basins*, Boulder: Geological Society of America.

Steiner, J. (1977), 'An Expanding Earth on the Basis of Sea-Floor Spreading and Subduction Rates', *Geology* **5**, 313–318.

Turekian, K. K. (1976), *Oceans, 2nd edn.*, Englewood Cliffs: Prentice-Hall.

Van Andel, T. H. (1985), *New Views on an Old Planet*, London: Cambridge University Press.

Van Hilten, D. (1963), 'Palaeomagnetic Indications of an Increase in the Earth's Radius', *Nature* **200**, 1277–1279.

Ward, M. A. (1963), 'On Detecting Changes in the Earth's Radius', *Geophysical Journal* **10**, 217–225.

Wegener, A. (1929), *Die Entstehung der Kontinente und Ozeane, Fourth Edition*, Braunschweig: F. Vieweg. The first edition appeared in 1915; the third edition (1922) was translated into English in 1924; the fourth (and final) edition was also translated: *The Origin of Continents and Oceans*, New York: Dover (1966).

Wilson, J. T. (1966), 'Did the Atlantic Close and then Reopen?', *Nature* **211**, 676–681.

5. 20th-CENTURY PHYSICS

JOHN M. NICHOLAS

PLANCK'S QUANTUM CRISIS AND SHIFTS IN GUIDING ASSUMPTIONS

1. INTRODUCTION

Why do scientists lose confidence in the fundamental doctrines of their sciences? It is my conviction that they regard them as gravely problematic when compelled to by alternatives which exhibit success or great promise of it, and that they are given up only when there are available alternatives. In this paper I will address the bearing of the emergence of the quantum theory out of the problems of late-19th-century physics for our understanding of how guiding assumptions of a science are called into question in the context of empirical failures. The three 'Canonical Theses' that will receive particular attention are (GA2.5), (GA2.7) and (GA3).

These three theses jointly (with a small qualification) constitute a doctrine which can be found, for example, in the account of scientific crisis which Thomas Kuhn has promoted in his *Structure of Scientific Revolutions* (Kuhn, 1970). Normal Science, which constrains scientists to an exclusive disciplinary matrix, including fundamental commitments like guiding assumptions, prohibits the intrusion of novel fundamental viewpoints until protracted problem solving failure forces a loosening of the constraints and promotes fundamental innovation. Such crisis is typically a prerequisite to the emergence of novel fundamental theory which, consequently, occurs only when guiding assumptions face profound difficulties.

This doctrine of empirical failures impinging on theoretical fundamentals is of considerable interest. It seems to fly in the face of plausible arguments, notably deriving from Pierre Duhem, that empirical discrepancies should not reasonably be able to provide means for discriminating between fundamentals and peripheral assumptions in the diagnosis of the failure. As I am convinced that the only reasonable basis for the solution to the Duhemian puzzle is to appeal to the success of novel fundamental doctrines for guidance in laying blame, the view opposed by Kuhn and rendered in the theses above takes on a particular salience. If, for fundamentals to be compellingly challenged by empirical

317

discrepancy, a role has to be found for their novel, prospective replacements, then there must be a way for the novelties to gain the light of day without crisis.

In what follows I shall briefly sketch: How the plausibility of the theses (and Kuhn's theory of crisis for the case study at hand) has hitherto derived in great part from a set of historiographical artefacts of a process dubbed by Paul Forman "defining the problem situation by projecting the solution backwards" (Forman, 1968, p. 150), and how the critical diagnosis of the failure of classical theory came in the second half of the decade after the invention of Planck's revolutionary thesis. I shall remark on some aspects of the emergence of the Planck hypothesis, and shall claim that the crisis and compelling diagnosis of failure for classical physics only developed after and in the light of the development of Planck's theory.

2. ARTIFACTS AND ANOMALIES

In a brief encapsulation of the discovery of quantum theory and a diagnosis of what problems had precipitated it, Planck wrote to R. W. Wood of Trinity College, Cambridge. What he had done, he reported:

can be described as simply an act of desperation. By nature I am peacefully inclined and reject all doubtful ventures. But by then I had been wrestling unsuccessfully for six years (since 1894) with the problem of equilibrium between radiation and matter and I knew that this problem was of fundamental importance to physics. I also knew the formula that expresses the energy distribution in normal spectra. A theoretical interpretation therefore had to be found at any cost, no matter how high. It was clear to me that classical physics could offer no solution to this problem and would have meant that all energy would transfer from matter to radiation. In order to prevent this, a new constant is required to assure that energy does not disintegrate. But the only way to recognize how this can be done is to start from a definite point of view. This approach was opened to me by maintaining the two laws of thermodynamics . . . they must, it seems to me, be upheld in all circumstances. For the rest, I was ready to sacrifice every one of my previous convictions about physical laws. Boltzmann had explained how thermodynamic equilibrium is established by means of a statistical equilibrium, and if such an approach is applied to the equilibrium between matter and radiation, one finds that the continuous loss of energy into radiation can be prevented by assuming that energy is forced, at the onset, to remain together in certain quanta. This was a purely formal assumption and I really did not give it much thought except that no matter what the cost, I must bring about a positive result. (Quoted in Hermann, 1971, p. 23.)

The picture Planck here painted is very reminiscent of conventional textbook and grapevine accounts of the collapse of classical physics, but although it gives an important insight into Planck's perception of the crisis and especially of the quantum postulate as a formal constraint less important than the move into the combinatorial conception of entropy, it reinforces the dangers of hindsight vision on the events to which even the participating scientists themselves fall prey. For Planck described his motivation as the escape from what was to become the ultraviolet catastrophe for the equipartition theorem of classical physics. Planck had, however, rewritten his own story since the equipartition theorem, and the problem of dissipation of energy into high frequencies did not figure in his development of the energy element theory of black body radiation (although they might well have been prominent in the mind of Wood whose enquiry would have been informed by a more characteristically British set of preoccupations which only surfaced widely in the international forum in the second half of the decade after Planck's discovery). In fact, Planck was moved instead by what might be called an 'infrared catastrophe' to adopt the Boltzmannian strategy which led him to his historic invention.

The creation of historical artifacts such as Planck's linkage of the motivation for the quantum theory with equipartition problems and the failure at the ultraviolet frequencies presents a pervasive problem for forming an accurate image of revolutionary transformations. It very often suggests, I believe, an illusory conformity of the transformations to the three theses given above. This process of 'defining the problem configuration by projecting the solution backwards', whether it involves the influence of textbooks, grapevine histories within the scientific profession, or even the recollections of the creative scientists themselves, as in this case of Planck, contributes significantly to the plausibility of the view that substantial revisions of theory at a fundamental level are driven by long standing, merely empirical discrepancies, in contexts which reflect a grave reluctance for fundamental theoretical innovation.

Consider this representative doctrine from *The Structure of Scientific Revolutions*:

> . . . quantum mechanics was born from a variety of difficulties surrounding black body radiation, specific heats and the photo-electric effect . . . the awareness of anomaly had lasted so long and penetrated so deep that one can appropriately describe the field affected by it as in a state of growing crisis (Kuhn, 1970, p. 86).

Can we realistically take these three classes of problems to have had the role described in our three theses? That is to say, did the research context involve appeal to only one set of guiding assumptions?; did innovation at fundamental levels come only after prolonged perception of the critical character of the problems?; was prolonged perception of anomaly sufficient for the recognition of the failure of guiding assumptions? Overall, I think the answer is essentially in the negative. Let us consider first the black-body radiation problem.

Black-Body Radiation: The radiation problem was not one of the major anomalous problems of physics at the turn of the century. Planck himself was in all probability attracted to the problem not on account of its being the driving anomaly in that domain of physics, but rather on account of the extraordinary physical significance of the fact that the properties of equilibrium radiation, cavity radiation in its experimental approximation, were independent of the particular materials of the enclosure walls with which radiation was put into equilibrium. This remarkable characteristic was immensely attractive to a specialist in thermodynamics of Planck's stripe. Planck's major intent was to find an electromagnetic basis for the irreversibility of thermal processes as they were presented by the phenomenological thermodynamics which he had adopted from Clausius. Thermodynamics offered the most promising theory of principle in physics at that time, perhaps almost universally applicable but minimizing the need for special assumptions about the micro-constituents of matter. Appeal to such hypotheses appeared premature if not gratuitous to many physicists and physical chemists. The independence of the radiative properties of matter from the particular constituents of the systems involved must have caught that sense of 'the Absolute' which Planck ascribed to himself when he looked back on those years (Klein, 1962, p. 460). Martin Klein provides us with a nice encapsulation of the Planck program, begun in the middle 1890s:

The ultimate goal of the programme was the explanation of irreversibility for strictly conservative systems, with no appeal to statistical considerations, and a valuable by-product would be the determination of the spectral distribution of black-body radiation (Klein, 1970, p. 220).

Planck began his program by attempting to exploit radiative scattering as the basis of equilibrium in the 'black-body' cavity. From March 1895 through to February 1897, Planck developed his conception through to the exploitation of radiation damping without conversion to heat, thereby promising to explain the irreversibility of thermal processes on the basis of conservative mechanisms. However, by 1897 this approach had a major set-back in that Boltzmann pressed home to Planck that the equations of electrodynamics are invariant under time-reversal, that is, electrodynamics is no better off than rational mechanics in guaranteeing irreversibility. By 1898 Planck had made the first major concession to the generally statistical approach to the thermodynamic problems he was attacking. By analogy with the assumption Boltzmann had made concerning the 'molecular chaos', Planck proposed a randomly constituted 'natural radiation'. By November, 1899, Planck defined expressions for the entropy of an oscillating dipole, and established a basic function for the spectral distribution, in his attempts to demonstrate the irreversibility of the tendency of a system of such oscillators in radiation towards equilibrium and to generate Wien's law. This success and the recognition of the progress of the research program certainly gave increased prominence to the problem. For example, Joseph Larmor, in his 1900 British Association for the Advancement of Science report, identified the investigation as "perhaps the most fundamental and interesting problem now outstanding in the general theory of the relation of radiation to temperature" (Larmor, 1900, p. 221). Already, however, there was a suspicion that the Wien law did, in fact, break down under some circumstances. This was of special concern to Planck at this time, since he had come to the conclusion that the Wien law and the second law of thermodynamics shared the same scope of validity, although he had severed this connection by the time he had submitted his next paper in the series, in March 1900.

By October of that year the threat of the breakdown of Wien's law for the spectral distribution had been realized. Clear-cut exceptions to the law were found at long wave lengths, and Planck now found himself compromised not only with having a half-baked statistical approach to the question of radiation equilibrium through the introduction of the essentially statistical assumption of 'natural radiation', but also having lost Wien's law as a fixed point of reference for his program of rationalizing irreversibility in these situations. Planck was reduced at this stage to constructing arbitrary, *ad hoc* expressions in a disorganized search for some guidelines to follow.

Employing an interpolation technique, Planck generated a law which appeared to provide an accurate characterization of the energy spectrum throughout all the wave lengths in question (Planck, 1900, p. 202). It was this law, the now well-known Planck Distribution Law, to which he referred in the letter to Wood when he said that he "knew the formula that expresses the energy distribution in normal spectra. A theoretical interpretation . . . had to be found at all cost" (Quoted in Hermann, 1971, p. 23).

The theoretical interpretation to which Planck appealed was one which must have, initially, caused him some considerable pain. For having fought all along to preserve the 'classical' thermodynamical interpretation of the second law, he was now forced to appeal to the Boltzmannian point of view, importing in particular Boltzmann's probabilistic interpretation of the entropy function, $s \propto \log W$. As a consequence, although Planck could claim, as he did in the Wood letter, that he had preserved 'the two laws of thermodynamics', in fact, he had only been able to keep the second law at the expense of the introduction of the statistical interpretation which had been an anathema for him at the outset. There were rewards for this painful transition apart from grounding his new distribution law. For one thing, his successful treatment of the radiation problem had seemed to permit a determination of the Avogadro number.

There was of course a catch in all this. But if we are to believe Planck in this respect it was not too serious a problem; it was only a question of "a purely formal assumption" to which he "really did not give . . . much thought" as the Wood letter put it. In partially imitating the approach which Boltzmann had used in his 1877 memoir, Planck had exploited a device for treating as discontinuous the elements constituting possibly continuous energy distributions. Boltzmann had been able to apply combinatorial methods of a relatively simple kind, by treating a continuous distribution as discontinuous, and after the appropriate manipulations, had generated the continuous case by taking the mathematical limit of the small discrete quantities involved. The method was essentially a formal device which had originally been invoked in order to treat continuous vibrating string problems as problems in point particle mechanics. The singular aspect of its use by Planck was that the limiting process at the end of the combinatorial manipulation was no longer needed. For, assuming these terms to depend in a simple way upon oscillator frequency, the entropy function for the oscillator emerged as that required to generate the Planck radiation law (Planck, 1900, p. 273ff.). In this way the quantum hypothesis was introduced as a constraint on the constitution of the energy of the oscillator. Planck did not give it 'much thought', regarding it as 'purely formal assumption'.

In scrutinizing the Planck case for its relevance to the theses, we should attempt to be clear about what parts of the theoretical context we take to have the role of the guiding assumptions. With hindsight we can say that a Principle of Continuity was surrendered with the importation of quantum ideas into the account of energy exchange and transmission. However, in the debates about the adequacy of conventional theories which provided the context from which the quantum theory sprang, Continuity was not the issue. There was no set of empirical difficulties which faced explicit applications of Continuity. It was only when the quantum theory was on the table as a plausible candidate for the solution to some puzzles that the possibility of Discontinuity appeared as a significant option. The theory which was eventually adopted turned out to be the arbiter of what the original difficulties perhaps should have been!

Even if the problems which confronted conventional theories prior to the emergence of the quantum theory did not point to issues of Continuity, there were partial diagnoses of the recognized discrepancies between theory and evidence. In Planck's own work, the failure of the Wien radiation law at lower frequencies, which provoked the steps to the quantum (he had an infrared rather than ultraviolet 'catastrophe'), left him without a clear foundation for reworking his hitherto successful synthesis of electromagnetic and thermodynamic devices and with a temporary worry that the Second Law had perhaps been challenged. It is noteworthy, in reflecting on the applicability of (GA2.5), that the consensus that Wien's law failed at the lower frequencies formed in the middle months of 1900, Planck's substitute radiation law appeared in October of that year, Planck defected even further into the Boltzmannian techniques, and he relinquished his austere conception of the second law of thermodynamics in favour of the combinatorial representation of entropy over the next few weeks. At the end of a mere few months after the recognition of the anomaly he had formulated the quantum theory. Planck's period of anomaly, as it were, was relatively brief, certainly by the standards of the specific heat anomalies which had faced gas theory for forty years and more, and, looking further afield, of classic anomalies in gravitational theory which lasted for many decades without material loss of confidence in the advocates of the respective theories.

One difficulty in regarding this move of Planck's into statistical entropy as an exception to (GA2.5), broadly understood, where persistence of the anomaly seems not to be particularly lengthy, is that Planck was not working in a vacuum shielded from the alternative research strategies which Boltzmann's programme offered. It was perfectly clear to Planck that Boltzmann's rival approach to entropy was available from the outset (indeed, he had already taken a step towards it in 1897 with the introduction of randomness in the radiation modelled on Boltzmann's 'molecular chaos'). What would have happened in the absence of a rival theoretical problem-solving strategy is not at all obvious, and the uncertainty in the evaluation of (GA2.5) reflects the ambiguity of the thesis with respect to whether it is supposed to apply regardless of the availability of rival theoretical approaches. This seems crucially important if one is concerned with the role which success with the rival theory might have in determining how empirical difficulties are to be mapped into the allocation of blame in the traditional theories. Taken strictly to the letter, however, (GA2.5) is disqualified because, in Planck's own work, the guiding assumption of his very rigid interpretation of the Second Law of Thermodynamics was set aside without there being a long period of recognized anomaly.

The presence of Boltzmann's alternative theoretical orientation has clear relevance for some versions of (GA2.7) and (GA3). The first part of (GA2.7) and (GA3) are refuted, if we take 'scientists' collectively rather than individually, for the situation Planck faced was not a Kuhnian monotheoretic edifice but a decisively pluralistic one. Well prior to Planck's work, 'scientists' had developed a rival set of guiding assumptions to those of a strictly phenomenological thermodynamics. It did not take mere anomalies to generate innovative new approaches.

To a certain extent, the Planck case is ambiguous as to which elements of the situation should be taken to be the novel guiding assumptions. If we attend to the issue of statistical versus non-statistical conceptions of entropy, then we may say that Planck was strongly motivated (albeit very rapidly, *contra* (GA2.5), by an empirical discrepancy to move to the Boltzmannian camp, even if he did not develop the new approach himself. On the other hand, if we think of the quantum constraint on the oscillators in the cavity wall as the thin end of the wedge of a new set of guiding assumptions involving discontinuity and in the long run the relinquishing of Hamilton's equations, we see Planck developing his own new approach. ((GA3) would then be illustrated). However, considerable doubt has been raised as to whether Planck really saw that quantum step as more than a mere artifice which happened to be the locus for the introduction of a new physical constant (Kuhn, 1978).

Perhaps the case involves a serendipitous and unintended development of new guiding assumptions through Planck's infelicitous mimicking of Boltzmann's procedures. It would seem reasonable, however, to interpret (GA2.7) and (GA3) as relevant to intended rather than accidental innovations.

Specific Heats: Although, as the Quantum Revolution proceeded, the Principle of Continuity was eventually discerned to be a major guiding assumption which had to be replaced, it was through the eyeglass of the discontinuous quantum theory itself that the issue was identified. At a level rather less surprising than that of the grand Principle of Continuity, the quantum theory also identified the failure of the Equipartition Theorem of statistical mechanics in crucial applications which had thrown up seeming discrepancies, and these had received much attention prior to the turn of the century. Before the quantum theory emerged and explicitly branded Equipartition, the theorem was vigorously debated and the failure of its implications counted as one of two "Nineteenth Century Clouds over the Dynamical Theory of Heat and Light", as Lord Kelvin characterized it (Kelvin, 1970, p. 358). To what extent was the Equipartition Theorem really a parochial problem? Was it a guiding assumption or at least did it entrain a guiding assumption of dynamical physics prior to the development of quantum theory? It seems to me to not be such a guiding assumption, although one leading protagonist in the debates in question (Rayleigh) deemed its failure to point to the rejection of some methods of classical mechanics. Let us look more carefully at the collapse of Equipartition and what it betokened.

Discrepancies for the gas theory had been recognized from its first mature appearance, and so by the time Planck was producing the quantum innovation, these empirical problems had been about in one form or another for forty years (Brush, 1965; Garber, 1978; Pais, 1982). Maxwell had been so impressed with the difficulties that he was skeptical of the gas theory, although that did not stop him working on it and making major contributions. In 1875 he had identified as 'the greatest difficulty yet encountered by the molecular theory' the tensions between assigning large numbers of degrees of freedom to atoms and molecules in order to account for spectroscopic phenomena on the one hand, and few degrees of freedom to address specific heat phenomena on the other. He had also anticipated problems with treating atoms as perfectly elastic solids which would swallow up available energy in high frequency vibrations, which prefigured later preoccupations of Kelvin and Rayleigh with such foundational difficulties for the elastic solid model. In 1876 Equipartition rapidly scored a considerable success for the specific heats of monatomic mercury vapour, and Boltzmann added a theoretical model of diatoms as rigid dumbbells which promised more than a little success. However, the anomalies were pervasive and some fairly radical proposals were made in order to cope with them, including, from Boltzmann, the possibility that molecules and the ether were not in thermal equilibrium.

Gas theory then faced serious anomalies from its inception, although there was significant progress on many fronts. By the turn of the century, these difficulties had been recognized in one form or another for more than a few decades. The difficulties were not perceived by its advocates fundamentally to undermine the promise of the programme. In part this was because there was not universal consensus on the applicability of the Equipartition theorem. Kelvin, for celebrated example, had been skeptical of the derivation of the theorem from early on, and in the last fifteen years of the century had played a prominent role in the vigorous debates in English journals about the theorem. When Kelvin described his Equipartition 'cloud over 19th-century physics,' he was not a proponent of the theorem bruised by the prolonged intractability of the empirical puzzles. With the longstanding scope of his profound doubts about the cogency of proofs for the theorem, his turn of the century address might have sustained an 'I told you so'.

Despite the difficult puzzles which faced gas theory, Kelvin, writing in 1884, embraced the theory as "a true step on one of the roads" converging on a "Kinetic Theory of Matter" (Kelvin, 1884, p. 617). Throughout the period of maturation of the gas theory from the early work of Maxwell and Boltzmann, Kelvin had been convinced that the elastic solid conception of gas atoms was in error, even if it was a sufficient approximation for some purposes. He held that the solid model was liable to the difficulty (put to him by Maxwell) that eventually all translational energy would be dissipated into 'shriller and shriller' vibrations. He was more sanguine however that the vortex atom – "a configuration of motion in a continuous all-pervading liquid" (p. 616) – would escape this dissipation into high-frequency vibrations. Kelvin wanted to avoid explaining elasticity in gases by a deeper level of elasticity of solids and to that end constructed models which combined rigid parts with gyroscopic components to yield elastic systems.

Intriguingly, such models might have offered the possibility of ethers which did not have infinitely many degrees of freedom, and even if Kelvin has been convinced of the applicability of the Equipartition Theorem, he might not have agreed to Rayleigh's later standing wave analysis of cavity radiation. A volume of ether with finite numbers of degrees of freedom would not be a sink for all available energy, and a maximum might have been feasible for the intensity distribution on classical grounds. However, Kelvin was completely unconvinced by proofs of Equipartition.

As a consequence, it was not the Equipartition Theorem's failure with gas specific heats which led Kelvin to give up a guiding assumption. He thought that parts of gas theory were adequate. He did, however, see the matter theory in an idiosyncratic light and had his vortex theory as a programmatic commitment. That commitment might not count as a guiding assumption as defined in the theses, since it was in great part merely programmatic, but it is noteworthy that such a program existed side by side with other programs, so that the potential for the emergence of new guiding assumptions was there throughout, without mere discrepancies being the agent of change. At the level of guiding assumptions as strictly defined, however, it can scarcely be said that Kelvin saw the fundamentals of his science called into question by empirical puzzles. His celebrated characterization of the problems of Equipartition as a 'Cloud' over 19th century physics takes on a severity which is plausible if we view it from the perspective of the Quantum Theory, but not so from the viewpoint of his own prior skepticism of the theorem. The millennial tone of his review resonates unavoidably with our hindsight contrast of quantum and 19th-century physics. What sort of picture might Kelvin have painted if, other things being equal, the date had been 1875? In any case, it makes little sense to represent Kelvin's doubts as propelled by empirical failures such as anomalies in specific heats (whether of gases or solids), when the impact on guiding assumptions would have to be channelled by a theorem to which he did not subscribe. Even the advocates of the theorem recognized that the scope of applicability of the theorem was very much an unsettled issue. Kelvin's doubts were much more profound and theoretically charged and had been around for more than a few decades.

At the end of his 'Clouds' address, Kelvin quoted at length from Lord Rayleigh's paper on "The Law of Partition of Energy", and despite his own skepticism about the Equipartition Theorem, gave new visibility to Rayleigh's judgement that the theorem was general in its application.

Predictive failure pointed not to revisions in gas theory but in 'General Dynamics'. Rayleigh was unique in understanding the empirical puzzles in specific heats of gases to require a transformation at the level of guiding assumptions (as strictly defined), such as the basic principles of Lagrangian dynamics. Where Kelvin had sought to improve the situation by removing the impediment of elasticity where it made trouble, he did so by adapting the particular physical models by which general dynamical principles expressed themselves. The issue was diagnosed by Rayleigh as one of the idealized representation of constraints, that is to say, conditions in a problem whose violation would presuppose extremely high potential energies. This aspect of the Lagrangian approach was pinpointed by attempts to 'freeze', as Jeans was to put it, vibrational degrees of freedom by making connections effectively rigid. But as Rayleigh observed:

However great may be the energy required to alter the distance of the two atoms in a diatomic molecule, practical rigidity is never secured, and the kinetic energy of the relative motion in the line of junction is the same as if the tie were of the feeblest (Rayleigh, 1900a, p. 451).

The difficulties of the gas theory, especially in the conventional form supposing elastic solid atoms, were strategic because they required an articulation of the role of elasticity. This had really been apparent from the outset with, for example, Maxwell's pressing the inevitable dissipation of translational energy into the infinity of high frequency vibrations which perfect elasticity required. Rayleigh's 1900 conviction that Equipartition would hold for low frequencies, but be suspended for the higher modes in the continuous elastic ether of his version of cavity radiation, has a decades-old familiarity to it. It is a simple analog of the dissipation problem which Maxwell, Kelvin, and others recognized many years before.

This familiarity is important if we are to see Rayleigh's 1900 radiation law in perspective. It is all too easy to see his paper as the first version of the Ultraviolet Catastrophe spelling the doom of classical physics and heralding the demand for novelty like Planck's Quantum Theory. I think, however, this view should be resisted, because it suggests an illusory conformity to (GA2.5). In any case, Rayleigh's proposal of the radiation law with proportionality to temperature was not communicated with a view to refuting classical theory, but rather to showing where it might be successful within the legitimate bounds of applicability of the Equipartition Theorem. Further, he expressly threw the question of the success or failure of the proposal upon the 'distinguished experimentalists' who were only at that time consolidating the aggregate picture of the empirically given data across a wide range of temperatures and frequencies (Rayleigh, 1900b, p. 485). Indeed, the data for what I called above Planck's 'infra-red catastrophe', confirming the temperature proportionality of the Rayleigh proposal, were only just established in the summer of 1900 (Kangro, 1976, pp. 181–187).

In sum, in England the black-body problem confirmed what was illustrated by specific heats anomalies and the merely partial success with cavity radiation: the elastic solid theory of atomic matter and ethereal matter required fundamental adaptation at the level of representation of the elasticity assumed as primitive, and efforts had been underway for many years to address the strategic worry. Partial success with capturing a phenomenon may be thought of like the proverbial half-empty bottle; it is a matter of perspective whether one emphasizes the full half or the empty half. Rayleigh, like Lorentz in rendering the low frequency end of the distribution on electron theory grounds a few years later (Lorentz, 1952, sects. 59–76, esp. pp. 80–81), was noting the full half of the bottle, not the ultraviolet catastrophe: classical theory had promising suggestions about the likely distribution, at least in the range of frequencies at which it was not liable to the long diagnosed difficulty of dissipation into high frequencies. Rayleigh did, however, distinguish himself in targeting the apparatus for the employment of general mechanical procedures as such, rather than reconceptualizing the elasticity at the level of physical mechanism, but it was the same theoretical nexus that was targeted. By the second half of the decade after Planck's quantum appeared, however, the partial success of the half-full bottle was to have become the catastrophe of the empty half.

British physicists had been very willing to articulate a range of new candidates for guiding assumptions without prolonged persistence of anomaly (Kelvin and others had very rapidly responded to the demonstration of stability of Helmholtz's vortices of the late 1850s), and further, some of the derivative models from these rival approaches did much to set into relief the loci of fundamental difficulty of the elastic solid theory. Much the same role was exercised by the new Quantum Theory itself in the first decade after its appearance. However, failures of the gas theory with specific heats of more complex molecules than monatomic ones did not come to precipitate a crisis by virtue of their vintage. The kinetic theory's fundamental assumptions had been known to have foundational difficulties for too long; progress was still being made, however.

Photoelectric Effect: Bruce Wheaton has noted that:

Lenard's work on the photoeffect . . . is supposed to have been instrumental in the genesis of the light quantum hypothesis. It is frequently claimed that, following Lenard's work, the photoelectric effect constituted one of several "difficulties" for the wave theory of light . . . The claim would have greatly surprised Lenard (Wheaton, 1978, pp. 299–300).

The photoelectric effect had received much attention through the 1890s, not so much as a basic contradiction for classical theories but rather as an unexplained phenomenon, indeed an only partially analyzed phenomenon. Clearly there was no influence from that set of issues on Planck, any more than from Equipartition problems about either specific heats or cavity radiation. However, Einstein did consider the photoelectric effect 'fair game' for his photon hypothesis, published in 1905 (Einstein, 1905), and it was he who set this process of hindsight projection going. In a section of that paper, entitled "On the production of cathode rays by illumination of solids", Einstein wrote:

The usual ideas that the energy of light is continuously distributed over the space through which it travels meets with *especially great difficulties* when one tries to explain photo-electric phenomena, as was shown in the pioneering paper by Mr. Lenard (ter Haar, 1967, p. 104; emphasis added).

Here Einstein manages to give the impression that the classical wave-theory was recognized by Lenard to founder upon this phenomenon. In particular Lenard drew attention to the fact that the velocity of electrons escaping from an illuminated metal surface was independent of the intensity of the illuminating radiation, which seemed to indicate that there was not a simple dependency between the amplitude of the illuminating radiation and the ejection of the electron from the metal. Lenard drew the conclusion that the significant parameter in the emission of slow cathode rays under such circumstances was not the energy of the radiation, but rather energies that were triggered in the atoms of the surface; what were called 'storage atoms'. As J. J. Thomson wrote, in analyzing Lenard's result:

It proves that the velocity of the corpuscles is not due to the direct action upon the corpuscle of the electric force, which according to the Electro-magnetic Theory of Light occurs in the incident beam of light. It suggests that the action of the incident light is to make some system, whether an atom, a molecule or a group of molecules, in the illuminated system unstable, and that the systems rearrange themselves in configurations in which they have a much smaller amount of potential energy. The diminution in the potential energy will be accompanied by a corresponding increase in the kinetic, and in consequence of the increased kinetic energy corpuscles may be projected from the atoms in the illuminated substance (Thomson, 1906, p. 278).

Clearly the basic problem raised by Lenard's paper was not one which was thought to controvert the continuous wave theoretical approach, for what it did was to force the recognition that the action of radiation was not direct but involved intermediary processes, so that the character of the radiation was fundamentally determined by the atomic or molecular processes which led to the ejection of the electrons. This relegation of the basic effect to parameters in the atomic–molecular constitution was preserved in later resonance theories of photoelectric emission from illuminated solids. *What Einstein had done, was to use the 'eye-glass' of a novel hypothesis in such a way as to see confirmation of that hypothesis, which differed radically from the classical theory, as counter-evidence to the classical theory.* In this way Einstein himself contributed to the hindsight reappraisal of the problem-complex which physicists faced, and 'projected the solution backwards'.

Specific Heats Again: The temperature dependency of specific heats of solids came to have a very special status in the development of quantum theory. For Einstein came to realize that the adoption of Planck's quantum postulate for the proposed oscillators in the black-body would have repercussions for other properties than simply the character of the equilibrium which could be set up with radiation. In particular, Einstein saw the natural domain of extension to be the oscillators invoked in the molecular theory of heat. He set himself, therefore, to finding other contradictions between experiment and the kinetic theory, and found them in the deviations from the Dulong and Petit law which ascribed a single atomic heat to all solids, based on the principle that the atoms of all simple bodies have exactly the same heat capacity. While the Dulong–Petit principle had general success, among lighter elements there were significant, recognized deviations. By 1900 it had come to be recognized too that these light elements had heat capacities which varied with temperature and also at higher temperatures approached a value in accord with the Dulong–Petit law. Einstein adopted a quantum theoretical expression for the mean energy of an oscillator, and together with certain simple assumptions, was able to explain why in particular the high frequencies of oscillation to be expected for light elements would lead to 'anomalous' heat capacities at normal temperatures. The quantum expression introduced a fundamental dependency of the energy and the specific heat on frequency. At greater temperatures, or at lower frequencies (characteristic of the heavier elements), the heat capacity would approach the value to be expected on the basis of the equipartition theorem.

Einstein's theory led to a striking prediction which found a very special place in the interpretation given to the quantum theory by Walter Nernst. For there was a striking consilience between Einstein's theory and the proposals to which Nernst had been led in the elaboration of his Third Law of Thermodynamics. By April of 1910, Nernst had become optimistic that his work required that the specific heats of solids tended to zero, or close to it, as their temperatures approached absolute zero, which was exactly the prediction of their behaviour on the basis of Einstein's theory. Despite his initial disinclination to the quantum theory, Nernst became one of the staunchest supporters of it, and both Nernst and Lindemann, his student and collaborator, devoted considerable effort to the articulation of the theory within their sphere of interest, including even a quantal approach to the problem of specific heats in gases. By June 1911 the quantum problem had crystallized out in the form of a proposal by Ernest Solvay, a wealthy industrialist, to support a conference on the principle of energy quanta. The driving force behind this conference was Nernst. He had, in fact, approached Planck about the possibility of such a conference about a year before. In the light of the interpretation advocated here for the development of the crisis in quantum theory, it is worth quoting as significant section of Planck's reply. Planck did not think the conference feasible at the stage. He wrote:

The fundamental assumption for calling the conference is that the present state of theory, predicated on the radiation laws, specific heat etc., is full of holes and utterly intolerable for any true theoretician and that this deplorable situation demands a joint effort towards a solution, as you rightly emphasized in your program. Now my experience has shown that this consciousness of an urgent necessity for reform is shared by fewer than half of the participants envisioned by you . . . even among the younger physicists the urgency and importance of these questions is still not fully recognized. I believe that out of the long list of those you named, only Einstein, Lorentz, Wien, and Larmor are seriously interested in this matter, besides ourselves (Quoted in Hermann, 1971, p. 137).

Nearly eleven years after the development of the quantum postulate, it was hard to find, in Planck's experience, more than a handful of physicists who felt the crisis for the classical physics! By mid-1911 these doubts had been overcome, not because the pressure of finding a classical solution to the specific heats problem had grown intolerable, but rather because the experimental confirmation had been realized of a radical non-classical theory whose acceptance, as Nernst put it, "will no doubt involve an extensive reformation of our present fundamental points of view" (Hermann, p. 138). Nernst was correct in this diagnosis, and there can be little doubt that the conference had an immense impact throughout the community of physicists.

The role of the specific heat problems throughout this seems to have been that of yielding the first major confirmation of the quantum postulate outside radiation problems. Although the problem of heat capacity variation with temperature had come to be recognized as unexplained by 19th-century science, it seems to me to be an extreme concession to hindsight vision to see those problems as other than *puzzles* prior to the emergence of the quantum theory. *It is exactly because they became paradigmatic applications of the burgeoning quantum theory that they gained the methodological significance which is ascribed to them after the fact.*

3. CONCLUSION

My main preoccupation in this paper has been briefly to highlight some of the historical artifacts which prompt an easy acquiescence to our three theses. A more complete and satisfactory account of the Planck case would, no doubt, permit a clearer resolution to their evaluation. Part of the clue, however, lies in the last sentence of (GA2.7), and highlights the role which novel theory has in targeting blame within the conventional framework. I have noted above that the success of a quantum model with the specific heats of solids, in conjunction with the plausible, if not unproblematic, grounding of a black-body distribution law adequate at both high and low frequencies, gave a major impetus to interest in some puzzles which had not been thought of as tearing the heart out of 19th-century physics. With the merely partial success of electron theory as developed by Lorentz and Jeans to apply to temperature radiation, inevitably eyes fell upon Planck's theory and its ambiguous novelty, the quantum. Penetrating analyses of Planck's contribution by Einstein and Ehrenfest, largely ignored until Lorentz's ditching of the Jean's model in 1908, had revealed only that Planck's theory involved a genuine novelty. The impact of the Solvay Conference was an intellectual explosion, which precipitated out the decisive recognition that fundamental principles in conventional physics had to be sacrificed. It was not the lack of success of classical physics which dealt the real blow to guiding assumptions, but rather success when apparently violating them.

Of the three theses addressed in this paper, (GA2.7) has the greatest plausibility, at least in its concluding assertion that difficulties become acute when rival theories explain them. That mere persistence of intractability should be able to direct the force of empirical discrepancies onto fundamentals or guiding assumptions is *a priori* implausible, and despite Rayleigh's isolated conviction in 1900 that Equipartition held true, so that discrepancies meant a failure at the level of dynamics, it is transparent that the quantum diagnosis was decisive. Poincaré's post-Solvay linking of Planck's distribution with Discontinuity was the magisterial diagnosis, and it blamed the classical failures on a Principle of Continuity which prior to the quantum theory had not even been negotiable.

REFERENCES

Einstein, A. (1905), 'On a heuristic viewpoint about the creation and conversion of light', *Annalen der Physik* **17**, 132 ff., in English translation in ter Haar, pp. 91–107.

Forman, Paul: (1968), 'The Doublet Riddle and Atomic Physics *circa* 1924', *Isis* **59**, 158.

Garber, Elizabeth (1978), 'Molecular Science in Late Nineteenth-Century Britain', *Historical Studies in the Physical Sciences* **9**, 265–297.

Harman, Peter (1982), *Energy, Force, and Matter*, Cambridge: Cambridge U.P.

Hermann, Armin (1971), *The Genesis of the Quantum Theory (1899–1913).*, Cambridge, Mass.: MIT Press.

Kangro, Hans (1976), *Early History of Planck's Radiation Law*, London: Taylor and Francis.

Kelvin, Lord (1970), 'Nineteenth Century Clouds Over the Dynamical Theory of Heat and Light', *The Royal Institution Library of Science: Physical Sciences*, vol. 5, New York: Elsevier.

Klein, M. (1962), 'Max Planck and the Beginnings of Quantum Theory', *Archive for the History of the Exact Sciences* 1, 459–477.

Klein, M. (1965), 'Einstein, Specific Heats and the Early Quantum Theory', *Science* 148, 175.

Klein, M. (1970), *Paul Ehrenfest*, vol. 1, Amsterdam: North Holland.

Kuhn, Thomas S. (1970), *The Structure of Scientific Revolutions*, 2nd edn., Chicago: Chicago U.P.

Kuhn, Thomas S. (1978), *Black Body Theory and the Quantum Discontinuity, 1894–1912*, Oxford: Clarendon Press.

Larmor, Joseph (1900), Presidential Report, Section A, *Report of the British Association for the Advancement of Science 1900*, London: John Murray, pp. 613–628.

Lorentz, H. A. (1952), *Theory of Electrons*, New York: Dover Publications.

McGucken, William (1969), *Nineteenth-Century Spectroscopy*, Baltimore: The Johns Hopkins Press.

Pais, Abraham (1982), *'Subtle Is The Lord . . .': The Science and the Life of Albert Einstein*, Oxford: Oxford U.P.

Planck, M. (1900), 'On the Theory of the Energy Distribution Law of the Normal Spectrum', *Verh. d. deutsch. phys. Ges.* 2, p. 273ff, reprinted in ter Haar, pp. 82–90.

Planck, M. (1906), *Vorlesungen über die Theorie Wärmestrahlung*, Leipzig: J. A. Barth.

Planck, M. (1949), *Scientific Autobiography*, New York.

Planck, M. (1945), *Treatise on Thermodynamics*, New York: Dover Publications.

Rayleigh, Lord (1900a), 'The Law of Partition of Energy', *Philosophical Magazine* 49, 98–118, reprinted in Rayleigh (1903), pp. 433–451.

Rayleigh, Lord (1900b), 'Remarks upon the Law of Complete Radiation', Philosophical *Magazine* 49, 539–540, reprinted in Rayleigh (1903), pp. 483–485.

Rayleigh, Lord (1903), *Scientific Papers*, vol. IV, Cambridge: Cambridge U.P.

ter Haar, D. (ed.) (1967), *The Old Quantum Theory*, Oxford: Pergamon Press, pp. 79–81.

Thomson, J.J. (1906), *The Conduction of Electricity Through Gases*, 2nd edn., Cambridge: Cambridge U.P.

Wheaton, Bruce R. (1978), 'Philipp Lenard and the Photoelectric Effect, 1889-1911', *Historical Studies in the Physical Sciences* 9, 299–322.

HENK ZANDVOORT

NUCLEAR MAGNETIC RESONANCE AND THE ACCEPTABILITY OF GUIDING ASSUMPTIONS

1. INTRODUCTION

The present contribution is based on a detailed case study of the development of nuclear magnetic resonance (NMR) in physics and chemistry after World War II, described in my book (Zandvoort 1986). I claim that my case is typical for a major portion – but not all – of modern physics and chemistry and that the results to be discussed in this paper can be generalized to large parts of modern physics and chemistry. To be more precise, my case can serve as a paradigm example of that part of scientific development that the physicist Weisskopf has dubbed 'extensive' (see Zandvoort 1986, ch. VII.4).

The results that I will present here are both factual and methodological in nature. On the factual side, there is a negative result as well as a positive result. The *negative result* is that novel predictions in the sense of the Popper–Lakatos tradition are *unimportant* for the factual success of a research programme like NMR. My case disconfirms (GA1.3), which states that the acceptability of a set of guiding assumptions is judged largely on the basis of the success of its associated theories at making novel predictions. On the other hand, my case *confirms* (GA1.1) and (GA1.2), which assert the relevance of empirical accuracy and solved problems for the assessment of a set of guiding assumptions, provided that one interprets these two theses as *necessary but not sufficient* conditions for the positive appraisal of sets of guiding assumptions (that is, research programmes). That is to say that, judging from my case, there must be satisfied still something more than requirements like (GA1.1) or (GA1.2) in order for a set of guiding assumptions to be appraised positively.

This 'something more' constitutes the *positive factual result*. It amounts to the thesis that, in order to be factually successful, the *results* of a research programme – which are essentially specific theories – should be *useful* to developments in neighboring fields of science. In my book I have used this as a starting point for a new model of scientific development that does more justice to my case and similar extensive

337

developments in modern natural science than the standard models from which the canonical theses stem. In addition, this new model sheds light on the structure of interdisciplinary research in general, an issue which is too much neglected by philosophers of science. But since a requirement of the kind embodied by the 'positive factual result' stated above does *not* figure in the canonical set of theses here under test, I will deal below, in the final section, only very superficially with that result and the model based on it.

In order to obtain the central negative result of this paper – i.e. the non-importance of novel predictions for the appraisal of the development of NMR – several *methodological problems* had to be overcome. The discussion of these problems and their tentative solutions leads to the methodological results announced above. In fact, the bulk of my paper will deal with the methodological issue of how one can actually arrive at empirical tests of a thesis like (GA1.3) on the alleged importance of novel predictions. In this respect my paper is an illustration of the general point stressed in Laudan *et al.* (1986) that, even if it is (considered to be) clear what are the theses to be tested, 'large issues will still remain regarding the design and execution of empirical tests for them". I hope that my paper will make clear what kind of enterprise is in any event needed to *solve* such issues: namely clarification, or explication, of the concepts in terms of which those claims are being stated.

Section 2 of the paper is devoted to the definition of the technical term 'research programme' and its relation with the term 'set of guiding assumptions' coined in Laudan *et al.* (1986). Section 3 introduces a set of 'possible' criteria for the success – or acceptability – of research programmes, ordered in a series of increasing strength. This series I use as a *scale* onto which the views of (e.g.) Lakatos, Kuhn, and Laudan on the topic under discussion can be mapped. Section 4 in turn establishes the connection between the elements of this scale and a number of theses from several clusters under examination. Sections 5 and 6 are devoted to the problems and results of testing the adequacy of the criteria for successful research programmes for my particular case, i.e., the development of NMR in physics and chemistry. In these sections, the above-mentioned negative factual result is being generated. Section 7 summarizes the 'positive factual result', and the main virtues of the model of scientific development alluded to above.

2. RESEARCH PROGRAMMES AND GUIDING ASSUMPTIONS

Recent philosophers of science such as Lakatos, Kuhn, Laudan, and Sneed distinguish between two levels of scientific knowledge: one level of *specific theories*, and another level of *families of such specific theories*, united by one basic or global theory. As was explained in Laudan, *et al.* (1986), these philosophers have each devised their own, sometimes mutually conflicting technical definitions of the entities of the latter, more global level. I will build on Lakatos's definition according to which a research programme is a *series* of specific theories $T_l, \ldots, T_i, T_j, \ldots$ developed subsequently in time by extending one constant basic theory or *hard core* with a changing set of additional assumptions, sometimes called *protective belt*, with the overall aim of better explaining, on the basis of a given basic theory, a certain domain of phenomena. There is also associated with a research programme an *heuristic*, which offers suggestions as to how to modify or 'sophisticate' the protective belt in order to generate better specific theories. (I will not make very much use of this latter part of the definition, however.)

Whenever I want to refer to this particular definition I will use the term 'research programme'. However, Laudan, *et al.* (1986) also point out that there is *no* disagreement about the identification of the historical entities to which the various technical definitions intend to refer. In fact, it is for this class of intended referents of the various technical definitions – which apparently can be established independently of the specific form of those technical definitions – that Laudan, *et al.* (1986) coin the term '(set of) guiding assumptions'. Accordingly, the empirical claims of the various theories of scientific change – which are considered to be commensurable – are in the canonical set of theses under test being stated with this latter term 'guiding assumption'. Therefore, it would be adequate to use this latter term when I am referring directly, that is, 'pretheoretically', to my case, i.e., the development of NMR.

This paper deals with the factual criteria of success – or acceptability – of guiding assumptions. According to my basic theoretical choice, guiding assumptions are (to be represented as) research programmes, which according to the above technical definition, are *series* of specific theories. The latter induces the view that the success of an entire research programme must be evaluated in terms of the success of the individual theory transitions within the programme. The criteria to be presented below are structured according to this view.

3. CRITERIA FOR SUCCESSFUL RESEARCH PROGRAMMES

In this section I present four different *prima facie* candidates for a criterion for successful research programmes. Together they form a series of increasing strength, which can be generated by imposing increasingly stronger conditions upon what may be called the historical relationship between theory and evidence. For each element of the scale first a criterion for individual theory transitions within a research programme is being formulated (a), and next a criterion for the research programme as a whole (b). For a certain research programme RP the scale runs as follows:

ES.a. Theory transition $T_i \rightarrow T_j$ within RP is *explanatorily successful* (ES) iff
  i. at the time of the transition, T_j successfully explains more laws than T_i does.
ES.b. The development of RP is ES in a time interval $\triangle T$ iff in that time interval there are theory transitions all of which are ES.
WHS.a. Theory transition $T_i \rightarrow T_j$ within RP is *weakly heuristically successful* (WHS) iff
  i. idem ES.a.i
  ii. the hard core of RP was not formulated with the purpose of explaining all these laws. (Or: not all these laws formed part of the hard core's problem situation.)
WHS.b. The development of RP is WHS in a time interval $\triangle T$ iff in that time interval all its theory transitions are ES and some are WHS.
HS.a. Theory transition $T_i \rightarrow T_j$ within RP is *heuristically successful* (HS) iff
  i. idem Es.a.i
  ii. T_j was not formulated with the purpose of explaining all these laws. (Or: not all these laws formed part of T_j's problem situation.)
HS.b. The development of RP is HS in a time interval $\triangle T$ iff in that time interval all its theory transitions are ES and some are HS.
TS.a. Theory transition $T_i \rightarrow T_j$ within RP is *temporally successful* (TS) iff
  i. idem ES.a.i
  ii. some of these laws were unknown, hence not confirmed (though there may have been confirming instances available), at the time T_j was formulated.

TS.b. The development of RP is TS in a time interval $\triangle T$ iff in that time interval all its theory transitions are ES and some are TS.

Below, I will drop the suffix a or b from expressions like 'ES.a' or 'WHS.b' when from the context it is clear which version of a criterion (i.e., for theory transitions or for research programmes) is meant. 'Theory transition $T_i \rightarrow T_j$ is ES (WHS, etc)' can be shortened to 'theory T_j is ES (WHS, etc.)' if there is no ambiguity about what the predecessor T_i is. Also terminology like 'the transition $T_i \rightarrow T_j$ derives explanatory (weakly heuristic, . . .) success from consequence L'; or: 'T_j derives . . . success from L', will be obvious.

Criterion ES.a, in Zandvoort (1986), was derived from considerations of (observational) verisimilitude, in combination with the assumption that the aim of science is to develop theories that are getting ever closer to the truth. There I showed that if a theory T_j at a time t successfully explains more laws about the domain than T_i does, then it is not possible, given the assumption that the set $E(t)$ of empirical data about the domain accepted at time t is not mistaken, that T_i is closer to the truth than T_j whereas the opposite *is* possible. Here 'closer to the truth' is conceived in a particular technical explication due to T. Kuipers (see, e.g., Kuipers, 1982 or Kuipers, 1987), and 'law' is taken in a specific definition proceeding in the style of the structuralist conception of theories (see Zandvoort, 1986 or Zandvoort, 1987). On this type of *a priori*, epistemological considerations I concluded that ES.a would be a reasonable minimal requirement for accepting at time t a certain T_j at the cost of a T_i, whereas ES.b would be the corresponding minimal requirement for considering a research programme as a whole successful during a certain time interval $\triangle T$. (The latter presupposes the idealizing assumption that in the development of successful research programmes 'drawbacks' do not occur.)

For the present purposes we need only the intuitive reading of criterion ES. Two things *are* important in the present context, though. Firstly, criterion ES talks about the (successful) explanation of *laws*. This reflects the fact that what is important in science is to explain (or predict) *regularities* (or: phenomena, or: effects), not individual facts. Secondly, the criterion ES talks about the *structural* or *logical* relationship between a theory and the expressions describing the empirical evidence for the theory (the empirical laws), whereas the remaining criteria WHS, HS and TS arise out of ES by imposing increasingly stronger requirements on the *historical* relationship between the conception or construction of the theory and the body of known evidence $E(t)$ that the theory should explain. Stated alternatively, WHS, HS and TS are increasingly stronger explications of the notion of a *prediction*.

This scale I used to map the positions of the various authors that hold a guiding assumptions-view of scientific knowledge on the question of the proper criterion for *successful* guiding assumptions. This mapping procedure does not proceed entirely straightforwardly, not even for the case of Lakatos, although the scale is based on his particular definition of guiding assumptions-like entities. For details of the arguments that go into the mapping procedure I have to refer to Zandvoort (1986), ch. I.4. Here I will merely state the results.

For *Lakatos*, the problem is that he in fact considered any theory transition $T_i \rightarrow T_j$ which obeys criterion ES.a but that fails to make a novel prediction as unsuccessful, or even unscientific. I have argued however that, on Lakatos's own basic presuppositions, this cannot be defended, and I have accordingly repaired Lakatos's view on this particular point. Having done that, then the identification of Lakatos's view on the proper criterion for successful research programmes is straightforward: the 'earlier Lakatos' must be 'mapped' on criterion TS, whereas the 'later Lakatos', influenced by Zahar, defended criterion HS.

Laudan tries to capture sets of guiding assumptions with the technical term 'research traditions'. According to Laudan, the success of a research tradition is determined by how well it succeeds in *solving problems* which may be either *conceptual* or *empirical*. The criteria from my scale do not deal with conceptual, or theoretical problems, but only with empirical problems. Or rather, they do not deal *explicitly* with theoretical problems, for one may see criteria like these as efforts to deal with the issue of theoretical problems by formulating the *implications on the empirical level*. (In fact I think that this is how Popper saw the matter.) We may establish that according to Laudan a specific theory from a research tradition solves an empirical problem exactly when that theory successfully explains a law in my sense. In particular, Laudan explicitly denies that for the successful solution of an empirical problem any requirement of predictivity would be relevant (Laudan, 1977, pp. 114–7). To be short, if we restrict the discussion to what Laudan calls empirical problems, then we may conclude that Laudan defends our criterion ES, not merely as a necessary condition for cognitively successful research programmes (traditions), but also and especially as a sufficient one.

Turning now to *Kuhn*, the relevant question is under which conditions he considers *normal science* successful. Since Kuhn equates the task of normal science with *solving puzzles*, we must say that in his view normal science is *successful* as long as it succeeds in solving puzzles. I further assume that a solution to a Kuhnian puzzle is the same kind of entity as a specific theory within a research programme. From the various expositions in Kuhn (1970) one may next induce the following. On the one hand, Kuhn does not explicitly require of a specific theory, in order to successfully solve a puzzle, that it has any predictive merits. But on the other hand, he assumes that in typical cases the puzzles for a certain period of normal science are being generated by that normal scientific development itself, which is to say that according to Kuhn they have come into being only after the fundamentals of the development under consideration have settled down. From this, it appears that Kuhn's *criterion for successful normal science* may best be equated to our criterion WHS.

4. THESES FROM THE CANONICAL SET ADDRESSED IN THIS PAPER

The above exercise may be seen as an effort to make the acceptability cluster theses more precise in order to make them more amenable to empirical test. Thus, one may map (GA1.3) either on HS.b or on TS.b, depending on the way one wants to interpret the term 'novel prediction'. ES.b covers important aspects of (GA1.2) and (GA1.1). Finally, WHS.b provides one – though perhaps not the only – possible explication of (GA1.4).

From the fact that my criteria for successful *research programmes* are defined in terms of criteria for successful *theory transitions* (thus *inverting* (T2.3), it is clear that in my view the acceptability cluster is closely related to the inter-theory relations cluster and the appraisal cluster. Here, especially (T1.1), (T1.2), (T1.3) and (T2.2) are relevant. Apart from this, from my treatment of my case in Section 6 below it will transpire that (T2.1) in my view refers to 'guiding assumptions' rather than to specific theories. Finally, my material confirms (T2.5) whereas it disconfirms (T2.4), but these aspects will remain rather implicit in the analysis.

Let us then quickly go to that material itself. Section 5 gives an outline of my case, the development of NMR, and argues that this case conforms to the general structure of theory development that is implicit in the research programmes terminology. Section 6 asks which one of the criteria from the scale is actually being fulfilled in the case of NMR.

5. THE NMR PROGRAMME

In 1945 the research groups of two physicists, Felix Bloch and Edward Purcell, independently of each other performed an experiment for which they were to receive a Noble prize in 1953, and which we now know as the Nuclear Magnetic Resonance or NMR experiment. Disregarding certain differences between their experiments, the essentials were as follows (see for more details Zandvoort (1986), and also Rigden (1986)). A piece of material of which the nuclei possess a magnetic moment is placed in a magnetic field and is subjected to a beam of electromagnetic radiation from the radio region, of which the frequency is slowly varied. At a certain frequency, called the resonance frequency, the nuclei absorb radiation from the beam, which is detected as a fall in the intensity of the radiation. This phenomenon is called the *nuclear magnetic resonance phenomenon*.

The theory that explains this NMR phenomenon finds its basis in quantum theory. According to that theory, the magnetic moments $\vec{\mu}$ of the nuclei can take only a discrete number of orientations with respect to the direction of the external magnetic field. I will consider the often-occuring case (e.g., when one is dealing with hydrogen nuclei) that there are two such orientations, which are then called the 'spin down' state and the 'spin up' state. The two states are associated with different energies. Under the absorption or emission of (electro) magnetic radiation, transitions may be induced between states of different energy if the frequency v obeys Planck's equation $\Delta E = h v$. Here ΔE is the energy difference between the states and h Planck's constant. If the two states would be equally populated, then the number of transitions per unit of time would be the same for both types of possible transitions (i.e., up $\rightarrow$ down and down $\rightarrow$ up). In this case there would be no net absorption or emission of radiation. But when there is thermal equilibrium in the system, then Boltzmann's law says that this number is *not* the same, and this results in a net absorption from the beam, which is the experimental quantity that is being detected in the NMR experiment.

The above theory is actually not complete enough to have the NMR phenomenon as an empirical consequence. The point is that, as soon as the radiation induces transitions, the population difference tends as a consequence to disappear: this is called saturation. In order to make the NMR experiment really feasible, there should be a so-called *spin-lattice relaxation* mechanism tending to restore the population difference indicated by Boltzmann's law. Such a spin-lattice relaxation mechanism can be characterized by a characteristic time, the spin-lattice relaxation time T_1. The shorter T_1, the stronger the relaxation mechanism and the easier the NMR experiment can be performed. Therefore Bloch and Purcell had to know (the order of magnitude of) T_1 in order to judge the feasibility of their projected experiment. To this aim they combined an earlier theory on relaxation phenomena in the electronic case due to Waller to their basic theory of NMR outlined above.

In addition to T_1, the line width of the resonance signal is an important quantity for judging the feasibility of the experiment: the smaller the line width, the better. The theory outlined so far predicts a zero-line width, but this is obviously not realistic. In an effort to improve on this, both Bloch and Purcell developed a crude argument involving the small local magnetic fields of neighboring nuclei. These local fields would give a spread in the resonance frequency roughly of the order of the magnitude of the nuclear magnetic moments, or at least so Bloch and Purcell argued. This effect was called dipolar broadening.

Thus prepared with theoretical knowledge and specific expectations, the Bloch and Purcell groups performed their experiments in the fall of 1945. The results were twofold: (a) they *succeeded* in revealing the NMR phenomenon as such; but (b) the specific expectations about the particular details of the phenomenon, e.g., the spin-lattice relaxation time and the line width, were smashed by their experimental findings: the actual relaxation times turned out to be much shorter than theoretically expected, and the dipolar broadening turned out to be absent in many substances where Bloch and Purcell had certainly expected it, that is: the actual line widths turned out to be much smaller than had been anticipated theoretically.

The NMR experiment was motivated by the desire to detect the values of nuclear magnetic moments: for if the frequency at which absorption occurs has been measured, then from Planck's relation the energy difference $\triangle E$ between the states can be obtained, which in turn is coupled to the value of the magnetic moment μ by the relation

$\triangle E=2\mu H$, where H is the value of the external magnetic field. But the *dis*agreement of theoretical expectations with the outcome of the experiment actually triggered an entirely new *research programme*, which may be considered more or less apart from the above mentioned motivation for the NMR experiment. That programme's aim was to give an ever better account of the various details of what one may crudely refer to as 'the NMR phenomenon', but which might perhaps more adequately be called the *domain of NMR phenomena*: the various finer details that can be discerned in the signals forming the output of the NMR experiment. The theory outlined above on which Bloch and Purcell based their experiment constitutes the first specific theory from this NMR programme (the '1945 theory' from now on).

From 1945 onward a host of new specific theories was developed within the context of this programme. Originally, the emphasis was on the study of relaxation times and line widths under various conditions and in different types of substances. The main outcome of this concern was the now famous theory of Bloembergen, Purcell and Pound, published in 1948. This theory, usually referred to as the *BPP theory of nuclear magnetic relaxation*, explained the two major anomalies confronting the 1945 theory of the NMR phenomenon mentioned above, i.e., the unexpectedly small relaxation times and the absence of dipolar broadening in many cases where it had been expected.

The next task was to work on particular line shapes in solids where 'dipolar broadening' *does* occur; after a while work on resonance lines in liquids or gases became increasingly important. Here the resonance lines were very narrow, indeed much narrower than the resolving power of the first experimental setups. While that resolving power was increased, soon a host of unforeseen effects was discovered in this area, which in turn meant a challenge to the theoretical NMR programme as conceived here: the challenge to *explain* these effects. One of the first of these effects was the so-called Knight shift in metals and metal-salts, reported by Knight in 1949 and explained theoretically by Townes *et al.* Another shift is the so called *chemical shift*, discovered by Dickinson and independently by Proctor and Yu in 1950, and theoretically explained by Ramsey, at about the same time. A further effect that was encountered in liquids is the *hyperfine splitting*, reported for the first time by Proctor and Yu in 1951, and explained, after a number of unsuccessful efforts by others, in 1952 by Ramsey and Purcell.

The development of this programme dealing with the discovery and explanation of NMR phenomena, which I followed during a crucial period lasting until 1953, was factually considered very successful. This appears from the positive appraisal of 'the scientific community', expressed, e.g., in a Nobel prize for Bloch and Purcell for their contributions to this programme. The statement might also be vindicated in terms of the large number of papers that came out and the large amount of manpower that went in the programme. Therefore, if a criterion X.b from my scale is to be empirically adequate for the NMR programme, then that NMR programme should score positive on that criterion. In order to find that out, the success of all specific theory transitions in the NMR programme within the time interval considered must be rated on the corresponding criterion X.a. From the results we then can conclude whether the criterion X.b in question has been satisfied or not. If so, we have no clash with historical reality; if not – and if moreover the time interval considered was long enough: see Zandvoort (1986), pp. 150 and 73 – then the criterion in question has been falsified.

The next section deals with (what initially appears as) just *one* theory transition within the NMR programme outlined above, i.e., the transition from the original, 1945 theory of the NMR phenomenon to the BPP theory. This will allow me to go in some depth into my case. It is also the best way to discuss the *methodological problems* announced in the introduction. After having arrived, in spite of these problems, at my factual results for this particular theory transition – which actually turns out to be more than just *one* theory transition – I will then generalize these results to the other theory transitions in the time interval considered, referring to my book for substantiation. This then results in conclusions concerning the appraisal, not merely of isolated theory transitions, but of entire research programmes.

6. THE EXAMPLE OF THE BPP THEORY

6.1. *The BPP Theory and its Problem Situation*

In the face of criterion HS especially, it is important to know what were the problems for which a certain theory was intended as a solution: I will call this the theory's *problem situation*. What was the problem situation of the BPP theory? Now the task that its authors had set themselves right after the first successful NMR experiments was to formulate a theory that would be capable of explaining the two anomalies confronting the

1945 theory of the NMR phenomenom: i.e., (a) the too short actual spin-lattice relaxation times, and (b) the too small resonance line widths. These two anomalies will, therefore, be major constituents of the problem situation of the BPP theory.

As was explained in Section 5 above, the 1945 theoretical expectations of long spin-lattice relaxation times were adopted from an earlier relaxation theory of Waller. These expectations had turned out to be incorrect. Apparently therefore there was something wrong with Waller's theory. Now the general problem of formulating a theory of spin-lattice relaxation amounts to finding local interactions with the magnetic moments that fluctuate over time. Against this background, Waller's theory contained two central assumptions. The first one identified the source of the required interactions as the local magnetic fields of the neighboring nuclei. The second one postulated that the *fluctuations* of these fields arose from the lattice vibrations. Bloembergen *et al.* stuck (at least initially, see below) to the first assumption, but replaced Waller's source of motion by the random thermal Brownian motion of the atoms or molecules in which the spins are contained. The local field arising from this motion they described by a random function $H(t)$ characterized by a correlation time τ_c. They subsequently calculated the intensity of the Fourier transform of $H(t)$ at the resonance frequency v_0, and from this they induced the rate of spin transitions, i.e., the spin-lattice relaxation rate. In this way they obtained expressions that relate the spin-lattice relaxation time T_1 to the correlation time τ_c and to the resonance frequency v_0.

According to the BPP theory also the line narrowing in liquids and other substances has its origin in this thermal Brownian motion. Therefore, there is in this theory a close connection between T_1 and the line width, which Bloembergen *et al.* represented by a parameter called T_2. The connection is shown in Figure 1, which gives a plot of T_1 and T_2 as a function of the correlation time, for two different values of the resonance frequency v_0.

Actually the expressions for T_1 and T_2 plotted in this figure are themselves not yet *observational* consequences of the BPP theory, because of the occurrence of τ_c which cannot be evaluated experimentally without further ado. But with the help of a formula which Debeye had derived in the context of his theory of dielectric dispersion, Bloembergen *et al.* were led to an expression relating τ_c to the experimental quantity η/T. Eliminating τ_c through this expression from the theoretical

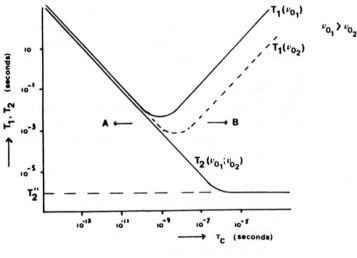

Fig. 1.

formula plotted in Figure 1 finally yielded genuine observational consequences relating the experimental quantities T_1 and T_2 to the experimental quantity η/T.

6.2. *Differentiating Between Explanatory and Predictive Merits of the BPP Theory: a First Methodological Problem*

There can be no doubt that, under any reasonable construal of 'successful explanation', the BPP theory successfully explains the short relaxation times and the narrow line widths that bothered the 1945 theory of
the NMR phenomenon. The theory transition under consideration, i.e.,
the transition from the 1945 theory to the BPP theory, is in other words
at least explanatory successful, i.e., obeys criterion ES.a from the scale.
It almost immediately follows that the transition is weakly heuristically
successful as well, since (a) the effects explained were disclosed by the
NMR experiment, and (b) irrespective of how the *precise* characterization of the hard core of the NMR programme should run, it is clear that
this hard core was there *before* the NMR experiment had been performed for the first time. In fact, this consideration will virtually always
apply in the case of theory transitions within the programme of NMR.

The question whether the BPP theory was successful in *explaining* certain phenomena has thus been answered straightforwardly. But in addition we want to know whether the BPP theory made successful *novel predictions* in the sense of Lakatos and his school, i.e., whether it is heuristically and temporally successful as well. In order to find possible candidates for such novel predictions, let us have a look at the theoretical curves of Figure 1, but now with $c \cdot \eta / T$, where c is an arbitrary constant, substituted for τ_c, thus making the curves (qualitative) *observational* consequences of the BPP theory. From that figure it is obvious that at least some of the empirical regularities derivable from the theory transcend in a non-trivial way the empirical evidence that was available at the time Bloembergen *et al.* started their research. As examples the following regularities can be mentioned:

(a) for sufficiently small η / T (that is, in region A), T_1 and T_2 are equal, and depend linearly on η / T;

(b) there is a minimum for the value of T_1, and this minimum depends on the resonance frequency in a specific way.

(You may convince yourself of the fact that we deal with *regularities* by thinking in front of the above statements universal quantifiers of the type 'for all liquids (and for a certain class of solids) it holds that'.)

Now the empirical evidence as it was presented together with the theory is largely in agreement with regularities like these. Figure 2, giving a plot of observed values of T_1 and T_2 versus η / T for the proton resonance in glycerol, is an example of such evidence. The (transition to) the BPP theory therefore again derives explanatory success, and *a fortiori* weak heuristic success, from the type of observational consequences presently under discussion.

But does this transition derive *temporal* success as well from these consequences? (The need to consider also the possibility of heuristic success will be seen to dissolve automatically as we proceed.) In order to answer this question we must decide between the following two possibilities: (a) was the BPP theory (together with the observational consequences under discussion) formulated first, and were the experiments subsequently done?; (b) or were the experiments performed first, and was the theory proposed only afterward? But the moment we try to do so we are facing the problem that theory formulation and acquisition of experimental evidence are thoroughly intertwined in the present case.

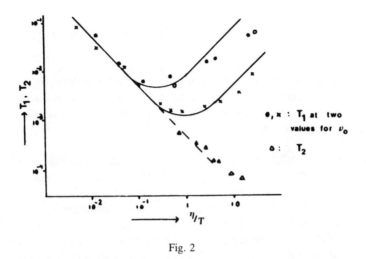

Fig. 2

This makes the distinction between the above two possibilities too simplistic as it stands.

A possible way out would be to dissect the research that led to the 1948 paper into a number of subsequent theoretical stages. For each transition between these stages we should then establish the type of success obtained, in order to come in this way to an overall rating. To a certain extent Bloembergen *et al.* themselves have made something like this possible. For as it happens they published in 1947 (before the 1948 paper was finished) a brief account of their work in *Nature,* and we may conceive the accounts given in these two papers as two subsequent stages in the development of the theory. There are indeed certain quantitative differences between the 1947 and 1948 versions of the BPP theory. It is plausible though (see Zandvoort, 1986, p. 93–94) that it is exactly the lack of quantitative fit of the 1947 theory with experiment that motivated the alterations leading to the 1948 theory, such that from these quantitative differences the transition 1947 → 1948 cannot derive heuristic or temporal success.

One might try to dissect the development leading to the BPP theory into still more subsequent stages, now by taking recourse to less public material, for instance that contained in private notebooks, private correspondence, or obtained from interviews. I have the following *justification* for not having done this. We want to test the correctness of a criterion like HS or TS, taken as (part of) an empirical theory of scientific development. More precisely, we want to test a claim of the following type:

Empirical claim: exactly those research programmes satisfying criterion X factually receive positive appraisal and as a consequence tend to be prolonged, whereas those obeying a criterion weaker than X are appraised negatively and tend to disappear.

Such an empirical claim may be read in two ways, depending on whether we interpret the claim as a description of certain *actions in accordance with a rule* or as a description of certain actions *following a rule*. In the first reading the claim says that research programmes are being pursued and abandoned, and therefore persist and disappear, *as if* the scientists were evaluating the programmes's progressiveness or non-progressiveness according to the criterion X in question, and *as if* they made their choices on which programme to work accordingly. In the second reading, the *as if* clause is dropped, and it is assumed that the scientists are *actually* evaluating research programmes against criterion X (and act accordingly).

If we do not want to devaluate a theory of scientific development based on such a claim to a *black box* theory, reproducing at most the regularities observed in the development of science without providing a deeper explanation, we should stick to the action-following-a-rule interpretation. From this it appears to follow that the material with which the success of a given theory transition is evaluated must be generally accessible to the members of the scientific community. But then the relevance of information obtained from notebooks, etc., appears to be negligible, since it will have been largely hidden from the scientific community which is responsible for the evaluation of the success of the development in question. From this it might be argued that the proper source of data to be used in testing empirical claims of the present type is the published literature. (See on this point also Gardner (1982), pp. 6 and 8, who draws a slightly different conclusion from it.)

My conclusion is this. Since the BPP theory and the evidence presently under discussion were presented in one paper, the historical relationship between the two was hidden from the scientific community. Because of this, that evidence is irrelevant for establishing the heuristic or temporal success of the theory transition in question. This peculiarity I call the *first methodological problem with testing claims like HS and TS.*

6.3. *Two Further Methodological Problems in Establishing Predictive Success*

In their 1948 paper, Bloembergen *et al.* showed that their theory not only accounted for relaxation in *liquids,* but also dealt rather satisfactorily with relaxation in at least certain solids and in at least one gas (i.e., hydrogen gas), as well. Moreover they showed that their theory could quantitatively explain the enhancement of spin-lattice relaxation in liquids to which paramagnetic ions had been added. Also in the case of these further applications of the BPP theory the relevant observational evidence had in part been published in advance, and in part was presented together with the theoretical treatment, such that the first methodological problem recurs. But now we must in addition consider the possibility that the phenomena addressed in the present applications did not belong to the *problem situation* in which the theory was developed. If so, we would *not* merely be dealing with a case of 'merely' explanatory, and therefore weakly heuristic, success as above, but moreover with a case of genuine heuristic success.

But now we immediately find a *second methodological problem* in our way. For we now must decide what the exact problem situation of the BPP theory was, where there are in fact no clear criteria on which such a decision can be made. Did that problem situation contain *all* known anomalies to the original theory of NMR on which Bloch and Purcell had based their first experiments? Or should we include only those anomalies of that theory that Bloembergen *et al.* took as their particular starting point for constructing their theory? I will adopt the latter option, although there are in fact no convincing intuitive or theoretical reasons why this is the relevant one in the context of a criterion like HS. Under this decision the phenomena that we presently consider did *not* form part of the problem situation, such that it seems that here we may indeed assign some heuristic success to the BPP theory. However, this assignment tacitly assumes that, with the additional applications to solids, gases, and liquids doped with paramagnetic ions, we are *still dealing with the same theory.* But in fact this assumption is a very problematic one. For in order to explain, e.g., the enhanced relaxation times in liquids containing paramagnetic ions, Waller's first assumption

mentioned in Section 6.1 above, which formed part of the original BPP
theory, had in fact to be replaced by the assumption that the local fields
are due, not to the surrounding nuclei, but to (the electronic moments
of) paramagnetic ions instead. (See for details Zandvoort (1986), p. 97.)

For this reason we are *in a certain sense* dealing with a different
theory. On the other hand a central idea from the BPP theory, namely
to calculate the fluctuations in terms of Brownian motion, is left un-
touched such that there is also ground for a claim that actually we still
have the same theory. The same holds for the other applications of the
BPP theory mentioned above.

What we are missing here is a clear *criterion of identity* for theories
within research programmes in order to decide whether we are dealing
with one and the same theory, or with two different elaborations of the
original BPP theory instead. In the former case, we have heuristic
success for the transition 1945 → BPP; in the latter case we have, in all
probability, merely weak heuristic success for the transition from one
version of the BPP theory to another. This missing criterion of identity
for specific theories constitutes the *third methodological problem* with
testing theses on the importance of successful predictions for the appraisal of research programmes. Neither Lakatos nor any of his followers
has paid explicit attention to it. This is very surprising, for very often
these authors have to rely on the notion of heuristic, rather than
temporal, success in order to make their cases. And as we may see from
the above examples, the problem of distinguishing heuristic success
(i.e., success in the sense of Lakatos) from *weak* heuristic success (which
is *non*-success in the sense of Lakatos) is then likely to occur. And
indeed the problem *does* occur, and is causing rather severe problems,
e.g., in the case studies presented in Howson (1976): see Zandvoort
(1986), note 26 to ch. IV.

Someone who *did* have a strong explicit opinion about the matter is
Popper, i.e., Lakatos's 'predecessor'. According to him, whenever the
slightest modification or reinterpretation has been introduced in a theory,
the result should be considered as an entirely new theory and its merits
should be judged accordingly, so as not to fall prey to a 'conventionalist
strategy'. It is therefore in the spirit of this precursor for those defending
criterion HS to take a stringent position on the identity criteria for
specific theories. This leads inevitably, however, to the conclusion that
in the present examples we are dealing with *different specific theories*,

and to the subsequent conclusion that the transitions occurring among them are at most WHS.

Still the fact of the family resemblance of the variants of the BPP theory remains. This can be reflected in the description of the case (as was done in Zandvoort 1986) by considering the elaboration of the BPP theory into various specific directions as the elaboration of a *subprogramme* or *programmita* contained within the larger programme of NMR.

We therewith must not consider anymore *the* BPP theory, but instead the theories of the BPP programmita. Of course, I have not been exhaustive in my treatment of the empirical consequences of the theories of this programmita discussed so far. In my book I have argued that the performance of additional consequences was usually poor upon subsequent empirical investigation. This in turn triggered new members of the 'family of BPP theories', i.e., of the BPP programmita, but again the fate of these was essentially similar to that of their predecessors. To be brief, in my book I have tried to make the case that essentially all theory transitions within the BPP programmita were explanatorily successful, and also weakly heuristically successful, but not something stronger.

6.4. *Conclusions and their Generalizability*

The conclusions reached above, which do not just hold for the part of the NMR programme considered above but for the entire NMR programme, can be summarized as follows. *Firstly*, trying to test criteria of success of the type of my scale provokes a host of difficult methodological problems, having to do with (a) the fact that very often a specific theory and its evidence are presented at the same time, (b) the vagueness of the notion of the problem situation of a specific theory, and (c) the absence of a criterion of identity for specific theories. In a sense, all three issues have to do with the vagueness, or the lack of a sufficiently precise explication, of the notion of a novel, or genuine prediction. *Secondly*, within a research programme like NMR we should distinguish between major theory transitions, leading to major new specific theories like the BPP theory, and minor theory transitions resulting from further elaborations of major new specific theories, giving rise to the development of subprogrammes or programmita's within the programme.

Thirdly, as far as the methodological problems allow for such a conclusion, both the major and the minor theory transitions show weak heuristic success, but obey no stronger criterion from my scale. And that means, *finally*, that the NMR programme as a whole also exhibits weak heuristic success, but not something stronger.

The final conclusion, together with the fact that the NMR programme was *factually* very successful, implies that, as far as NMR and similar developments are concerned, Lakatos is wrong on the criterion for successful research programmes. My material does not allow one to infer whether the NMR programme would factually have been unsuccessful if it had obeyed ES but not WHS, such that, in the face of the evidence provided by the NMR case, both Kuhn and Laudan *may* still be right (but note the remarks in the next section on the necessary but not sufficient character of criteria like ES and WHS for my case).

The core result is that (GA1.3) is disconfirmed. Further, (GA1.2) and – provided we accept the translations suggested in Section 4 – (GA1.1) and (GA1.4) were fulfilled by the NMR programme. But although the NMR programme *was* judged as acceptable, this of course does not imply that judgement was based *largely* on those criteria. Instead, so it will be argued in the next section, a criterion like (GA1.2) should be seen as a *necessary* condition for success/acceptability, not as a sufficient one. Under the same proviso also (T1.2) is supported by my case. Finally, (T2.5) is definitely supported by my case – although I have not explicitly argued for that – whereas (T2.4) certainly is disconfirmed.

In the introduction I claimed that my results can be generalized to a type of modern natural science that, although abundant, does not coincide with *all* scientific research. Thus, while for *this* type of research – which was earlier called 'extensive research' by Weisskopf – (GA1.3) is (claimed to be) false, it might still be true for other types of research – e.g., for the 'intensive' type which Weisskopf contrasted with extensive research. Roughly, I think that Lakatos and his followers mainly addressed the latter, intensive type. For this type of research (GA1.3) *does* seem to contain some truth, even though the case studies of these authors are at crucial places unconvincing because of insufficient awareness of the methodological problems presented above. My 1986 contains the material for an *explanation* of this divergence in the factual rules for the acceptability of theories or guiding assumptions. This explanation, the structure of which strongly resembles Laudan's 'reticulated model' (Laudan 1984), refers both to differences in the complexity of the empirical situations being investigated and to differences in the cognitive aims governing the research in question. I will go into this explanation on another occasion.

7. THE IMPORTANCE OF EXTRINSIC SUCCESS

My case study cannot force a further decision between criteria ES and WHS. Especially ES might be considered as a *necessary* condition for success for a programme like NMR. But this criterion is not *sufficient* for bringing a programme like NMR the success that it actually had. In addition, the NMR programme fulfilled another requirement which, although not discussed by the philosophers of science addressed up till now was pivotal for its factual success. This condition is this: (some of) the theories from a programme like NMR must be *useful* as theories of measurement for revealing phenomena from domains that *other* research programmes, localized in neighboring fields of natural science, try to explain. In the case of the NMR programme (which derived its original motivation from nuclear physics), these other programmes were mainly found in chemistry. For it turned out that, once theories like the BPP theory, the theory of chemical shift, or the theory of hyperfine splitting, each of which explains certain details of the NMR resonance signal, had been developed, such theories could subsequently be *used* in making inferences concerning the structure and dynamics of molecules from the signal. The latter are phenomena from domains that *chemical* research programmes try to explain or predict. In order to explain such phenomena, one must be in the position to empirically measure or observe them: and this is exactly what the theories from the NMR programme enable one to do. I have called this type of success, which was obtained by the NMR programme, extrinsic, to distinguish it from the intrinsic success that the criteria discussed up till now were dealing with. I have argued that this criterion of extrinsic success was very influential, not merely in the appraisal of specific theories within the NMR programme, but also in determining the directions in which the entire development proceeded. Lack of space forces me to refer at this point to Zandvoort (1986). The analysis in that book culminates, in chapter VII, in a unifying treatment of Lakatos's model of scientific development – including the criterion HS – and the so-called finalization model of Böhme *et al.* (1983). The resulting *model of intrinsic and extrinsic success of research programmes* incorporates these two pre-decessors as special cases, and is more successful than either in dealing with developments from the bulk of modern physical science. My model sheds light on a type of *interdisciplinary relationship* which is extremely important for the progress of much of modern natural science, and which might in addition be of value for understanding, and possibly improving, those parts of science oriented explicitly to the solution of societal problems.

REFERENCES

Böhme, G., Van den Daele, W., Hohlfeld, R., Krohn, W., and Schäfer, W. (1983), *Finalization in Science*, (Boston Stud. in the Phil. of Sci. 77), Dordrecht: D. Reidel.

Gardner, M.R. (1982), 'Predicting novel facts', *Brit. J. Phil. Sci.* **33**, 1–15.

Howson, C. (ed.) (1976), *Method and Appraisal in the Physical Sciences*, Cambridge: Cambridge U.P.

Kuhn, T. S. (1970), *The Structure of Scientific Revolutions*, 2nd edn., Chicago: University of Chicago Press.

Kuipers, T. A. F. (1982), 'Approaching Descriptive and Theoretical Truth', *Erkenntnis* **18**, 343–378.

Kuipers, T. A. F., (ed.) (1987), *What is Closer-to-the-Truth?*, (Poznan Stud. in the Phil. of the Sci. and the Mum. 10), Amsterdam: Rodopi

Lakatos, I. (1978), *The Methodology of Scientific Research Programmes* (Philosophical papers vol. I, edited by J. Worrall and G. Curry), Cambridge: Cambridge U.P.

Laudan, L. (1977), *Progress and Its Problems*, Berkeley: University of California Press.

Laudan, L. (1984), *Science and Values*, Berkeley: University of California Press.

Laudan, L., Donovan, A., Laudan, R., Barker, P., Brown, H., Leplin, J., Thagard, P., and Wykstra, S. (1986), "Testing theories of scientific change", *Synthese* **69**, 141–223.

Rigden, J. (1986), 'Quantum States and Precession: the Two Discoveries of NMR', *Rev. Mod. Phys.* **58**, 433–448.

Zandvoort, H. (1986), *Models of Scientific Development and the Case of NMR*, (Synthese Library, 184), Dordrecht: D. Reidel.

Zandvoort, H. (1987); 'Verisimilitude and Novel Facts', in Kuipers (1987), pp. 229–251.

ELECTROWEAK UNIFICATION AND THE APPRAISAL OF THEORIES

1. INTRODUCTION

The period from 1970 to 1973 was one of very intense and productive activity in theoretical and experimental research in high energy physics (HEP). In particular in the area of weak-interaction physics enormous strides were taken and one of the major results was the unification of the electromagnetic and the weak interaction, the so-called electroweak unification. The community of HEP researchers working in this area was sufficiently small and clearly identified so that the complete historical and scientific reconstruction of what occurred during that period was possible. Such reconstructions have been given both by historians and physicists alike.[1] Thus, this episode offers a very fruitful as well as a contemporary case study to which one can apply and test the many philosophical theses associated with theory appraisal.

This episode in the recent history of HEP offers the student of science who is particularly interested in theory change examples of the following aspects: the choice between competing theories, the role of methodological and axiological criteria in theory choice, the role of anomalies as well as the prediction of heretofore unexpected phenomena, the modification of theories in the face of empirical difficulties, and the complex interrelationship between theory and experiment. The picture that emerges is far more complex than any simple text-book model[2] of scientific change and progress. As is clearly stated in the introductory essay in this volume, the purpose of this project is to emulate the very methods of science itself in order to understand and describe the process of scientific change. Usually in science, one must initially simplify and idealize the description of the phenomenon in order to withdraw the more salient features. The preliminary hypothesis or theory must be made to confront the empirical data that it purports to explain. This is usually accomplished by making measurements and performing experiments. Appropriate modifications or adjustments in the theory may be necessary and the theory becomes more mature or sophisticated as more of the complexities of the phenomenon are taken into account.

This paper then will be a first step in the long process of understanding and describing theory appraisal. The theses concerning theory appraisal represent this initial simplification and idealization of the system to be studied. The system under study in this project is a proposed but not yet completely specified model of scientific change and the sources of the empirical data are the historical reconstructions of the scientific events that exhibit scientific change. This episode in the recent history of HEP is one data point and does not settle the issue of the validity of any of these theses. Neither falsification nor verification of any thesis is absolute. Certainly, one must learn this from the history of science itself. The construction of a successful theory is a complex dynamical process that involves not only testing the theory by means of experiments but also using the empirical data as a guide for appropriate modifications. Certainly then the construction of a theory of scientific change must be an analogous process.

Specifically, the theses that will be considered and evaluated in this paper are the following: (T1.1), (T2.1), (T2.11), (T2.5), (T2.9) and (T2.10).

While these theses do not exhaust those associated with theory appraisal, they do lend themselves directly to evaluation using the historical/scientific details of the four-year period in HEP history under scrutiny. Furthermore, as the evaluation is made, certain sub-groupings of these theses will emerge and will provide direction for future consideration and modification. In order then to evaluate these theses, the historical events preceding the period in question will only be briefly sketched. The assumption here is that while many of the theoretical and experimental developments in the preceding twenty-year period were important (and may also in the future be used to test and evaluate other theses of scientific change), the specific developments from 1970 to 1973 will be more than adequate for empirical testing. Not only are the scientific works well documented but most of the principal scientists are and have been available for personal interviews.[3]

2. BRIEF HISTORICAL SUMMARY

The theory used to describe the weak interaction, which is exemplified by such phenomena as the decay of the free neutron and the positron or electron decay of unstable nuclei, was initially proposed in 1934 by

E. Fermi (1934) (he actually formalized a suggestion made by W. Pauli in 1930). Fermi modelled his theory on the new and successful theory of quantum electrodynamics (QED) which is the quantized-field-theory version of Maxwell's electrodynamics. However, in contrast to the electromagnetic interaction which is essentially of infinite range, the weak-interaction is very short ranged. Thus, while QED requires a massless particle, the photon, to propagate the quantized electromagnetic field, the quantized-field theory of the weak interaction requires a massive particle to propagate the weak-interaction field. These particles were appropriately given the symbol W, and because of their mass and quantum numbers are referred to as the Intermediate Vector Bosons (IVBs).[4]

By the late 1960s, the accepted working theory of weak interactions was the so-called VA theory.[5] The particles that served as the propagators of the weak-interaction field were the two charged states of the IVB, the $W^\pm$. The neutral state of the IVB, the Z^0, was not required to describe the weak interaction. Typical weak-interaction phenomena could be described using conventional Feynman diagrams[6] by means of the exchange between interaction vertices of either a positive or a negative IVB. It should be noted that no IVBs were experimentally observed because their hypothetical mass, of the order of 50 GeV, was not within the range of the energies of the contemporary accelerators. Some typical Feynman diagrams for some typical weak-interaction scattering phenomena are shown in Figure 1.

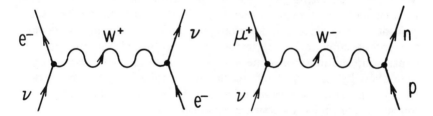

Fig. 1.   (a) lepton-lepton scattering,   (b) lepton-hadron scattering.

Events of the type that might involve a neutral IVB were not considered in the VA model. Thus, scattering events of the type shown below would not be expected to be observed experimentally. Indeed, this was the situation at the end of the 1960s. The upper limit on the observed fraction of scattering events that involved neutral IVBs, measured relative to the sample of those involving charged IVBs, was given as 10% by a major experiment performed at CERN (Cundy *et al.*, 1970). This was considered to be consistent with the non-existence of such events as described by the VA theory (see Figure 2).

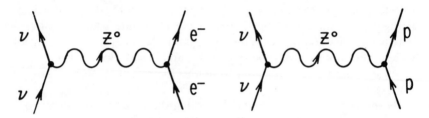

Fig. 2.   (a) lepton-lepton scattering,   (b) lepton-hadron scattering.

Within the HEP community these scattering events became known as charged-weak-current (CWC) and neutral-weak-current (NWC) events. This is in reference to the treatment of the scattering particles as leptonic or hadronic particle currents (in analogy to electrical currents) and the exchange of either a charged or a neutral IVB between the interaction vertices.

As mentioned earlier, the theory of weak interactions was strongly based upon the formalism and style of QED. However, one of the major problems facing QED was the apparent inability of the formalism to calculate exactly higher-order terms in the perturbation theory expansion of the Lagrangian describing the interaction.[7] Thus to calculate the cross-section for $e^- e^-$ scattering using this perturbation theory expansion, one must calculate a series of integrals, with each integral multiplied by the expansion parameter α raised to the appropriate power. The numerical value of α is much less than one.[8] Thus each integral is multiplied by a successively smaller fraction. Except for the first term in the expansion, each succeeding integral diverges, that is its numerical value cannot be analytically calculated. QED theorists thus had to be content with using only the first term in the expansion. This greatly limited the precision of their calculations and hindered their efforts to compare their results with the very precise experimental data.

However, in 1947 a group of HEP theorists outlined the solution to this problem.[9] They essentially showed that by calculating the divergent integrals and inserting the physical mass and charge of the electron at the end of the calculation, the divergences would disappear. This procedure, referred to as renormalization, can be applied to any desired degree of accuracy in the QED calculation. However, for those theorists working in weak-interaction physics, this procedure of renormalization was not applicable. Thus, the VA theory was useful only for first-order calculations. In addition to this problem, the higher-order contributions did not decrease with, but rather increased rapidly because of the massive exchange particles involved in the propagation of the weak-interaction field. The theorists ignored this in their calculations, but it certainly weakened their confidence in the model. Also, since the masses of the IVBs were experimentally unknown, the theorists essentially had to guess as to what values to use in their calculations. The VA theory then was at best a heuristic guide and a phenomenological model rather than a fundamental theory used to describe weak interactions.

In 1971 a Dutch graduate student, G. 't Hooft (1971a, 1971b) proved that another class of theories used in HEP could also be renormalized. This group of theories, the Yang–Mills gauge-invariant theories, [10] had been first introduced by Yang and Mills (1954). Although modelled after QED, these gauge-invariant theories were not renormalizable and thus were not immediately useful in weak-interaction theory. However, many physicists studied and developed various gauge-invariant models of both weak and strong interactions. In 1961 S. Glashow (1961) proposed a gauge-invariant model that in principle indicated the procedure for the unification of the weak and the electromagnetic interactions. However, the masses of the exchange particles had to be arbitrarily inserted into the calculations. A. Salam, J. C. Ward, and P. Higgs worked on methods for determining the mass and other properties of the exchange particles. In particular, Higgs developed a mechanism that generated the masses of the exchange particles in a consistent fashion, free of ad-hoc or arbitrary assumptions. In 1967 S. Weinberg (1967) proposed a gauge-invariant model of the electroweak unification that employed the Higgs mechanism. Independently of Weinberg's work, in 1968 A. Salam presented a similar model. Thus by the late 1960s, the Weinberg–Salam (WS) model, although only one of several different gauge-invariant theories, was a serious candidate for providing the description of a long-sought-after goal in HEP, the unification of the weak and the electromagnetic interactions. The basic stumbling block was still the problem of renormalization. Thus the work of 't Hooft in 1971, while seemingly out of

required fewer free parameters and it was relatively easy to interpret physically. Finally and most importantly it provided the means of unifying the electromagnetic and the weak interactions. HEP experimentalists through their interactions with the theorists recognized the importance of this theoretical advance and set about to search for these NWC events.

Prior to 1972, the neutrino-scattering experiments provided no evidence for NWC events. Only those scattering events that involved the exchange of a charged IVB ($W^\pm$) were observed. The measured cross-sections were in good agreement with the VA theory. This agreement was not unexpected since the VA theory used experimentally-measured parameters to calculate the various quantities of interest. In a very real sense, the NWC events were not observed because, one could argue, they simply were not needed to complete the description of weak interactions given by the extant theory.

HEP experimentalists at both CERN and FNAL[13] were preparing experiments to study lepton-lepton and lepton-hadron scattering. CERN was to employ a huge bubble-chamber analysis system while FNAL was to use spark-chamber analysis techniques. Both groups modified their experimental arrangements to specifically search for the NWC events. It must be emphasized that the experimental modifications were made using accepted experimental methodological and axiological criteria.[14] After a considerable amount of effort that involved six universities in the CERN collaboration and three universities in the FNAL collaboration, plus the research-lab personnel at both national labs, the positive evidence was presented to the scientific community. The CERN group first announced its results (F. J. Hasert *et al.*, 1973) and shortly thereafter, the FNAL group announced its results (B. Aubert *et al.*, 1973). Thus, by late 1973, the combined experimental evidence along with the fruitful modification of the WS model by means of the GIM mechanism, led to the general acceptance of the electroweak-unification theory. The WS model had not only replaced the VA theory as the primary description of weak interactions but had also emerged from among the many gauge-invariant theories as the primary description of electroweak unification.

3. DATA ANALYSIS

The process of theory appraisal and choice involved not only the choice of the WS model over the VA model but also the choice of the WS model (with the GIM modification) from among the many competing gauge-invariant theories. Of immediate importance was the proof, by 't Hooft, of the renormalizability of the gauge-invariant theories of the Yang–Mills type. With this, the analogy to QED as desired by Yang and Mills was complete. By the late 1960s, QED was certainly viewed by HEP researchers as the fundamental theory of HEP. In the language of this volume, QED constituted the set of guiding assumptions for HEP. The choice of the WS model over the VA theory was consistent within the broader context of QED since the WS model was modelled after QED. Even though the VA theory was also analogous to QED in many respects, it was not renormalizable. In summary, the WS model offered improved quantitative descriptive ability, because it was renormalizable. It offered improved simplicity, because it united two major interactions under the exchange mechanism that united the IVBs and the photon using the appropriate group theory techniques. It exhibited fertility or the capacity for expansion of its explanatory scope by the introduction of the GIM mechanism. Finally, one of the goals of HEP research, both theoretical and experimental, has been to unify the four basic interactions in nature. Thus, the WS model is consistent with the basic goals, beliefs and cognitive values of the HEP community. This belief is evident in a comment made in a letter from the HPWF[15] group to R. Wilson, the director of research at FNAL:

> There has been an increasing awareness of the need of more sensitive searches for neutral-weak currents and neutral-weak intermediate bosons. The existence of a neutral-weak current or a neutral-weak propagator would cast additional light on the connection between weak and electromagnetic interactions. As the center of mass energy, $S^{1/2}$, available to experiments increases, and GS moves closer in magnitude to α, the possibility of finding such a connection becomes more realistic. We might now stand in a position analogous to that of Oersted, Ampère, and Faraday 150 years ago as they attempted to elicit the connection between electricity and magnetism (Benvenuti *et al.*, 1972).

A final comment on this particular phase of theory appraisal is important before moving on to the role of empirical criteria. The choice of the WS model, with the appropriation of the GIM mechanism, over the VA theory and its subsequent emergence from among the other competing gauge-invariant theories as the dominant model, is more complex than the previous discussion has indicated. While the actual emergence of the WS model of electroweak unification appears to have been rapid (over the period from 1970 to 1974) the actual mechanics took several decades to develop. The proposal of the Yang–Mills type of gauge-invariant theories in 1954, the work of Glashow, Ward, Salam and Higgs in the early 1960s, the independent work of Weinberg and Salam in the late 1960s, Glashow, Iliopolous and Maiani in 1970, and the proof of renormalization by 't Hooft for Yang–Mills-type theories in 1971, all form an important series of interconnected events. The importance of internal consistency, clarity and coherency of concepts, and consistency within the broader context of QED as the set of guiding assumptions for HEP, were extremely important in each of these stages. For example, the problem of arbitrarily assigning exchange-particle properties had to be resolved and was done by Higgs and Weinberg independently. Symmetry considerations guided Glashow and the other HEP theorists in 1964 in the introduction of a new quantum number, charm, which was then later applied to the WS model in 1970. The paradox of the association of the massive neutral IVB with the massless photon was resolved by spontaneous symmetry breaking (SSB)[16]. Finally, renormalization provided the key. It put the gauge-invariant theories squarely in line with QED. The point is this: the assessment did not occur over a period of a few months or years but rather over a period of almost two decades.

The role of empirical criteria in the emergence of the WS model is a much more complicated process than it appears to be initially. In general HEP experiments are not merely tests of the epistemological status of a theory. HEP experiments (as well as others) are quite often designed without any specific theoretical motivation. In this case the HEP experimentalists and theorists worked pretty much independently of each other until the new theory was taken seriously by the theoretical community. If we focus on the SNC decays, as exemplified by the decay

$$K^0 \rightarrow \mu^+ + \mu^-$$

and on the NWC scattering as exemplified by the lepton-hadron scattering,

$$\nu + p \rightarrow \nu + p$$

we can develop a preliminary sense of the applicability of various theses (T1.1, T2.1, T2.11, T2.5, T2.9 and T2.10). The NWC events were not an empirical problem for the VA theory, so they did not represent any sort of anomaly (T1.1). The WS model without the GIM mechanism predicted significant measurable cross sections for both strangeness-conserving and SNC events, and so the latter posed an empirical problem for this model. However, with the modification proposed by the GIM mechanism, an apparent counter example became a success-fully solved empirical problem (T2.1). As discussed earlier, the WS model did not pose any conceptual difficulties and at the time(1971–1973), it solved the empirical problems that it confronted (T2.11). Thus in this case, (T1.1) seems to be inapplicable while (T2.1) and (T2.11) seem to be appropriate for the description of what occurred.

The experiments performed at CERN and FNAL were certainly motivated by the enthusiasm and interest in the WS model. In an interview with P. Galison in 1980, C. Rubbia[17] remarked;

Steve Weinberg was the one who, with rare insistence . . was chasing me and many other people (to do the neutral-current search). I learned all these things (about gauge-theories and neutral currents) from him directly. I remember I was down in the old cyclotron at 44 Oxford Street. He called me up in the beginning I thought, my God, what (is) he asking me? (Then) I realized how beautiful things were (Galison, 1983, p.494)

However, the assumptions used to perform these experiments were not drawn directly from the model. The WS model predicted measur-able cross-sections for NWC events as well as CWC events. The details of the theory concern the properties of the exchange particles, the techniques for calculating the cross-sections and the like. In contrast to this, the experimental set up is concerned with maximizing the condi-tions that are appropriate to observing the scattering events while at the same time not introducing any favorable or unfavorable bias. Such problems as background scattering events, energy distribution, beam purity and the like are experimental problems to be solved using established HEP experimental procedures. Thesis (T2.5) is definitely appropriate in this case. This thesis touches upon the problem of theory-laden observations and draws a rather clear distinction. Those observations which rely upon some background theory, and do not rely upon the theory under study, do not offer any philosophical problems. On the other hand, observations which are closely intertwined with the very theory in question would certainly raise serious doubts about the validity of the results. In the present case, the NWC experiments fall into the former category. Thesis (T2.5) is very important and in a certain sense is both a descriptive as well as a normative principle.[18]

It must be emphasized that while the HEP experimentalists were motivated and guided by their theoretical counterparts, their experiments were designed and performed within the well-defined limits of their established methodology. As pointed out earlier, the basic goal of electroweak unification was clearly understood by the experimentalists. However, their detached and unbiased approach is evident in these comments by A.K. Mann and D. Cline (see note 17):

> You can say, well, we came to the conclusion immediately that we had seen weak neutral currents. But you'd be surprised, that was the last conclusion we came to. Our first conclusion was that we were making some mistake and that these muons were somehow escaping the apparatus or being missed by us in some way and that no effect of that magnitude could exist (Galison, 1983, p. 494, interview with A. K. Mann).

> Three pieces of evidence now in hand point to the distinct possibility that a μ-less signal of order 10% is showing up in the data. At present I don't see how to make these effects go away (Galison, 1983, internal FNAL memo from D. Cline)

In short, the experimentalists were influenced by the theorists in a completely unbiased fashion. The experimentalists knew what was at stake but would not allow any bias or prejudice to influence their work because of their strong commitment to their methodological values and criteria.

The last two theses (T2.9) and (T2.10) are closely related. The former identifies those experiments which are crucial for a theory in a singular sense, i.e., there is no other theory in direct competition and thus an 'either or' situation does not develop. The latter describes the situation in which an experiment does allow one to choose one theory over its rivals. The observation of the NWC scattering events clearly identified the WS model as the preferable theory over the VA theory. Thesis (2.10) is appropriate for this aspect. However, (T2.9) is also appropriate in the sense that with the modification of the WS by means of the GIM mechanism, the agreement between the predicted and measured cross-section for the decay process

$$K^0 \rightarrow \mu^+ + \mu^-$$

served as a successful crucial test. Even though (T2.9) and (T2.10) may seem to be mutually exclusive, the circumstances may be such that both are appropriate. The WS model therefore faced two distinct types of tests. The first involved a direct comparison of the WS model with the

VA theory. The second involved a direct comparison of the WS model with the data, and no direct comparison with any other theory.

An interesting point arises when one considers the significance of these crucial tests, be they singular or plural in their importance to theories. Recall that the WS model did not initially describe correctly the SNC decays. This did not result in the rejection of the model but rather stimulated further research and led to the incorporation of the GIM mechanism. The basic simplicity, coherence and potential for growth that the WS model exhibited encouraged the HEP theorists to correct the deficiency. Once corrected, the crucial test was crucial no more. Initial failure led to ultimate success and great improvement. Secondly, the observation of NWC scattering events was taken as a confirmation of the power of the WS model. However, the VA theory really said nothing about these events. It is not that they were forbidden, but rather that they were simply not needed to describe the weak-interaction phenomena known and understood up to that point in time. The experiments did not report any significant evidence for them and the theory did not need them. So, upon close inspection, (T2.9) and (T2.10) really only assume a descriptive role and serve very little purpose when interpreted as normative principles. Historically speaking, most crucial experiments seem to be crucial only after the fact and then quite often many years after the fact.

4. CONCLUSIONS

Before drawing any conclusions it should be emphasized that the conclusions are based only upon this particular historical episode and that the theses under study represent preliminary idealizations. Just as in the case of a scientific experiment, the conclusions that one can draw are limited to that experiment and are limited by the degree of simplification that is used to set up the experiment. Continuing with this thought, experiments are rarely repeated in exact form, they are refined and updated. Succeeding experiments build upon their predecessors. Thus this philosophical experiment should be viewed in the same manner.

Now let us move on to draw some first-round conclusions. Unsolved empirical problems seem to play a surprisingly small role in the process of theory assessment (let us distinguish between the assessment of a single theory and the comparison of two or more theories). Theorists seem to view these as merely temporary problems. Thus the K^O decay did not result in the dismissal of the WS model. Rather it led to a very fruitful modification, the GIM mechanism with the introduction of the quantum-number charm. Woven into this was the role that symmetry and internal coherence played in the introduction of a fourth quark, associated with charm, by Glashow and others in 1964. Before the K^O decay presented problems to the WS model, charm had been introduced into the theoretical scheme of HEP. GIM successfully applied it then to the WS model in 1970 and resolved the empirical problem. The important point here is that the empirical problem did not influence the development of this concept.

After a review of the various theses associated with theory appraisal it seems that by delineating the various aspects, the true nature of the appraisal process has been lost. Scientists do not consider the idea of turning counter examples into solved problems in isolation from other factors. As outlined above, the temporal aspect (charm was invented independently of the K^0 decay problem) as well as the formal aspects (symmetry, predictive accuracy, etc.) are dynamically related. The WS model was accepted as the working model of electroweak unification for a number of factors, all of which occurred in a complicated network of interrelated influences. No single factor carried the argument in favor of the WS model.

Consider the empirical evaluation of the theory. First of all it was not independent of the theoretical evaluation (consistency, coherency, etc.). The absence of NWC scattering events was not an anomaly for the VA theory. What motivated the interest in the WS model as well as other gauge-invariant theories was the proof of renormalizability. They were shown to be consistent with the general principles of QED. The prediction of NWC scattering events by the WS model was viewed as a crucial test not so much because the VA theory did not require them but because of the manner in which they were predicted by the WS model. The neutral IVB is an integral part of the structure of the WS model. There is nothing arbitrary in the predicted cross-sections for the various NWC scattering events. Furthermore, the modification of this model introduced by the GIM mechanism was done in a coherent, consistent fashion rather than as a simple *ad-hoc* procedure to remove anomaly. The empirical assessment then does not constitute a simple verification or falsification of the theory. The two aspects of the scientific endeavor, theory and experiment, actually proceed in a more complicated fashion.

The relationship is symbiotic, each feeding off the other. They motivate and justify each other. Thesis (T2.5) therefore is extremely important. It must be stressed that the experiments that are performed (and the use of the term 'test' is too close in connotation to falsification or verification) are sometimes guided and motivated by specific theories but are not based in principle upon those theories. As mentioned earlier, this thesis is not only descriptive of good science but normative as well. Theses (T2.9) and (T2.10) then could be treated as corollaries of (T2.5). Experiments do allow for a crucial assesment of a theory either in a singular sense or in a plural sense as discussed in the previous section.

In conclusion, based upon this one historical episode, the assessment or appraisal of the WS model involved several interlocking aspects. One can split the assessment into a formal and an empirical assessment. The former pertains to theoretical criteria such as consistency (internal and external), coherency, capacity for expansion and so forth. It is impossible to separate these, or weight them or organize them into some hierarchical structure. They all influence the evaluation and they will assume different relative degrees of importance depending upon the empirical problems. The other aspect, the empirical assessment, is also a more dynamical mechanism. It is motivated by theory yet guided by its own methodological and even sometimes axiological criteria.

The theses that have been considered in this paper may be summarized in the form of a conclusion and a corollary.[19]

CONCLUSION: *The empirical appraisal of a theory is based upon phenomena which can be detected and measured without using assumptions drawn from the theory under evaluation* (T2.5) *and the first part of* (T2.11).

COROLLARY: *Under certain circumstances scientists can identify certain phenomena as empirical problems that can be used to test a single theory or as a test to choose between two or more rival theories* (T1.1, T2.1, T2.9 and T2.10).

NOTES

[1]The recent work of P.Galison (1983) and A. Pickering (1984) provide an excellent historical reconstruction of this period in HEP. The primary sources such as the paper by

S. Weinberg (1967) or the two papers by G. 't Hooft (1971a, 1971b) are listed in the bibliographies of both of these works. Because such a complete list of primary sources would be very lengthy, only those to which direct reference is made will be listed in the bibliography of this paper. The interested reader may refer to the very extensive bibliography in Pickering's book.

[2]By the phrase 'text-book model' is meant that model as described by T.S. Kuhn in the introductory section of *The Structure of Scientific Revolutions*.

[3]Refer to Galison (1983) for a number of enlightening interviews with physicists such as C. Rubbia and S. Weinberg. Pickering (1984) has also made ample use of such first-person interviews and recollections.

[4]These are particles of spin one and negative parity.

[5]This is short-hand notation for vector current minus axial-vector current. In the interaction equation, in addition to a vector term there is also an axial-vector term (spin one and positive parity). The construction of this equation involves the subtraction of the axial-vector current from the vector current term. R. Feynman and M. Gell Mann (1958) and independently R. Marshak and E. Sudarshan (1958) introduced this formalism.

[6]Feynman diagrams such as those in Figures 1 and 2 represent one of the several techniques that can be employed in the calculation of the terms in the perturbation-theory expansion of the interaction equation (referred to as the Lagrangian, refer to note 7).

[7]The equation that is written down is referred to as the Lagrangian. It is a quantized version of the classical expression which represents the kinetic and potential energies of the particles involved. It is the form of the potential energy function that specifically describes the dynamics of the interaction.

[8]α is the so-called fine-structure constant. It is a dimensionless quantity with a numerical value of 1/137. It appears in many of the QED calculations.

[9]A conference on the foundations of quantum mechanics was held in June 1947 on Shelter Island, New York. Theoretical discussions concerning the recent experimental work of W. Lamb and R.C. Retherford on the hydrogen spectrum led to the initial outline of the solution of the renormalization problem. In the Spring of 1948 a second conference was held at Pocono Manor in Pennsylvania at which J. Schwinger and R. Feynman presented their approaches to the solution of the renormalization problem. S.I. Tomanaga in Japan had independently developed a solution to this problem as well and all three published their work in 1947 and 1948. F. Dyson in 1949 showed that their different approaches were equivalent. Feynman's approach entailed the use of diagrams such as those in Figures 1 and 2. These have the most recognizable form of the technique of renormalization.

[10]In classical electromagnetic theory, certain transformations of the potential energy functions can be performed without affecting the physical predictions of the theory. This gauge (measurement) invariance carried over into QED. Thus in constructing a theory to describe the weak or the strong interaction, Yang and Mills were guided by QED to construct analogous gauge-invariant theories.

[11]Refer to Weinberg's (1974) general review of the state-of-the-art of gauge-invariant theories in which he discusses natural and realistic theories.

[12]In 1964 J.D. Bjorken and S.L. Glashow (1964), along with Y. Hara (1964), Z. Maki and Y. Ohnuki (1964), and D. Amati, H. Bacry, J. Nuyts and J. Prentki (1964) published work that was important in the development of the concept of charm.

[13]CERN is the acronym for the *Organisation Européenne Pour La Recherche Nucleaire* and FNAL is the acronym for the Fermi National Accelerator Laboratory.

[14]Refer to A. Pickering (1984) and M. J. Hones (1987) for two different perspectives concerning the role of methodology and axiology in these HEP experiments.

[15]HPWF refers to collaboration of Harvard University, The University of Pennsylvania, The University of Wisconsin and FNAL. This group performed the spark-chamber experiments at FNAL.

[16]Basically SSB involves the understanding that the symmetry present in the Lagrangian for a particular interaction might not be present in the physical system itself. Thus the physical spectrum of particles might not correspond to those simply read-off from the Lagrangian. Nature in a sense spontaneously breaks the mathematical symmetry of the Lagrangian. Superconductivity and ferromagnetism are examples of this phenomenon. Specifically in the case of the WS model, SSB enabled one to pair the massless photon with the massive neutral state of the IVBs, the Z^0, in a particular group-theory representation. The spontaneous breaking of the particular symmetry of this group chosen by WS leads to two neutral particles, one massless and the other massive, and both with specific quantum numbers.

[17]C. Rubbia, A. K. Mann and D. Cline were the organizers of the HPWF collaboration.

[18]D. Shapere (1982) has discussed this problem in detail. His clear distinction between background theory and the theory under study is very helpful in understanding this thesis.

[19]This attempt to summarize the philosophical analysis of the past fifteen pages is an excellent example of the physicist's penchant for expressing his or her results in the form of a few simple equations or theorems. This is known more formally as the virtue or cognitive value of simplicity.

REFERENCES

Amati, D., Bacry, H., Nuyts, J., and Prentki, J. (1964), 'SU$_4$ and Strong Interactions', *Phys. Lett.* **11**, 190–192.

Aubert, B. *et al.* (1974), 'Further Observations of Muonless Neutrino-Induced Inelastic Interactions', *Phys. Rev. Lett.* **32**, 1454–1457.

Benvenuti, A., Cline, D., Imlay, R. L., Ford, W., Mann, A. K., Pilcher, J. E., Reeder, D. D., Rubbia, C., and Sulak, L. R. (1972), letter to R. R. Wilson, 14 March 1972.

Bjorken, J. D. and Glashow, S. L. (1964), 'Elementary Particles and SU$_4$', *Phys. Lett.* **11**, 255–257.

Cundy, D. B. *et al.* (1970), 'Upper Limits for Diagonal and Neutral Current Couplings in the CERN Neutrino Experiments', *Phys. Lett.* **31B**, 478–480.

Fermi, E. (1934), 'Versuch Einer Theorie der β-Strahlen', *Zeit. Phys.* **88**, 161–171.

Feynman, R. P. and Gell-Mann, M. (1958), 'Theory of the Fermi Interaction', *Phys. Rev.* **109**, 193–198.

Galison, P. (1983), 'How the First Neutral-Weak-Current Experiments Ended', *Rev. Modern Phys.* **55**, 477–509.

Glashow, S. L. (1961), 'Partial Symmetries of Weak Interactions', *Nuclear Phys.* **22**, 579–588.

Glashow, S. L., Iliopoulos, J., and Maiani, L. (1970), 'Weak Interactions with Lepton-Hadron Symmetry', *Phys. Rev.* **D2**, 1285–1292.

Hara, Y. (1964), 'Unitary Triplets and the Eightfold Way', *Phys. Rev.* **134B**, 701–704.

Hasert, F. J. *et al.* (1973), 'Search for Elastic Muon-Neutrino Electron Scattering', *Phys. Lett.* **46B**, 121–124.

Hones, M. J. (1987), 'The Neutral-Weak-Current Experiments: A Philosophical Perspective', *Studies in History and Philosophy of Science*, vol. 18, No. 2, pp. 221–251.

Kuhn, T. S. (1970), *The Structure of Scientific Revolutions*, Chicago: University of Chicago Press.

Kuhn, T. S. (1977), 'Objectivity, Value Judgments, and Theory Choice', in *The Essential Tension*, Chicago: University of Chicago Press, pp. 320–340.

Laudan, L. (1977), *Progress and Its Problems*, Berkeley: University of California Press.

Laudan, L. (1984), *Science and Values*, Berkeley: University of California Press.

Maki, Z. and Ohnuki, Y. (1964), 'Quartet Scheme for Elementary Particles', *Prog. Theoret. Phys.* **32**, 144–158.

Pickering, A. (1984), 'Against Putting the Phenomena First: The Discovery of the Weak-Neutral Current', *Studies in History and Philosophy of Science*, vol. 15, No. 2, pp. 85–117.

Pickering, A. (1984), *Constructing Quarks*, Chicago: University of Chicago Press.

Salam, A. (1968), 'Weak and Electromagnetic Interactions', in *Elementary Particle Physics* (edited by N. Svartholm); Stockholm: Almquist and Wiksills.

Shapere, D. (1982), 'The Concept of Observation in Science and Philosophy', *Philosophy of Science* **49**, 485–525.

Sudarshan, E. C. G. and Marshak, R. E. (1958), 'Chirality Invariance and the Universal Fermi Interaction', *Phys. Rev.* **109**, 1860–1862.

't Hooft, G. (1971a), 'Renormalizable Lagrangians for Massless Yang–Mills Fields', *Nuclear Phys.* **B33**, 173–199.

't Hooft, G. (1971b), 'Renormalizable Lagrangians for Massive Yang–Mills Fields', *Nuclear Phys.* **B35**, 167–188.

Weinberg, S. (1967), 'A Model of Leptons', *Phys. Rev. Lett.* **19**, 1264–1266.

Weinberg, S. (1974), 'Recent Progress in Gauge Theories of the Weak, Electromagnetic and Strong Interactions', *Rev. Mod. Phys.* **46**, 255–277.

Yang, C. N. and Mills, R. (1954a), 'Isotopic Spin Conservation and a Generalized Gauge Invariance', *Phys. Rev.* **95**, 631.

Yang, C. N. and Mills, R. (1954b), 'Conservation of Isotopic Spin and Isotopic Gauge Invariance', *Phys. Rev.* **96**, 191–195.

TABULAR SUMMARY OF RESULTS

Introductory Note: In the following table each theoretical claim is preceded by a letter indicating its standing after being examined in one or more of the historical case studies included in this book. **C** indicates Confirmed, **R** indicates Refuted, and **N** indicates No Consensus was reached or the topic was not addressed. These results are presented in a different form in part one, pp. 29–30 and 39–40.

Although we use the terms 'confirmed' and 'refuted', we realize our results are indicative rather than definitive. Elsewhere (pp. 14–40) we have given our reasons for reaching the conclusions reported and have cited the evidence we found persuasive. Other scholars are encouraged to re-examine these preliminary judgments by bringing additional empirical evidence to bear.

Readers interested in locating evidence and arguments for and against individual theses should consult the 'Index of Thesis Citations and Evaluations'.

THESES ABOUT GUIDING ASSUMPTIONS

Guiding Assumptions 1: Acceptability
GA1. The acceptability of a set of guiding assumptions is judged largely on the basis of:
    N 1.1 Empirical accuracy;
    C 1.2 The success of its associated theories at solving problems;
    R 1.3 The success of its associated theories at making novel predictions;
    C 1.4 Its ability to solve problems outside the domain of its initial success;
    R 1.5 Its ability to make successful predictions using its central assumptions rather than assumptions invented for the purpose at hand.

Guiding Assumptions 2: Anomalies
GA2. When a set of guiding assumptions runs into empirical difficulties:
    N 2.1 Scientists believe this reflects adversely on their skill rather than on inadequacies in the guiding assumptions;
    R 2.2 Scientists are prepared to leave the difficulties unresolved for years [the cases examined suggest this thesis should be modified];
    R 2.3 Scientists often refuse to change those assumptions;
    R 2.4 Scientists ignore the difficulties as long as the guiding assumptions continue

to anticipate novel phenomena successfully [the cases examined suggest this thesis should be modified];

N 2.5 Scientists believe those difficulties become grounds for rejecting the guiding assumptions only if they persistently resist solution;

N 2.6 Scientists introduce hypotheses which are not testable in order to save the guiding assumptions;

N 2.7 Those difficulties become acute only if a rival theory explains them.

Guiding Assumptions 3: Innovation

R GA3. New sets of guiding assumptions are introduced only when the adequacy of the prevailing set has already been brought into question.

Guiding Assumptions 4: Revolutions

GA4. During a change in guiding assumptions (i.e., a scientific revolution):

R 4.1 Scientists associated with rival guiding assumptions fail to communicate;

N 4.2 A few scientists accept the new guiding assumptions, which fosters rapid change, but resistance intensifies when change appears imminent;

R 4.3 Guiding assumptions change abruptly and totally;

R 4.4 The entire scientific community changes its allegiance to the new guiding assumptions;

N 4.5 Younger scientists are the first to shift and then conversion proceeds rapidly until only a few elderly holdouts exist.

THESES ABOUT THEORIES

Theories 1: Inter-theory relations

T1. When confronted with rival theories, scientists prefer a theory which:

C 1.1 Can solve some of the empirical difficulties confronting its rivals;

C 1.2 Can solve problems not solved by its predecessors;

R 1.3 Can solve all the problems solved by its predecessors plus some new problems.

Theories 2: Appraisal

T2. The appraisal of a theory:

C 2.1 Depends on its ability to turn apparent counter-examples into solved problems;

C 2.2 Depends on its ability to solve problems it was not invented to solve;

N 2.3 Is based on the success of the guiding assumptions with which the theory is associated;

R 2.4 Is based entirely on those phenomena gathered for the express purpose of testing the theory and which would not be recognized but for that theory;

C 2.5 Is based on phenomena which can be detected and measured without using assumptions drawn from the theory under evaluation [the cases examined suggest that this thesis is confirmed, but with some qualifications];

R 2.6 Is usually based on only a very few experiments, even when those experiments become the grounds for abandoning the theory;

N 2.7 Is sometime favorable even when scientists do not fully believe the theory (specifically when the theory shows a high rate of solving problems);

C 2.8 Is relative to prevailing doctrines of theory assessment and to rival theories in the field;

C 2.9 Occurs in circumstances in which scientists can usually give reasons for identifying certain problems as crucial for testing a theory;

C 2.10 Depends on certain tests regarded as 'crucial' because their outcome permits a clear choice between contending theories;

N 2.11 Depends on its ability to solve the largest number of empirical problems while generating the fewest important anomalies and conceptual difficulties.

INDEX OF THESIS CITATIONS AND EVALUATIONS

THESES ABOUT GUIDING ASSUMPTIONS

Guiding assumptions 1: Acceptability

GA1.   The acceptability of a set of guiding assumptions is judged largely on the basis of:
1.1 empirical accuracy;
1.2 the success of its associated theories at solving problems;
1.3 the success of its associated theories at making novel predictions;
1.4 its ability to solve problems outside the domain of its initial success;
1.5 its ability to make successful predictions using its central assumptions rather than assumptions invented for the purpose at hand.

Guiding assumptions 2: Anomalies

GA2.   When a set of guiding assumptions runs into empirical difficulties:
2.1 scientists believe this reflects adversely on their skill rather than on inadequacies in the guiding assumptions;
2.2 scientists are prepared to leave the difficulties unresolved for years;
2.3 scientists often refuse to change those assumptions;
2.4 scientists ignore the difficulties as long as the guiding assumptions continue to anticipate novel phenomena successfully;
2.5 scientists believe those difficulties become grounds for rejecting the guiding assumptions only if they persistently resist solution;
2.6 scientists introduce hypotheses which are not testable in order to save the guiding assumptions;
2.7 those difficulties become acute only if a rival theory explains them.

Guiding assumptions 3: Innovation

GA3. New sets of guiding assumptions are introduced only when the adequacy of the prevailing set has already been brought into question.

Guiding assumptions 4: Revolutions

GA4.   During a change in guiding assumptions (i.e., a scientific revolution):
4.1 scientists associated with rival guiding assumptions fail to communicate;
4.2 a few scientists accept the new guiding assumptions, which fosters rapid change, but resistance intensifies when change appears imminent;
4.3 guiding assumptions change abruptly and totally;
4.4 the entire scientific community changes its allegiance to the new guiding assumptions;
4.5 younger scientists are the first to shift and then conversion proceeds rapidly until only a few elderly holdouts exist.

THESES ABOUT THEORIES

Theories 1: Inter-theory relations
T1. When confronted with rival theories, scientists prefer a theory which:
    1.1 can solve some of the empirical difficulties confronting its rivals;
    1.2 can solve problems not solved by its predecessors;
    1.3 can solve all the problems solved by its predecessors plus some new problems.

Theories 2: Appraisal
T2. The appraisal of a theory:
    2.1 depends on its ability to turn apparent counter-examples into solved problems;
    2.2 depends on its ability to solve problems it was not invented to solve;
    2.3 is based on the success of the guiding assumptions with which the theory is associated;
    2.4 is based entirely on those phenomena gathered for the express purpose of testing the theory and which would not be recognized but for that theory;
    2.5 is based on phenomena which can be detected and measured without using assumptions drawn from the theory under evaluation;
    2.6 is usually based on only a very few experiments, even when those experiments become the grounds for abandoning the theory;
    2.7 is sometime favorable even when scientists do not fully believe the theory (specifically when the theory shows a high rate of solving problems);
    2.8 is relative to prevailing doctrines of theory assessment and to rival theories in the field;
    2.9 occurs in circumstances in which scientists can usually give reasons for identifying certain problems as crucial for testing a theory;
    2.10 depends on certain tests regarded as 'crucial' because their outcome permits a clear choice between contending theories;
    2.11 depends on its ability to solve the largest number of empirical problems while generating the fewest important anomalies and conceptual difficulties.

CONTRIBUTORS

Brian S. Baigrie is a philosopher of science. His recent publications include 'Relativism, truth, and progress', *Transactions of the Royal Society of Canada*, ser. 5, vol. 4 (1990), 9–19; and 'The justification of Kepler's ellipse', *Studies in History and Philosophy of Science* **21** (1990), 633–644. Address: Institute for History and Philosophy of Science and Technology, Room 313, Victoria College, University of Toronto, Toronto, Canada M5S 1K7.

William Bechtel is primarily interested in the philosophical foundations of cognitive science and the history and philosophy of biochemistry and cell biology. Among his recent publications are *Philosophy of Science: An Overview for Cognitive Science* (Hillsdale, N.J.: Lawrence Erlbaum Associates, 1988); with A. Abrahamsen, *Connectionism and the Mind* (Oxford: Blackwell, 1991); and, with R. C. Richardson, *Discovering Complexity* (Princeton: Princeton University Press, 1992). Address: Department of Philosophy, Georgia State University, Atlanta, GA 30303-3080.

Arthur M. Diamond, Jr., is an economist with a special interest in the economics of science. He publications include 'The determinants of a scientist's choice of research projects', in Tamara Horowitz, Allen I. Janis and Gerald J. Massey (eds.), 'Scientific Failure' (in press); and 'Science as a rational enterprise', in *Theory and Decision* **24** (1988), 147–167. Address: Department of Economics, University of Nebraska, Omaha, NE 68182-0048.

Betty Jo Teeter Dobbs is an historian and Newton specialist who studies Renaissance and early modern cosmologies, with particular emphasis on the historical significance of alchemy. Her recent publications include *Alchemical Death and Resurrection: The Significance of Alchemy in the Age of Newton* (Washington, D.C.: Smithsonian Institution Libraries, 1990); and *The Janus Faces of Genuis: The Role of Alchemy in Newton's Thought* (Cambridge: Cambridge University Press, 1992). Address:

Department of History, Voorhies Hall, University of California, Davis, CA 95616.

Arthur Donovan is an historian of science and technology. His publications include *Philosophical Chemistry in the Scottish Enlightenment* (Edinburgh: Edinburgh University Press, 1975); 'Lavoisier — Science, Administration, and Public Life in Eighteenth-Century France' (in press); and 'Education, Industry, and the American University', in Robert Fox and Anna Guagnini (eds.), 'Education and Industrial Performance in Modern Europe' (in press). Address: Department of Humanities, U.S. Merchant Marine Academy, Kings Point, NY 11024.

Maurice Finocchiaro is a philosopher and historian of science. Some of his many publications are *Galileo and the Art of Reasoning* (Dordrecht: Kluwer, 1980); and *The Galileo Affair: A Documentary History* (Berkeley: University of California Press, 1989). Address: Department of Philosophy, University of Nevada, Las Vegas, NV 89154.

Henry Frankel is a philosopher and historian of science with a special interest in the modern earth sciences. His recent publications include 'From continental drift to plate tectonics', *Nature* **335** (1988), 122–130; and 'The development of plate tectonics by J. Morgan and D. McKenzie', *Terra Nova* **2** (1990), 202–214. Address: Department of Philosophy, University of Missouri, Kansas City, MO 64110.

James R. Hofmann concentrates on research in the history and philosophy of nineteenth- and twentieth-century physics. He is the author of 'How the models of chemistry vie', *Proceedings of the Philosophy of Science Association* **1** (1990), 405–419. He is currently working on a biography of Andre-Marie Ampere. Address: Philosophy Department, California State University, Fullerton, CA 92634.

Michael J. Hones is an experimental physicist primarily interested in the philosophy of science. His recent publications include 'Reproducibility as a methodological imperative in experimental research', *Proceedings of the Philosophy of Science Association* **1** (1990), 585–599; and 'Scientific realism and experimental practice in high energy physics', *Synthese* **86** (1991), 29–76. Address: Physics Department, Villanova University, Villanova, PA 19085.

Larry Laudan is a philosopher of science whose books include *Progress and Its Problems* (Berkeley: University of California Press, 1977); *Science and Hypothesis* (Dordrecht: D. Reidel, 1981); *Science and Values* (Berkeley: University of California Press, 1984); and *Science and Relativism* (Chicago: University of Chicago Press, 1990). Address: Department of Philosophy, University of Hawaii, Honolulu, HI 96822.

Rachel Laudan studies the history and philosophy of science and technology. Her most recent book is *From Mineralogy to Geology* (Chicago: University of Chicago Press, 1987), and she is currently completing a study tentatively titled 'Science, Progress and History'. Address: Department of General Science, University of Hawaii, Honolulu, HI 96822.

Ronald Laymon is a philosopher of science with a particular interest in the use of idealizations and approximations in science and law. His recent publications include 'Using Scott Domains to explicate the notions of approximate and idealized data', *Philosophy of Science* **54** (1987), 194–221; and 'Cartwright and the lying laws of physics', *Journal of Philosophy* **86** (1989), 353–372. He is currently working on a book entitled 'Not Quite Right: Approximations and Idealizations in Science'. Address: Department of Philosophy, 350 University Hall, Ohio State University, Columbus, OH 43210.

Seymour H. Mauskopf is an historian of science with a special interest in marginal sciences, such as parapsychology, and in the historical relations between science and technology, especially in the study and production of explosives and munitions. Among his publications are 'Marginal science', in R. C. Olby, *et al.* (eds.), *Companion to the History of Modern Science* (London: Routledge, 1990), pp. 869–885; and 'Chemistry and cannon: J.-L. Proust and gunpowder analysis', *Technology and Culture* **31** (1990), 398–426. Address: Department of History, Duke University, Durham, NC 27706.

Deborah G. Mayo is a philosopher of science who is especially interested in problems of statistical inference. With Rachelle Hollander, she has edited *Acceptable Evidence: Science and Values in Risk Management* (Oxford: Oxford University Press, 1991), and is the author of 'Novel evidence and severe tests', *Philosophy of Science* **58** (1991), 523–552.

She is currently working on a book entitled 'Error and the Growth of Experimental Knowledge'. Address: Department of Philosophy, Virginia Polytechnic Institute and State University, Blacksburg, VA 24061.

John M. Nicholas is an historian and philosopher of science. He has edited *Moral Priorities in Medical Research* (Toronto: Hannah Institute, 1988), and is the author of 'Realism for shopkeepers: behavioralist notes on constructive empiricism', in J. Brown, *et al.*, *An Intimate Relation* (Boston: Kluwer, 1989), pp. 459–476. He is currently working on a study of force concepts in seventeenth-century mechanics and optics. Address: Department of Philosophy, University of Western Ontario, London, Ontario, Canada N6A 3K7.

Richard Nunan is a philosopher with special interests in the philosophy of science, applied ethics, and the philosophy of law. Among his publications are 'Novel facts, Bayesian rationality, and the history of continental drift', *Studies in History and Philosophy of Science* 15 (1984), 289–314; and 'Heuristic novelty and the asymmetry problem in Bayesian confirmation theory', forthcoming in *British Journal for the Philosophy of Science*. He has recently interviewed a number of leading European geologists and is completing several articles on modern geological controversies. Address: Department of Philosophy, College of Charleston, Charleston, SC 29424.

C. E. Perrin, an historian of chemistry and a well-known and widely published expert on Antoine Lavoisier and the chemical revolution of the eighteenth century, died in 1988. Two of his many articles are 'The triumph of the antiphlogistonians', in Harry Woolf (ed.), *The Analytic Spirit: Essays in the History of Science in Honor of Henry Guerlac* (Ithaca, N.Y.: Cornell University Press, 1981), pp. 40–63, and 'Research traditions, Lavoisier, and the chemical revolution', in Arthur Donovan (ed.), *The Chemical Revolution* (*Osiris*, ser. 2, vol. 4, 1988), 53–81.

Alan J. Rocke is an historian of science primarily interested in nineteenth-century chemistry. His books include *Chemical Atomism in the Nineteenth Century: From Dalton to Cannizzaro* (Columbus: Ohio State University Press, 1984), and *The Quiet Revolution: Hermann Kolbe and the Science of Organic Chemistry* (Berkeley: University of California

Press, in press). Address: Program in History of Technology and Science, Case Western Reserve University, Cleveland, OH 44106.

Henk Zandvoort is affiliated with the Delft University of Technology and is primarily interested in the philosophy of the technical sciences. He has written *Models of Scientific Development and the Case of NMR* (Dordrecht: D. Reidel, 1986). Address: Ministry of Education and Science, P.O. Box 25000, 2700 LZ Zoetermeer, The Netherlands.